Lecture Notes in Physics

263

Quantum Chaos and Statistical Nuclear Physics

Proceedings of the 2nd International Conference on Quantum Chaos and the 4th International Colloquium on Statistical Nuclear Physics, Held at Cuernavaca, México, January 6–10, 1986

Edited by T.H. Seligman and H. Nishioka

Springer-Verlag
Berlin Heidelberg GmbH

Editors

T. H. Seligman
Instituto de Física, Laboratorio de Cuernavaca
Universidad Nacional Autónoma de México (UNAM)
Apdo. Post. 20364-01000 México D.F.

H. Nishioka
Max-Planck-Institut für Kernphysik
Postfach 10 39 80, 6900 Heidelberg, FRG

ISBN 978-3-662-13591-4 ISBN 978-3-540-47230-8 (eBook)
DOI 10.1007/978-3-540-47230-8

Originally published by Springer-Verlag Berlin Heidelberg New York in 1986
Softcover reprint of the hardcover 1st edition 1986

2153/3140-543210

TO PETER MOLDAUER

We have been deprived of the presence of Peter Moldauer at this meeting by his untimely death in an accident. He was one of the few who kept the subject of statistical nuclear physics alive during many years when interest in the subject was very limited. While he published a large number of papers in the field, I believe his friends remember him mainly as a critic, and this may well have been his most important contribution. He set down many erroneous ideas, and in many other instances, he forced us to rethink results and give more exact, more stringent motivations for ideas correctly developed, but inadequately or insufficiently founded. Thus, by being a foe to his friends, he advanced the subject in many ways. I believe it to be a great loss that we cannot take advantage of his critical mind at this time in which new ideas and methods are revolutionizing the field.

T.H. Seligman

INTERNATIONAL ADVISORY COMMITTEE

O. Bohigas (Orsay)
G. Casati (Como)
J. Flores (Mexico City)
E. Heller (Los Alamos)
J. Hüfner (University of Heidelberg)
P. Mello (Mexico City)
P. Pechukas (Columbia, New York)
T.H. Seligman (Cuernavaca, Mexico)
H. Weidenmüller (MPI Heidelberg)
A.S. Wightman (Princeton)

ORGANIZING COMMITTEE

T.H. Seligman (IFUNAM)
E. Blaisten-Barojas (IFUNAM
A. Dacal (IFUNAM)
B. Lesche (IFUNAM)
K.B. Wolf (IIMAS-UNAM)

PREFACE

The convergence in quantum chaos and statistical nuclear physics of methods and physical background made it seem desirable to hold a joint meeting on both subjects. This decision influenced to some degree the weight given to different aspects. The overlap is clearly centered around random-matrix methods, as may be seen from the first two sections of these proceedings. The fact that it was not possible to make a clear division of the subject of spectra and states in accordance with the two conference headings seems in itself to justify bringing nuclear statistics and quantum chaos together in one meeting.

Nevertheless, this should not distract from the importance of the other matters. At this point, statistical scattering still lacks a simple analogon, the exciting developments of time-dependent processes have not penetrated nuclear physics. Whether a relevant development in this respect will occur remains to be seen, but I hope that the contacts established will prove fruitful.

August 1986 The Editors

Table of Contents

RIEMANN'S ZETA FUNCTION: A MODEL FOR QUANTUM CHAOS?

M.V.Berry
H.H.Wills Physics Laboratory, Tyndall Avenue,
Britol BS8 1TL, U.K.

1 INTRODUCTION

The celebrated hypothesis of Riemann[1] is that all the complex zeros of his function $\zeta(z)$ have real part 1/2, so that the quantities $\{E_j\}$ defined by

$$\zeta\left(\tfrac{1}{2} - iE_j\right) = 0 \tag{1}$$

are all real. There is evidence supporting the hypothesis: the first few million E_j have been computed and are all real, and it has been proved that uncountably many E_j are real. My purpose in this speculative paper is to extend the old suggestion that the E_j are real because they are eigenvalues of some Hermitian operator $\hat{H}$. The extensions are that if $\hat{H}$ is regarded as the Hamiltonian of a quantum-mechanical system then

(i) $\hat{H}$ has a classical limit

(ii) the classical orbits are all chaotic (unstable)

(iii) the classical orbits do not possess time-reversal symmetry.

To make these assertions plausible I will combine two sorts of evidence. The first (section 2) concerns largely numerical results connecting $\{E_j\}$ with the spectra of infinite random complex Hermitian matrices. The second (section 3) concerns analogies between a (divergent) representation of the number of Riemann zeros with $0<E_j<E$ and an asymptotic formula expressing the number of quantum energy levels in any given interval as a sum over classical closed orbits. Finally (section 4), I will give a semiclassical interpretation of the Riemann-Siegel formula (the basis of a powerful method for computing the $\{E_j\}$[1]) leading to a conjectured generalization that would be a quantization formula for classically chaotic systems.

The most useful formulae for $\zeta(z)$ will be the familiar ones: as a product over primes p or a sum of inverse powers of integers n, namely

$$\zeta(z) = \prod_p \left(1-p^{-z}\right)^{-1} \qquad (\mathrm{Re}\, z>1) \tag{2a}$$

$$= \sum_{n=1}^{\infty} n^{-z} \qquad (\mathrm{Re}\, z>1) \tag{2b}$$

even though neither representation converges on the line Rez=1/2 where the zeros are. Of course there are many representations which are valid for Rez=1/2, such as this resummation of (2b):

$$\zeta(z) = \frac{1}{1-2^{1-z}} \sum_{n=1}^{\infty} (-1)^{n+1} n^{-z} \qquad (\mathrm{Re}\, z > 0) \tag{3}$$

The symmetry [1] relating $\zeta(z)$ and $\zeta(1-z)$ implies that each of the zeros (1) with real E_j has a counterpart with $-E_j$, and with this understanding we henceforth regard $\{E_j\}$ as a set of positive numbers with $E_1<E_2<E_3\ldots$.

2 RIEMANN ZEROS AND RANDOM MATRICES

A useful separation of the $\{E_j\}$ into an average part and a fluctuating part can be achieved by means of the Riemann staircase. This is defined as

$$N_R(E) \equiv \sum_{j=1}^{\infty} \Theta(E-E_j) \tag{4}$$

where Θ denotes the unit step function. $N_R(E)$ is simply the number of zeros with $E_j<E$. The average $\langle N_R(E)\rangle$ is a smooth approximation to the staircase, whose form is known [1] to be

$$\langle N_R(E)\rangle = \frac{E}{2\pi}\left(\ln\left\{\frac{E}{2\pi}\right\} - 1\right) + \frac{7}{8} . \tag{5}$$

Fig.1a shows just how close an approximation this is, even for small E. Figs 1b,c show that it is necessary to include the term 7/8.

Deviations from $\langle N_R(E)\rangle$ constitute the fluctuations in $\{E_j\}$. The statistics of these fluctuations can be studied numerically, and it is found that with high accuracy they coincide with the statistics of the eigenvalues of a typical member of the Gaussian unitary ensemble' (GUE)[2] of complex Hermitian matrices whose elements are Gauss -distributed in a way that is invariant under unitary transformations.

One such statistic is the probability distribution of the normalized spacings $\{S_j\}$ between adjacent zeros; these are defined by

$$S_j \equiv (E_{j+1}-E_j)\Big/\langle d_R((E_j+E_{j+1})/2)\rangle \tag{6}$$

where $\langle d_R(E)\rangle$ is the average density of zeros, given by (5) as

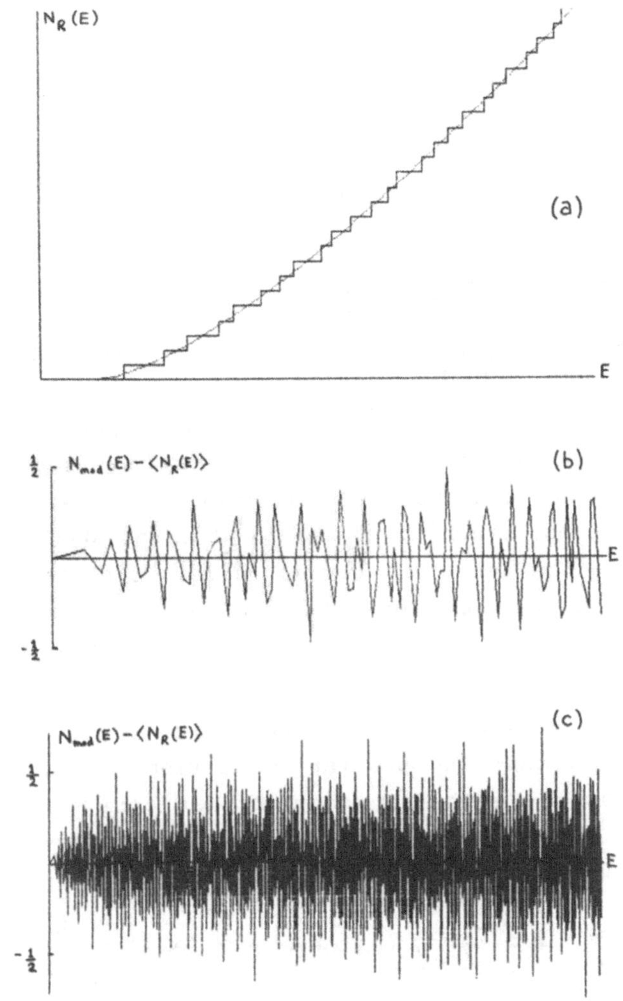

Fig.1 a) Riemann staircase $N_R(E)$, and (dotted) its average $\langle N_R(E)\rangle$, for the lowest 25 zeros; b) deviation $N_{mod}(E)-\langle N_R(E)\rangle$ for the lowest 100 zeros; c) $N_{mod}(E)-\langle N_R(E)\rangle$ for the lowest 1000 zeros ($N_{mod}(E)$ is $N_R(E)$ made continuous by replacing the N'th step by the straight line joining E_N, $N-1/2$ and E_{N+1}, $N+1/2$).

$$\langle d_R(E)\rangle = \frac{d}{dE}\langle N_R(E)\rangle = \frac{1}{2\pi}\ln\left\{\frac{E}{2\pi}\right\} \tag{7}$$

The spacings distribution P(S) is shown in fig.2, together with $P_{GUE}(S)$, for which random-matrix theory [2] gives the close approximation

$$P_{GUE}(S) = \frac{32}{\pi^2} S^2 \exp\left\{-4S^2/\pi\right\}. \tag{8}$$

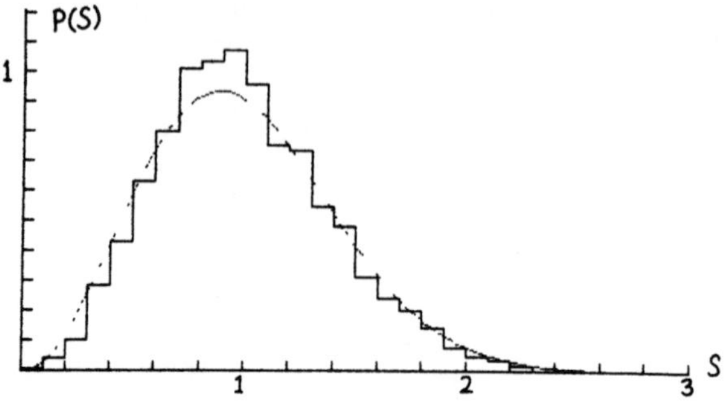

Fig.2 Histogram of distribution P(S) for the first 5000 Riemann zeros, together with $P_{GUE}(S)$ dotted.

Far more extensive computations have been performed by Odlyzko; these are unpublished, but preliminary reports have been given by Dyson [3] and Bohigas and Giannoni [4]. Odlyzko calculated sequences of up to 10^5 Riemann zeros, reaching to the 10^{11}th, and studied not only P(S) but also correlation functions betwen pairs, triplets and quartets of zeros. Apart from one apparent exception, to which we will return in section 3, his statistics are in excellent agreement with those of the GUE.

A theorem of Montgomery [5] supports the conclusion suggested by Odlyzko's results, that the statistics of Riemann zeros are precisely those of the GUE. The theorem concerns the form factor $K(\tau)$ of the zeros; this is the Fourier transform of the pair correlation function, and is defined as

$$K(\tau) = \lim_{M\to\infty} \left\{ \frac{1}{M} \sum_{j=1}^{M} \sum_{k=1}^{M} \exp\{2\pi i\tau(x_j - x_k)\} - \frac{\sin M\pi\tau}{\pi\tau} \right\} \tag{9}$$

where $\{x_j\}$ are the Riemann zeros, scaled so as to have unit mean spacing, that is

$$x_j \equiv \langle N_R(E_j) \rangle. \tag{10}$$

Montgomery proves that for $|\tau|<1$, $K(\tau)$ coincides with the GUE form factor [6]

$$\left. \begin{aligned} K_{GUE}(\tau) &= |\tau| \qquad (|\tau|<1) \\ &= 1 \qquad (|\tau|>1) \end{aligned} \right\}, \tag{11}$$

and conjectures that this agreement continues to hold when $|\tau|>1$.

Now, GUE statistics are distinctive [4]: they differ sharply, for example, from the Poisson statistics of numbers generated sequentially by a random process, and from the Gaussian orthogonal ensemble (GOE) statistics of eigenvalues of random real symmetric (as opposed to complex Hermitian) matrices. It would seem difficult to simulate GUE statistics by a process not involving eigenvalues of complex Hermitian matrices, so from now on I will regard the experimental observation that the Riemann zeros obey GUE statistics as evidence that there is indeed a nontrivial complex infinite Hermitian matrix $\hat{H}$ with eigenvalues E_j. ('Nontrivial' means that $\hat{H}$ must not contain the $\{E_j\}$ explicitly - as in the diagonal matrix $E_i \delta_{ij}$ or any simple unitary transform thereof.)

Among the class of operators $\hat{H}$ which are quantum Hamiltonians with classical limits, those with discrete energy spectra obeying GUE statistics have classical orbits which are <u>chaotic</u> and <u>without time--reversal symmetry</u>. This has been demonstrated numerically by Seligman and Verbaarschot [7], Berry and Robnik [8], and Robnik and Berry [9], and explained theoretically by Berry [10]. Both of the above conditions are necessary for GUE statistics: for systems whose orbits are not chaotic but integrable (with or without time-reversal symmetry), Berry and Tabor [11] showed the energies to be Poisson--distributed; and for systems which are chaotic but which do have time-reversal symmetry, Bohigas, Giannoni and Schmit [12] and others [13,4,10] have shown the energies to be GOE-distributed.

Of course the observation that GUE statistics are shared by both the Riemann zeros E_j and the eigenvalues of classically chaotic systems without time-reversal symmetry does not imply that the $\{E_j\}$ come from a classically chaotic $\hat{H}$, but we now turn to analytical evidence which strongly suggests that they do.

3 OUROBOROLOGY

Ouroboros [14] was the mythical snake that swallowed its tail, and serves to symbolize the constructive interference of quantum waves associated with classical orbits (fig.3). Such interference forms the basis of a technique for generating the energy spectrum, developed by Gutzwiller [15-17] and Balian and Bloch [18]; for an elementary review, see [19]. The technique has been used to explain why classically integrable systems have Poisson-distributed energy levels

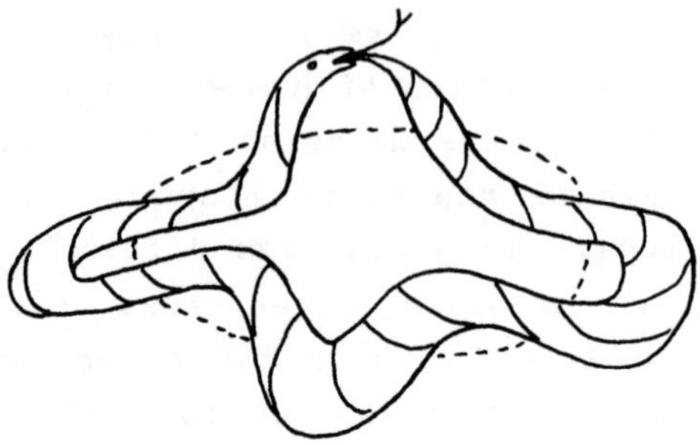

Fig.3 A quantum wave (Ouroboros) interfering constructively round a classical closed orbit.

[11] and why chaotic systems have GOE- or GUE- distributed levels [10].

Ouroborology is based on representing the spectral density as the imaginary part of the trace of the Fourier transform of the propagator (Green function of the time-dependent Schrödinger equation), which is expressed semiclassically (i.e. for small Planck's constant $\hbar$) as a sum over classical paths. The spectral staircase N(E), defined by (4) with $\{E_j\}$ now being the energy levels, is just the integral of the spectral density, so N(E) can similarly be expressed in terms of closed classical paths with energy E, the relation being that the fluctuating part of the staircase is

$$N_{osc}(E) \equiv N(E) - \langle N(E) \rangle = \mathrm{Im} \sum_{p} \sum_{m=1}^{\infty} B_{pm} \exp\{i S_{pm}(E)/\hbar + \phi_{pm}\}. \tag{12}$$

In this formula, ⟨N(E)⟩ is the average staircase (cf (5)) and will be discussed later. The double sum is over all closed orbits, each being an m-fold traversal of a primitive orbit labelled p. S_{pm} is the action of the orbit m, given in terms of the canonical phase-space variables q_μ, p_μ by

$$S_{pm}(E) = \oint p_\mu dq_\mu \quad (= m S_{p1}(E)). \tag{13}$$

The phases ϕ_{pm} and amplitudes B_{pm} depend on the focusing and stability of a bundle of (non-closed) orbits centred on the closed one. We require only the formulae for systems with two freedoms whose closed orbits are all isolated and unstable (making the dynamics chaotic) and without focal points. Then $\phi_{pm}=0$, and [15,16]

$$B_{pm} = \left[2\pi m \sinh\{m\lambda_p(E)/2\}\right], \tag{14}$$

where $\lambda_p(E)(>0)$ is the instability exponent of the primitive orbit p at energy E (i.e. $\exp\{\pm\lambda_p\}$ are the eigenvalues of the 2x2 matrix M_p of linearized phase-space deviations transverse to p, and

$$2\sinh\{m\lambda_p/2\} = \left[-\det\{M_p^m - 1\}\right]^{1/2}.$$

Thus for such systems the semiclassical asymptotic formula for the spectral fluctuations is

$$N_{osc}(E) = \frac{1}{2\pi}\sum_p \sum_{m=1}^{\infty} \frac{\sin\{m S_{p1}(E)/\hbar\}}{m \sinh\{m\lambda_p(E)/2\}}. \qquad (15)$$

Next, we note that $N_{osc}(E)$, as defined by (4) and the first member of (12), has unit discontinuities at each eigenvalue. Therefore the closed-orbit sum (15) can at best be conditionally convergent, with the discontinuities determined by the very long orbits. For these, $m\lambda_p/2$ is large, and so we can replace the series by

$$N_{osc}(E) \approx \frac{1}{\pi}\sum_p \sum_{m=1}^{\infty} \frac{1}{m}\exp\{-m\lambda_p(E)/2\}\sin\{m S_{p1}(E)/\hbar\}. \qquad (16)$$

This replacement will be reconsidered later.

Let us now turn to $\zeta(z)$ with $z=\frac{1}{2}-iE$. Just above the real E axis, the phase of decreases by as ReE passes each Riemann E_j. Moreover $\zeta(z)\to 1$ as $\mathrm{Im}(E)\to+\infty$ (i.e. as $\mathrm{Re}z\to+\infty$). Therefore the fluctuating part of the Riemann staircase is

$$N_{R,osc}(E) \equiv N_R(E) - \langle N_R(E)\rangle = -\frac{1}{\pi}\lim_{\eta\to 0}\mathrm{Im}\ln\zeta\left(\tfrac{1}{2} - i(E+i\eta)\right). \qquad (17)$$

Now pretend that the product formula (2a) can be used when Rez=1/2 (more about this later), substitute into (17) and expand the logarithms. This gives

$$N_{R,osc}(E) \approx -\frac{1}{\pi}\sum_p \sum_{m=1}^{\infty} \frac{\sin\{mE\ln p\}}{m\,p^{m/2}} \qquad (18)$$

which apart from a sign, to be discussed later, has the same form as the semiclassical expression (16) if the following identifications are made: the label p for primitive closed orbits denotes prime numbers; the actions of the closed orbits are

$$S_{pm} = mE\ln p\;; \qquad (19)$$

Planck's constant $\hbar$ is unity, so that the semiclassical limit is $E\to\infty$; and the instability exponents (independent of E) are

$$\lambda_p = \ln p\,. \qquad (20)$$

It is worth remarking that it follows from (19) that the periods of the closed orbits would be

$$T_{pm} = \frac{dS_{pm}}{dE} = m \ln p = \ln p^m. \tag{21}$$

There is thus a formal analogy between fluctuations of the Riemann staircase and fluctuations of the spectral staircase of a classically chaotic system. Because of the distinctive GUE statistics (and for other reasons [10]) the classical orbits must lack time-reversal symmetry. Mathematicians have noticed essentially the same analogy as that between (15) or (16) and (18), in the context of a special case for which a certain transform of (15) is exact rather than asymptotic. This is the Selberg trace formula, which equates a sum over eigenvalues of the Laplace-Beltrami operator (playing the role of $\hat{H}$)on a manifold of constant negative curvature to a sum over closed geodesics on this manifold (all unstable, i.e. chaotic). But despite extensive study (see McKean [20] and Hejhal [21]) this analogy has not led to the identification of the mysterious 'Riemann' classical system with the properties (19-21). The 'closest approach' has been the discovery by Pavlov and Faddeev [22] and Gutzwiller [23] of a scattering (rather than bound) system whose phaseshifts (rather than energy levels) are given by $\zeta(z)$ with Rez=1 (rather than 1/2).

An apparently anomalous outcome [3] of Odlyzko's computation of the form factor $K(\tau)$(eq.9) of the Riemann zeros gives further support to the semiclassical analogy. Although he finds good overall agreement with the GUE formula (11), close examination of the difference $K-K_{GUE}$ reveals a series of spikes for small τ . Such spikes are predicted by the semiclassical theory, because as I have shown elsewhere [10] the universality of the GUE statistics ceases to hold for large energy scales, that is short time scales. For $K(\tau)$ the semiclassicl formula for $|\tau|<1$, expressed in terms of the closed-orbit amplitudes and periods, is [10]

$$K(\tau) = \pi^2 \tau^2 \sum_p \sum_{m=1}^{\infty} B_{pm}^2 \, \delta(\tau - T_{pm}/2\pi\hbar\langle d\rangle). \tag{22}$$

Now as $\hbar \to 0$, $\hbar\langle d\rangle \sim \hbar^{-(D-1)}$ for a system with D freedoms (D>1), so that the spike associated with a given orbit slides towards $\tau=0$ as $\hbar \to 0$. Near any finite τ, then, spikes are semiclassically thickly clustered and it is their average [10] which gives $K \approx |\tau|$ as in (11). But an accurate evaluation of $K(\tau)$ should reveal at least the first few spikes. In the Riemann case, (21) shows that the spikes should occur at τ-values proportional to logarithms of powers of primes. If the first of Odlyzko's spikes occurs at $\tau = K\ln 2$, the others occur at positions which on his picture are indistinguishable from $K\ln 3$, $K\ln 4$, $K\ln 5$ and $K\ln 7$, but as expected there is no spike at $K\ln 6$ because 6 is not a power of a prime.

Four objections may be raised against the chaos analogy for the Riemann zeros.

Objection 1: the Riemann closed-orbit formula (18) depends on the product (2a) which does not converge when Rez=1/2. But the analogous semiclassical formulae (15) and (16) almost certainly do not converge either. The physical reason for this is that the number of closed orbits of a chaotic system proliferates exponentially as their period (or action) increases, overwhelming the effect of the instability exponents λ_p in making the amplitudes decay exponentially. The mathematical reason was explained to me by Dr.A.Voros in the context of the Selberg trace formula: the fundamental quantum object for which semiclassical techniques give an expression in terms of closed orbits is not $N_{osc}(E)$ (or its derivative which is the fluctuating part of the spectral density) but the trace of the resolvent

$$g(E) \equiv \sum_j \frac{1}{E - E_j} \tag{23}$$

for which the closed-orbit formula makes sense only when $\eta \equiv \mathrm{Im}E$ exceeds some finite value. This means that in the formula

$$N(E) = -\frac{1}{\pi} \lim_{\eta \to 0} \mathrm{Im} \int_0^{E+i\eta} dE' g(E') \tag{24}$$

the limit $\eta \to 0$ cannot be taken when using ourolorology for g(E). Retaining finite η has the effect of introducing further factors $\exp\{-i\eta T_{pm}(E)/\hbar\}$ into (15) and (16), making the sums converge but at the price of giving a staircase whose steps are smoothed by η, thereby frustrating attempts to discriminate individual eigenvalues. Objection 1 therefore disappears because both the Riemann and semiclassical closed-orbit formulae share the disadvantage of not converging for real E. One could argue that this shared disadvantage strengthens the analogy.

It is interesting to look numerically at the divergence of the product (2a), especially in view of earlier computations [10] of the spectral density (derivative of (18) which with small numbers of primes showed pronounced peaks at the lowest few zeros, nicely simulating the delta-functions that the exact spectral density must possess. It is simplest to calculate the truncated product

$$|\zeta_M(\tfrac{1}{2}-iE)| \equiv \prod_{p<M} \left|1 - \frac{e^{iE\ln p}}{p^{1/2}}\right|^{-1}. \qquad (25)$$

As fig.4a shows, very few factors suffice to discriminate the lowest zeros. As M increases, however, $|\zeta_M|$ oscillates increasingly fast between the zeros, in contrast to the exact $|\zeta|$ which has only one maximum between each pair of zeros (cf fig.6 later); and when M=10000 (fig.4b) the oscillations are threatening to obscure the first zero. That such

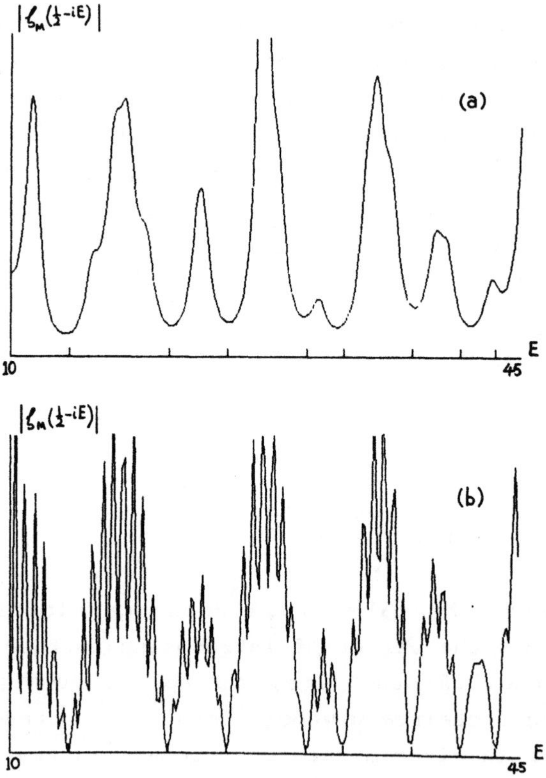

Fig.4a Truncated Riemann product $\zeta_M(\frac{1}{2}-iE)$ as a function of E for M=5 (three factors in (25)), with ticks marking the exact Riemann zeros E_j; b) as a) but with M=10000

obscuration will eventually occur is illustrated in fig.5, which shows $|\zeta_M|$ as a function of M evaluated at the exact positon of the first Riemann zero E_1=14.135... : at first $|\zeta_M|$ decreases, apparently indicating convergence onto E_1, but, when M exceeds about 2000, $|\zeta_M|$ begins to oscilalte with increasing amplitude. Rough asymptotics shows

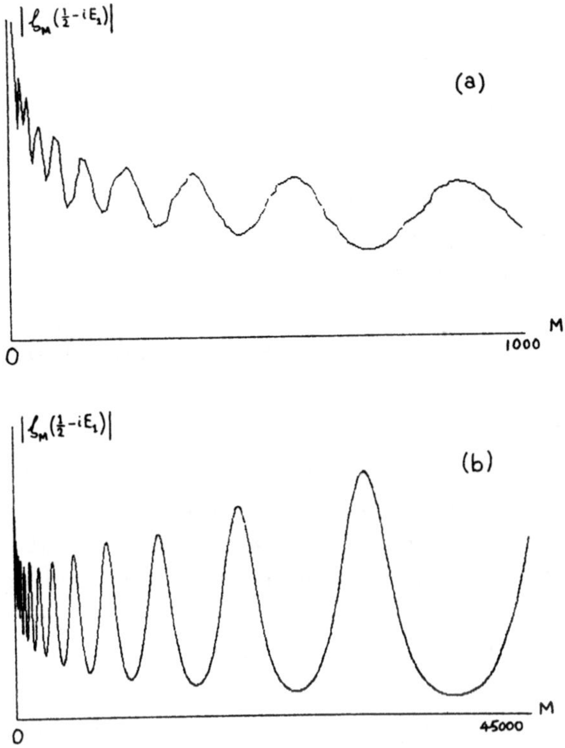

Fig.5 Truncated Riemann product $|\zeta_M(\frac{1}{2}-E_1)|$ as a function of M, evaluated at the lowest Riemann zero E_1, for a) M<1000; b) M<45000.

that $|\zeta_M|$ eventually diverges as

$$|\zeta_M| \sim \exp\{M^{1/2}\sin\{E\ln M\}/E\ln M\}. \tag{26}$$

The fact that (18) relies on evaluating the Riemann product on the line Rez=1/2, which is displaced by 1/2 from the nearest line on which it converges (z=1), suggests that E=1/2 is the smallest interval over which the Riemann zeros can be discriminated in this way, and hence that ouroborology might fail altogether when the mean separation of zeros is about 1/2. From (7), this occurs when $E_j \sim 2\pi\exp(4\pi)$ $2\mathrm{x}10^6$, i.e. $j \sim 3\mathrm{x}10^6$, and preliminary numerical exploration in this region indeed suggests the beginning of a failure of (18) to discriminate individual zeros.

Objection 2: the passage from (15) to (16), in which $\sinh(m\lambda/2)$ was replaced by $\exp(m\lambda/2)/2$, was a swindle, implying that for a proper analogy the Riemann formula (18) ought to involve not $p^{m/2}=\exp\{m\ln p/2\}$

but 2sinh{m ℓnp/2}. One answer lies in considering not ζ itself but the product

$$P(E) \equiv \prod_{k=0}^{\infty} \zeta\left(\tfrac{1}{2}+k-iE\right). \tag{27}$$

For real E this converges and has the same zeros as $\zeta(\frac{1}{2}-iE)$, and so can be employed instead of $\zeta(\frac{1}{2}-iE)$ to approximate the fluctuations of the Riemann staircase. An easy calculation (cf.(17-18)) gives

$$N_{P,osc}(E) = -\frac{1}{\pi}\lim_{\eta\to 0} \operatorname{Im}\ln P(E+i\eta) \approx -\frac{1}{2\pi}\sum_{p}\sum_{m=1}^{\infty}\frac{\sin\{mE\ln p\}}{m\sinh\{m\ln p/2\}} \tag{28}$$

which obviously is analogous to (15).

Objection 3: the average density of Riemann zeros (7) has a logarithmic form which is not easy to interpret as the average density <d> of eigenvalues of an operator $\hat{H}$ with a classical limit $H(q_\mu, p_\mu)$. For example, in a finite D-dimensional 'billiard' enclosure, $\langle d\rangle \sim E^{(D/2-1)}$; replacing the enclosure by a binding potential similarly fails to give a logarithm. However, Simon [24,25] draws attention to a class of Hamiltonians which are classically unbound but have discrete quantal energy spectra. Of these, a planar billiard with a channel reaching to infinity whilst narrowing hyperbolically does have a logarithmic average level density (and moreover displays intermittent chaos). This can be seen in an elementary way using the Weyl formula

$$\langle d(E)\rangle \sim \frac{\text{area}}{4\pi} \tag{29}$$

(defining energy as $E \equiv k^2$ where k is the de Broglie wavenumber), together with the idea that waves do not penetrate (except with exponential evanescence) where the channel is narrower than a wavelength, that is narrower than about k^{-1}. If the channel boundary has equation y=B/x, the area for waves with energy E is thus

$$\text{area} \sim B\int^{y\sim 1/k} y\,dx = B\int^{Bk} dx/x \sim B\ln B\sqrt{E} \sim \ln E \tag{30}$$

giving $\langle d\rangle \sim \ln E$ as claimed. This disposes of objection 3, although there is of course no suggestion that the Riemann $\hat{H}$ really is a billiard of this type (even with magnetic field, to break time-reversal symmetry).

Objection 4: the semiclassical and Riemann formulae (16) and (18) have opposite signs. I have no clear answer to this. It is possible to get negative signs for some of the B_{pm} in (14), when the unstable orbits have focal points [15,16]: for a primitive orbit with n_p focal points, the sign is $(-1)^{[mn_p+1)/2]}$, where [ ] denotes integer part. But no choice of n_p gives negative signs for all the B_{pm}. However, this determination of the sign implicitly assumes an $\hat{H}$ of the form 'Kinetic energy + potential energy', acting on scalar states; without these restrictions it might be possible to get all-negative amplitudes.

4 THE RIEMANN-SIEGEL FORMULA: A RULE FOR QUANTIZING CHAOS?

It follows from the functional equation for $\zeta(z)$ [1] that the following function Z(E) is real and even for real E:

$$Z(E) \equiv \exp\{-i\theta(E)\}\zeta(\tfrac{1}{2}-iE) \tag{31}$$

where

$$\begin{aligned}\theta(E) &= \arg\Gamma(\tfrac{1}{4}+\tfrac{E}{2}) - \tfrac{E}{2}\ln\pi \approx \tfrac{E}{2}\ln\tfrac{E}{2\pi} - \tfrac{E}{2} - \tfrac{\pi}{8} \\ &= \pi(\langle N_R(E)\rangle + 1)\end{aligned} \tag{32}$$

(cf.5). The Riemann-Siegel formula is an asymptotic representation of Z(E) for large E:

$$Z(E) = -2\sum_{n=1}^{Q(E)} \frac{\cos\{\pi\langle N_R(E)\rangle - E\ln n\}}{n^{1/2}} + R(E) \tag{33}$$

where

$$Q(E) \equiv [\sqrt{(E/2\pi)}] \tag{34}$$

and R(E) is a series of remainder terms [1] whose main effect is to cancel the discontinuities of the main sum arising from the E-dependence of the limit Q. Fig.6 shows how accurate the formula is, even for small E.

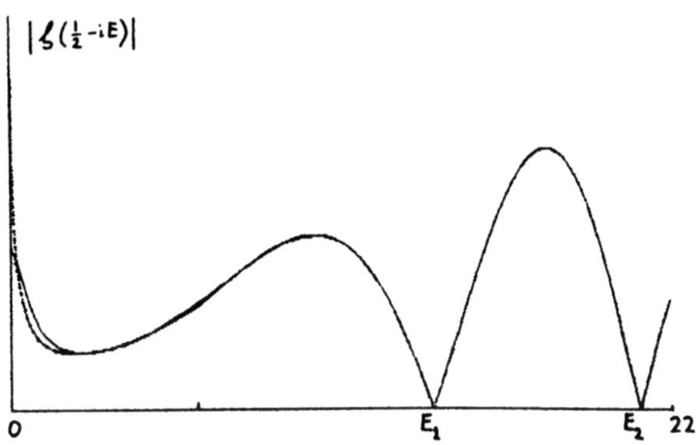

Fig.6. Comparison of $|\zeta(\frac{1}{2}-iE)|$ (full line) with Riemann-Siegel formula plus one correction term (dotted line). (The number Q(E) of terms in the sum (33) changes from zero to one at $E=2\pi$, indicated by a tick)

Now I will outline how the Riemann-siegel formula can be obtained from the series (2b) by an argument with a semiclassical interpretation suggesting a generalization. First note that in obtaining (2b) by expanding the products in (2a), use is made of the factorization theorem that for any integer n we can write

$$\ln n = \sum_p m_p \ln p \tag{35}$$

with a unique choice of the set of integers $\{m_p=0,1,2...\}$. On the semiclassical analogy, the sum index n can thus be interpreted as runing over all possible combinations of orbit periods (cf.21). Each such combination $\ln n$ will be called a pseudoperiod (pseudoperiods where all m_p except one are zero are periods of actual orbits).

Next, split the sum (2b) into two, with pseudoperiods with n less than and greater than some initially arbitrary value Q, and apply Poisson's summation formula to the second sum:

$$\begin{aligned} \zeta(\tfrac{1}{2}-iE) &= \sum_{n=1}^{Q} \frac{\exp\{iE\ln n\}}{n^{1/2}} + \sum_{n=Q+1}^{\infty} \frac{\exp\{iE\ln n\}}{n^{1/2}} \\ &= \sum_{n=1}^{Q} \frac{\exp\{iE\ln n\}}{n^{1/2}} + \sum_{m=-\infty}^{\infty} \int_{Q+\delta}^{\infty} dn \, \frac{\exp\{i(E\ln n - 2\pi mn)\}}{n^{1/2}} \end{aligned} \tag{36}$$

where $0<\delta<1$. For large E the integrals may be approximated by the method of stationary phase. The stationary point of the m'th integral is at $n=E/2\pi m$, which lies in the integration range only if $1 \leq m \leq [E/2\pi(Q+\delta)]$. The choice $Q=Q(E)$ (equation 34) and $\delta \to 0$ gives the same limits for the sums over n and m, and then stationary phase leads to

$$\zeta(\tfrac{1}{2}-iE) \approx \sum_{n=1}^{Q(E)} \left(\frac{\exp\{iE\ln n\}}{n^{1/2}} + \frac{\exp\{i(-\frac{\pi}{4} + E\ln\{E/2\pi m\} - E)\}}{n^{1/2}} \right) \tag{37}$$

which gives the Riemann-Siegel sum (33) when combined with the definitions (31) and (32).

What the Poisson technique has achieved in (37) is a resummation of the orbits with long pseudoperiods, to give a series of the same form as the sum over the short pseudoperiods, with precisely the correct phase relation to make $\zeta(\frac{1}{2}-iE)$ have the required analytic structure (31). To see what immense advantage this resummation has produced as compared with the naive ouroborology result (18), consider the oscillations of the terms in (33). The first term, with n=1 (corresponding to the zero pseudoperiod with all m_p equal to zero in 35), oscillates fastest. This term alone gives zeros

$$\langle N_R(E_j)\rangle = j - \tfrac{1}{2} \qquad (j=1,2,3\ldots), \tag{38}$$

with the correct density but which are (asymptotically) uniformly distributed instead of GUE-distributed. The terms with n>1 (nonzero pseudoperiods) oscillate more slowly, and the highest term n=Q(E) is almost constant:

$$\frac{d}{dE}\left(\pi\langle N_R(E)\rangle - E\ln Q\right) = \pi\langle d_R(E)\rangle - \ln Q(E)$$
$$= \ln\{\sqrt{(E/2\pi)}\} - \ln\{[\sqrt{(E/2\pi)}]\} \approx 0 \tag{39}$$

By contrast, the terms given by unresummed ouroborology oscillate ever faster to give the noise and divergence visible in figs. 4b and 5b.

It is natural to speculate that a similar resummation might be possible for semiclassical quantum chaos. For the simplest nontrivial case, embodied in equations (14-16), this speculation generates, by an argument to be outlined in a moment, a real function W(E) whose zeros would be the eigenvalues. The pseudoorbits are simpler than in the Riemann case, in that each primitive orbit contributes at most one traversal. Thus the actions, periods and instability exponents may be written

$$\left.\begin{array}{l} S^{(k)}(E) = \sum_p \ell_p S_{p1} \;;\; T^{(k)}(E) = \sum_p \ell_p\, dS_{p1}/dE \;;\; \lambda^{(k)}(E) = \sum_p \ell_p \lambda_p \\ \text{with } \{\ell_p = 0 \text{ or } 1\} \text{ labelled } k \text{ in order of increasing } T^{(k)} \end{array}\right\} \tag{40}$$

Then the analogue of the Riemann-Siegel formula (33) would be

$$W(E) = -2\sum_{k=1}^{k_{max}} (-1)^{g_k} \exp\{-\lambda^{(k)}(E)/2\}\cos\{\pi\langle N(E)\rangle - S^{(k)}(E)/\hbar\} \tag{41}$$

where g_k is the number of terms (nonzero ℓ_p's) in the sums (40), and k_{max} is given by the condition that the highest term is non-oscillatory:

$$T^{(k_{max})}(E) = \pi\hbar\,\langle d(E)\rangle \tag{42}$$

Because $\langle d\rangle \sim \hbar^{-D}$ for a system with D freedoms, the longest pseudoperiod is of order $\hbar^{-(D-1)}$, so the number of terms in (41) grows exponentially as $\hbar$ decreases or E increases; nevertheless (41) should at least give sensible results for individual eigenvalues in the semiclassical limit, unlike the ouroborology formula (16).

The conjectured Riemann-Siegel analogue (41) arises from considering the function

$$w(E) = \exp\left\{\int_0^E dE'\left(g(E') - \langle g(E')\rangle\right)\right\}, \tag{43}$$

where g(E) is the resolvent (23) and <g(E)> its average. For real E approaching the real axis from above, use of the average of (24) gives

$$\lim_{\eta\to 0} \omega(E+i\eta) = \exp\left\{-\int_{-\infty}^{\infty} dE' \ln\left|1-\frac{E}{E'}\right| \langle d(E')\rangle\right\} \exp\{i\pi\langle N(E)\rangle\} \prod_j \left(1-\frac{E}{E_j}\right) \tag{44}$$

Therefore $\omega(E)$ has zeros at the eigenvalues E_j. On the other hand, for sufficiently large Im E the fluctuations g-<g> can be approximated semiclassically by the analogue of (16) (which replaces the analogue of (15) by an argument similar to that centred on (27) and (28)). Thus

$$\omega(E) \approx \exp\left\{-\sum_p \sum_{m=1}^{\infty} \frac{\exp\{-m\lambda_p/2\}}{m} \exp\{imS_{p1}/\hbar\}\right\}$$

$$= \prod_p \left(1 - \exp\{-\lambda_p/2\}\exp\{iS_{p1}(E)/\hbar\}\right), \tag{45}$$

giving $\omega(E)$ as a product over primitive orbits, analogous to the product (2a) for $\zeta(z)$ with the difference that the factors do not appear as reciprocals (this originates in the mysterious sign difference described in objection 4 at the end of section 3).

Expanding the product in (45) we obtain an analogue of the sum (2a), involving the pseudoorbits (40). Then the conjectured Riemann-Siegel analogue (41), with W(E) identified from (44) as proportional to the product $\prod_j(1 - E/E_j)$, would follow, if only (!) one could resum the high pseudoorbits in a way similar to that which gave the true Riemann-Siegel formula (33).

Exactly the dynamical zeta function $\omega(E)$, in the form (45), has been written down by Gutzwiller [17]. Ruelle [26,27] introduced analogous zeta functions to study chaotic dissipative dynamical systems and these have been further investigated by Parry [28,29] (who introduced what I here call pseudoorbits) and Parry and Pollicott [30]; in these dissipative zeta functions, orbit actions and instability exponents are (by implication) considered as proportional to periods and entropy. None of these authors appear to have considered the possibility that a Riemann-Siegel formula might exist for dynamical zeta functions.

ACKNOWLEDGMENT No military agency supported this research.

REFERENCES

[1] H.M.Edwards, Riemann's Zeta Function (Academic Press: New York and London) (1974)

[2] C.E.Porter, Statistical theories of Spectra: Fluctuations (Academic Press: New York (1965)
[3] F.J.Dyson, in 'Symmetries in particle physics' (eds: I.Bars, A.Chodos and C-H Tze; Plenum: New York and London) pp 279-281 (1984)
[4] O.Bohigas and M.J.Giannoni, Chaotic motion and Random-matrix Theories in Mathematical and Computational Methods in Nuclear Physics eds.J.S.Dehesa, J.M.G.Gomez and A.Polls, Lecture Notes in Physics 209 (Springer-Verlag, N.Y.) pp.1-99 (1984)
[5] H.L.Montgomery, Proc.Symp.Pure Math. 24 181-193 (1973)
[6] M.L.Mehta, Random Matrices and the Statistical Theory of Energy Levels (Academic Press: New York and London) (1967)
[7] T.H.Seligman, and J.J.M.Verbaarschot, Phys.Lett. 108A 183-187 (1985)
[8] M.V.Berry and M.Robnik, J.Phys.A. (1986) In press
[9] M.Robnik and M.V.Berry, J.Phys.A. (1986) In press
[10] M.V.Berry, Proc.Roy.Soc.Lond. A400 229-251 (1985)
[11] M.V.Berry and M.Tabor, Proc.Roy.Soc.Lond. A356 375-394 (1977)
[12] O.Bohigas, M.J.Giannoni and C.Schmit, Phys.Rev.Lett. 52 1-4 (1984)
[13] M.V.Berry, Ann.Phys. (N.Y) 131 163-216 (1981)
[14] Ouroboros: see Encyclopaedia Brittanica (15th ed.) vol.9 p13 (1985)
[15] M.C.Gutzwiller, J.Math.Phys. 12 343-358 (1971)
[16] M.C.Gutzwiller, in 'Path Integrals and their Applications in Quantum, Statistical and Solid-State Physics' (Eds. G.J.Papadopoulos and J.T.Devreese) Plenum, N.Y. 163-200 (1978)
[17] M.C.Gutzwiller, in 'Stochastic Behaviour in Classical and Quantum Hamiltonian Systems' (eds. G.Casati and J.Ford) Lecture Notes in Physics 93 (Springer: Berlin)(1979)
[18] R.Balian and C Bloch, Ann.Phys.(N.Y) 69 76-160 (1972)
[19] M.V.Berry, Semiclassical Mechanics of Regular and Irregular Motion in Chaotic Behavior of Deterministic Systems (Les Houches Lectures XXXVI, eds.G.Iooss, R.H.G.Helleman and R.Stora (North-Holland: Amsterdam) pp171-271 (1983)
[20] H.P.McKean, Comm.Pure App. Math. 25 225-246 (1972)
[21] D.A.Hejhal, Duke Math.J. 43 441-482 (1976)
[22] B.S.Pavlov and L.D.Faddeev, J.Soviet Math 3 522-548 (1975)
[23] M.C.Gutzwiller, Physica 7D 341-355 (1983)
[24] B.Simon, Ann.Phys. 146 209-220 (1983)
[25] B.Simon, J.Funct.Anal. 53 84-98 (1983)
[26] D.Ruelle, Inventiones Math 34 231-242 (1976)
[27] D.Ruelle, Ergod.Th.and Dynam.Syst. 2 99-107 (1982)
[28] W.Parry Israel J.Math. 45 41-52 (1983)
[29] W.Parry Ergod Th.and Dynam.Syst. 4 117-134 (1984)
[30] W.Parry and M.Pollicott, Annals of Math. 118 573-591 (1983)

SPECTRAL FLUCTUATIONS OF CLASSICALLY CHAOTIC QUANTUM SYSTEMS

Oriol BOHIGAS, Marie-Joya GIANNONI and Charles SCHMIT
Division de Physique Théorique*, Institut de Physique Nucléaire,
F-91406 Orsay Cedex, France

I. Introduction

The previous Conference on Quantum Chaos was held at Como more than two years ago and many questions on spectral fluctuations presented there have been investigated since then [Ca-85]. Some suggested lines of inquiry are now becoming fully developed, some conjectures based on numerical experiments are being put on a firm theoretical basis and new theoretical as well as experimental progress is being made. It is the purpose of this contribution to give a brief account on some topics of this activity. The general theme will be : 'Given a system which is classically chaotic which are the manifestations in the corresponding quantum spectrum of its chaotic character ?' Emphasis will be given to works that are paving the way between the study of chaotic motion and random matrix physics, in line with the titles of the present joint Conferences, namely Quantum Chaos and Statistical Nuclear Physics (random matrix theories have developed mainly in connection with the latter). And it should be kept in mind that the fluctuation properties of nuclear spectra, although believed to be generic of systems whose classical counterpart is chaotic, continue to provide the only convincing example of spectral manifestation of chaoticity found in nature so far.

The material of this talk is organized as follows : In Section 2 we remind how spectral fluctuations are defined and characterized, and we briefly describe the more important models of spectral fluctuations, with emphasis on random matrix theory predictions, which are compared to existing data. In Section 3 we review the present status of computer experiments and of theories that have lead to the identification of several universality classes for spectral fluctuations of quantum systems. The basis of such classification, closely inspired by random matrix theories, has its origin

*Laboratoire associé au C.N.R.S.

in the regularity properties (integrability versus chaoticity) of the corresponding classical system. The role played by symmetries is emphasized and the transition regimes between the different universality classes is discussed. In Section 4 some recent results for the hydrogen atom in a strong magnetic field and on the spectrum of the Laplacian on a surface of negative curvature, problems both of current interest in this field, are discussed. We end up with some remarks and conclusions.

2. Fluctuation measures : models and predictions [BFF-81, BG-84]

We consider the discrete spectrum $E_1, E_2, E_3, \ldots$ of a d-dimensional quantum system (an atomic nucleus, an atom, a molecule or model systems like a particle in a potential, in particular a particle in a box) with Hamiltonian $H(\underset{\sim}{q},\underset{\sim}{p})$ and leading to the Schrödinger equation $(\Delta - V(\underset{\sim}{q}) + E)\,\Psi(\underset{\sim}{q}) = 0$ (1)

Let N(E) be the staircase function giving the number of points on the energy axis which are below or equal to E. The question now is to separate N(E) in a smooth part $N_{av}(E)$ and the remainder will define the fluctuating part $N_{f\ell}(E)$ of N(E)

$$N(E) = N_{av}(E) + N_{f\ell}(E) \tag{2}$$

For $N_{av}(E)$ one can obtain, by taking the simple semi-classical rule that each quantum state is associated with a phase-space volume h^d

$$N_{av}(E) \simeq \frac{1}{2^d \pi^{d/2}\,\Gamma(d/2+1)} \int_{V(\underset{\sim}{q}) \leq E} [E - V(\underset{\sim}{q})]^{d/2}\, d\underset{\sim}{q} \tag{3}$$

For the particular case of one particle in a box in two dimensions (billiard in the classical case), one has $N_{av}(E) \simeq SE/4\pi$, where S is the area of the two-dimensional domain. If one includes corrections, which are smooth terms of lower order in E, one obtains, with Dirichlet boundary conditions

$$N_{av}(E) \simeq (1/4\pi)(SE - \mathcal{L}\sqrt{E} + K) \tag{4}$$

In Eq.(4) $\mathcal{L}$ is the length of the parameter of the boundary whose area is S ; K is a constant containing complex information on the geometrical and topological properties of the domain.

In Fig.1 are compared N(E) and $N_{av}(E)$ for some particular cases. It can be seen that $N_{av}(E)$ indeed reproduces perfectly the average behaviour of N(E), starting from the bottom of the spectrum.

Before studying fluctuations one wants to get rid of $N_{av}(E)$ in order to characterize and compare fluctuation patterns of different systems whose corresponding average behaviours are not the same. For that purpose one can

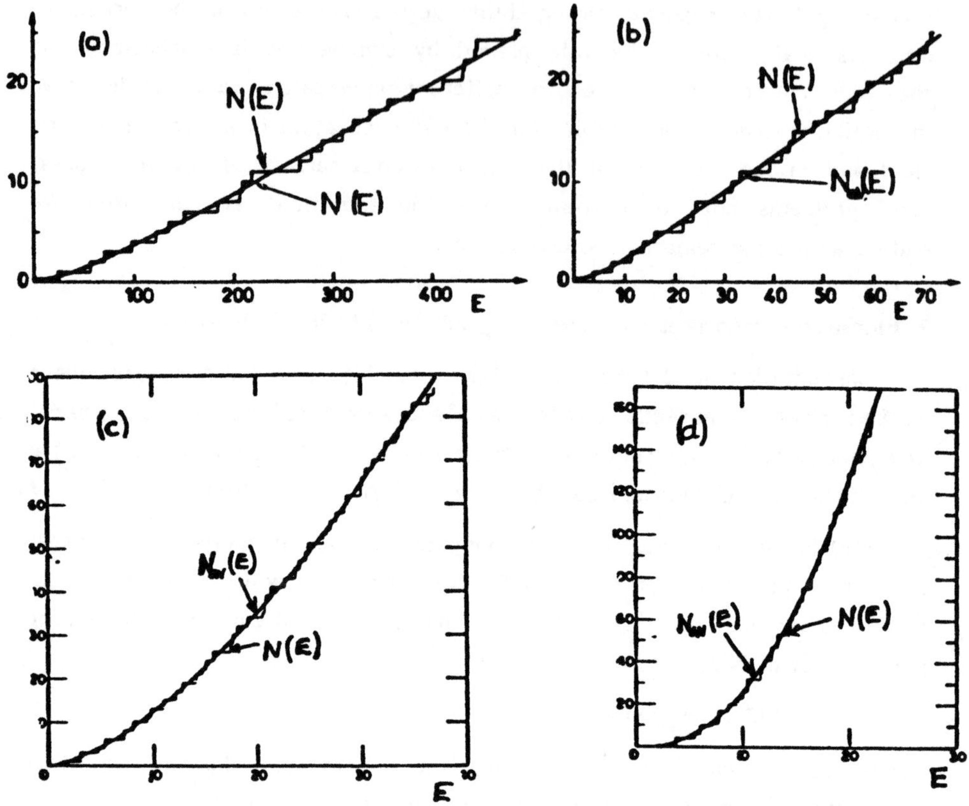

Fig.1 - Plots of N(E) and N_{av}(E) for : (a) a quarter of circle and Eq.(4) ; (b) a stadium (see Fig. 4) and Eq.(4) ; (c) $V(\underset{\sim}{q}) = q_1^4+q_2^4$ and Eq.(3), d = 2 ; (d) $V(\underset{\sim}{q}) = q_1^4 + q_2^4 + q_3^4$ and Eq.(3), d = 3 ((c) and (d) taken from [HMY-81]).

"unfold" the original spectrum $\{E_i\}$ through the mapping $E \mapsto x$

$$x_i = N_{av}(E_i) \qquad i = 1,2,3,... \tag{5}$$

The effect of (5) is that the sequence $\{x_i\}$ has on the average a constant mean spacing (or a constant density) equal to unity, irrespective of the particular form of the function N_{av}(E). We are now in position to study the statistical laws governing sequences $\{x_i\}$ having very different origins. By construction $x_i \simeq i-1/2$ (i=1,2,...) and the departures

$$\delta_i = x_i - (i - 1/2) \quad i = 1,2,... \tag{6}$$

of x_i from its average value i-1/2 are the level fluctuations.

To characterize level fluctuations in a systematic way one deals with the k-level correlation functions $R_k(x_1,...,x_k)$ ($R_k(x_1,...,x_k)dx_1...dx_k$ is the probability of finding one level within each of the intervals $[x_j,x_j+dx_j]$)

and measures derived from them. We shall only refer to

1/ The spacing distribution $p(x)$ between adjacent levels .

2/ Quantities directly related to the number statistic $n(L)$: given an interval $[\alpha, \alpha+L]$ of length L, it counts the number of levels contained in the interval. The average value of $n(L)$ is L (the mean spacing is unity). We shall consider higher moments or cumulants of $n(L)$:variance $\Sigma^2(L)$, skewness $\gamma_1(L)$ and excess $\gamma_2(L)$.

3/ The $\overline{\Delta}_3$ statistic of Dyson and Mehta. It measures, given an interval of length L, the least-square deviation of the staircase function from the best straight line fitting it. Its average value $\overline{\Delta}_3$ is related to the variance $\Sigma^2(L)$ of $n(L)$.

In general $\Sigma^2(L)$ and $\overline{\Delta}_3(L)$ can be expressed in terms of the 2-level correlation function, whereas $\gamma_1(L)$ ($\gamma_2(L)$) depends also on the 3-level (3- and 4-level) correlation function(s) [BHP-85].

Let us consider two trivial limiting cases

1/ Take a random variable s whose probability density $p(x)$ is $\exp(-x)$. Construct a sequence $\{x_i\}$ as follows

$$x_1 = 1/2,\ x_{i+1} = x_i + s_i \quad i = 1,2,3,..., \tag{7}$$

where s_i are outcomes of independent trials of the variable s. The resulting spectrum is what is called a Poisson spectrum (case of maximum randomness). One has $p(x) = e^{-x}$, $\Sigma^2(L) = L$, $\overline{\Delta}_3(L) = L/15$.

2/ A picked fence spectrum or spectrum of the harmonic oscillator in one dimension. It is the more ordered spectrum one can imagine, with no randomness. One has $p(x) = \delta(x-1)$, $\Sigma^2(L) = 0$, $\overline{\Delta}_3(L) = 1/12$.

The 'non trivial' models of spectral fluctuations are provided by random matrix theories (RMT) [Po-65,Me-67]. In RMT one considers the Hamiltonian matrix H as an N × N stochastic matrix (the matrix elements are random variables) ; the random matrix ensemble is specified by the probability density $\mathcal{P}(H)dH$. One is interested in asymptotic results valid for large N.

General underlying space-time symmetries obeyed by the system put important restrictions on the admissible matrix ensembles. We shall refer to results of the Gaussian Orthogonal Ensemble (GOE) and the Gaussian Unitary Ensemble (GUE) which correspond to ensembles of real symmetric matrices and of hermitian matrices respectively. In general, GOE applies when time-reversal invariance is a good symmetry and GUE when it is not (see next Section for some warnings). Besides these general symmetry considerations, no other property of the system is taken into account to define the GOE or GUE.

GOE and GUE eigenvalue fluctuations are presently well known. For GOE the spacing distribution is very well approximated by

$$p(x) \simeq (\pi/2) \times \exp(-(\pi/4)x^2) \tag{8}$$

and for GUE by

$$p(x) \simeq (32/\pi^2)x^2 \exp(-(4/\pi)x^2) \tag{9}$$

The spacing distributions in both cases show level repulsion, i.e. tendency to avoid level clustering or small probability of small spacings ($p(x)$ vanishes at the origin). The level repulsion is stronger for GUE than for GOE :near the origin $p(x) \sim (\pi^2/6)x$ for GOE whereas for GUE one has $p(x) \sim (\pi^2/3)x^2$. For GOE the number variance $\Sigma^2(L)$ for $L \gtrsim 1$ is given by

$$\Sigma^2(L) = (2/\pi^2)\, \ell nL + 0.44 \tag{10}$$

In (10) the logarithmic increase is to be compared to the linear increase with L for a Poisson spectrum. One can speak of the semicrystalline nature of the spectrum, showing long range order. The GUE spectrum is more rigid than the GOE spectrum : $\Sigma^2(L)$ is half the GOE-value plus 1/8. For large L, the average of $\Delta_3(L)$ is given for GOE by

$$\overline{\Delta}_3(L) \simeq (1/\pi^2)\, \ell nL - 0.007 \tag{11}$$

GUE shows also a logarithmic increase of $\overline{\Delta}_3(L)$ but with a smaller coefficient than in Eq.(11) (stronger rigidity).

How well do GOE predictions compare with experimental data ? We shall restrict to high resolution nuclear resonance data in the regime of well separed resonances (one example is given in Fig.2). The combined set of nuclear resonance - energy data of different nuclei- in short, the nuclear data ensemble (NDE) - has been treated as a sampling of eigenvalues of GOE matrices [HPB-82,BHP-83,BHP-85].

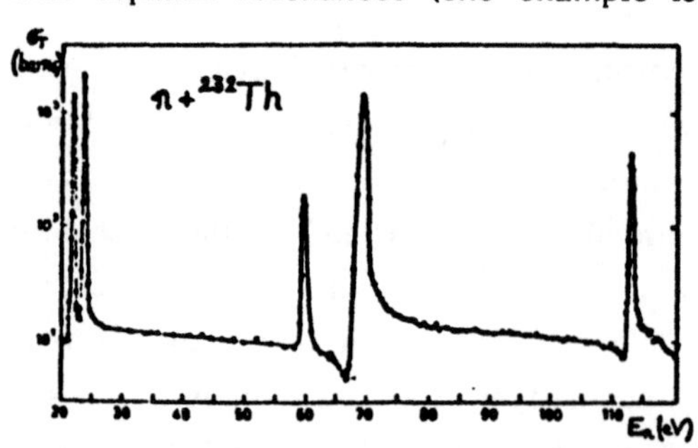

Fig.2 - Resonances for the reaction n+^{232}Th (taken from the compilation Neutron cross sections (1964)).

The data are the result of an impressive long term collective effort at Columbia University (neutron resonance spectroscopy) and Duke University (high resolution proton scattering experiments). Results are reproduced in Fig.3. As can be seen, all the fluctuation measures considered, which include a thorough study of 2-point measures and to some extent more than 2-point

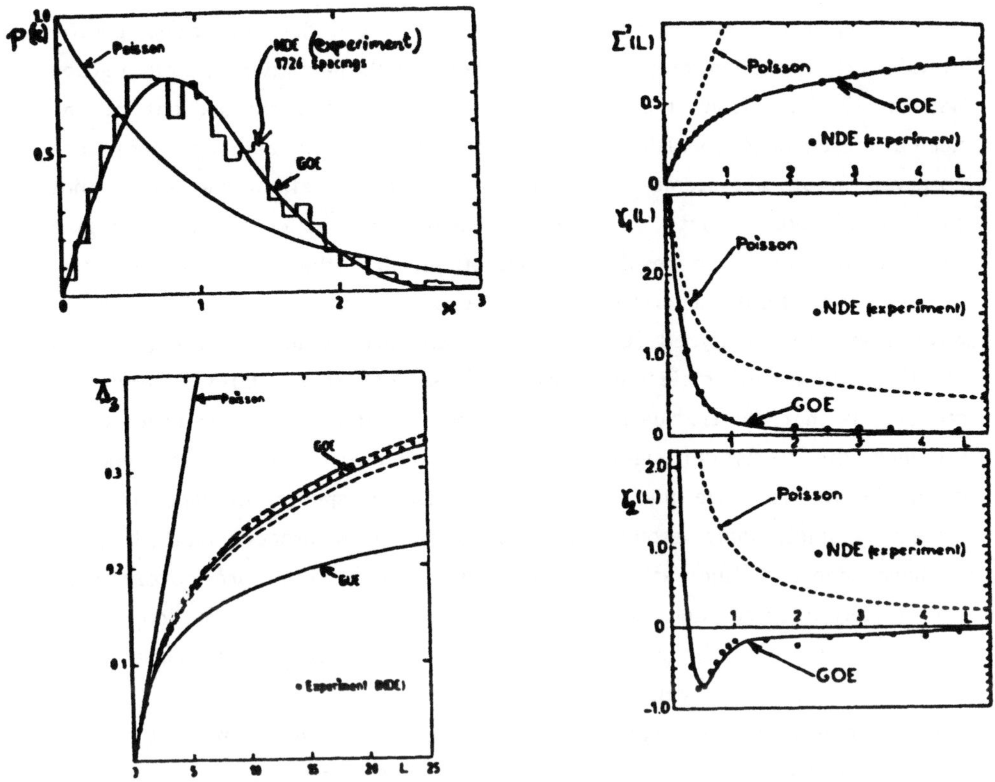

Fig.3 - Results of fluctuation measures of nuclear resonances. Poisson and GOE are given for comparison (taken from [HPB-82,BHP-83,BHP-85]).

measures as well, are fully consistent with GOE predictions.

Now, from the theoretical point of view, are GOE-fluctuations specific of GOE or, on the contrary, are there other models that give GOE fluctuation patterns ? And, from the experimental one, are the fluctuations of the nuclear resonances specific of nuclei or, on the contrary, are they also observed in other systems ? Concerning the first question, one knows that there is a whole variety of models of random matrix ensembles that give GOE-fluctuations (some results are analytical, the others are obtained from Monte Carlo calculations). Concerning the second one, some atomic spectra have also been analyzed [CG-83] and there are some pioneering studies of fluctuations of molecular spectra [HKC-83,MSP-84,LLJ-86]. And the outcome is that the fluctuations seem to be consistent with GOE predictions, although the statistical significance is much lower than for the nuclear case. We therefore see that, on the one hand, the spectra of very different systems (nuclei, some atoms and molecules), when properly scaled, seem to have identical fluctuation patterns, even though they are governed by very different force laws (short range interactions and Coulomb long range forces). On the other hand, these characteristic fluctuation patterns, although not specific of, are well reproduced by GOE, a parameter-free theory. Thus a simple picture emerges : the level fluctuation laws seem to be universal, as well from the experimental than from the theoretical point of view.

3. Spectral fluctuations and chaotic motion

To obtain some clues on the origin of the universality of level fluctuation laws and also to investigate how "complicated" a system must be in order to show RMT-fluctuations ([BG-84]) let us first discuss some results of numerical experiments on two-dimensional systems. The guiding lines we will try to follow are : i) the notion of simple or complicated as used in studying classical dynamical systems, namely integrable or chaotic respectively, ii) Einstein's precept "everything should be made as simple as possible, but not simpler".

Consider a particle in a box in two dimensions with boundaries as shown in Fig.4. The classical problem is a billiard problem, which is integrable

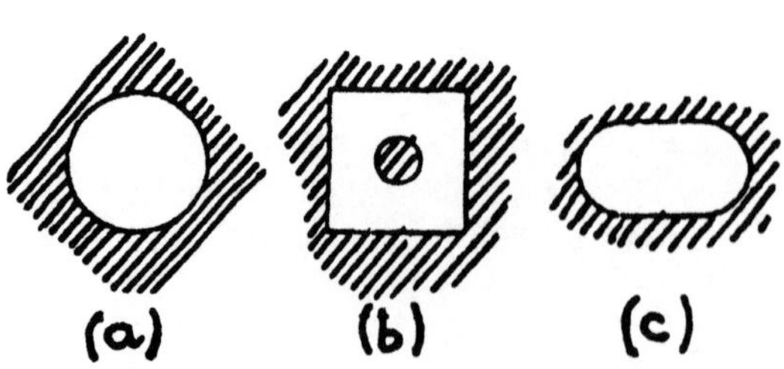

Fig.4 - Different boundaries in which a particle is enclosed : (a) circular billiard ; (b) Sinai's billiard ; (c) Bunimovich's stadium billiard.

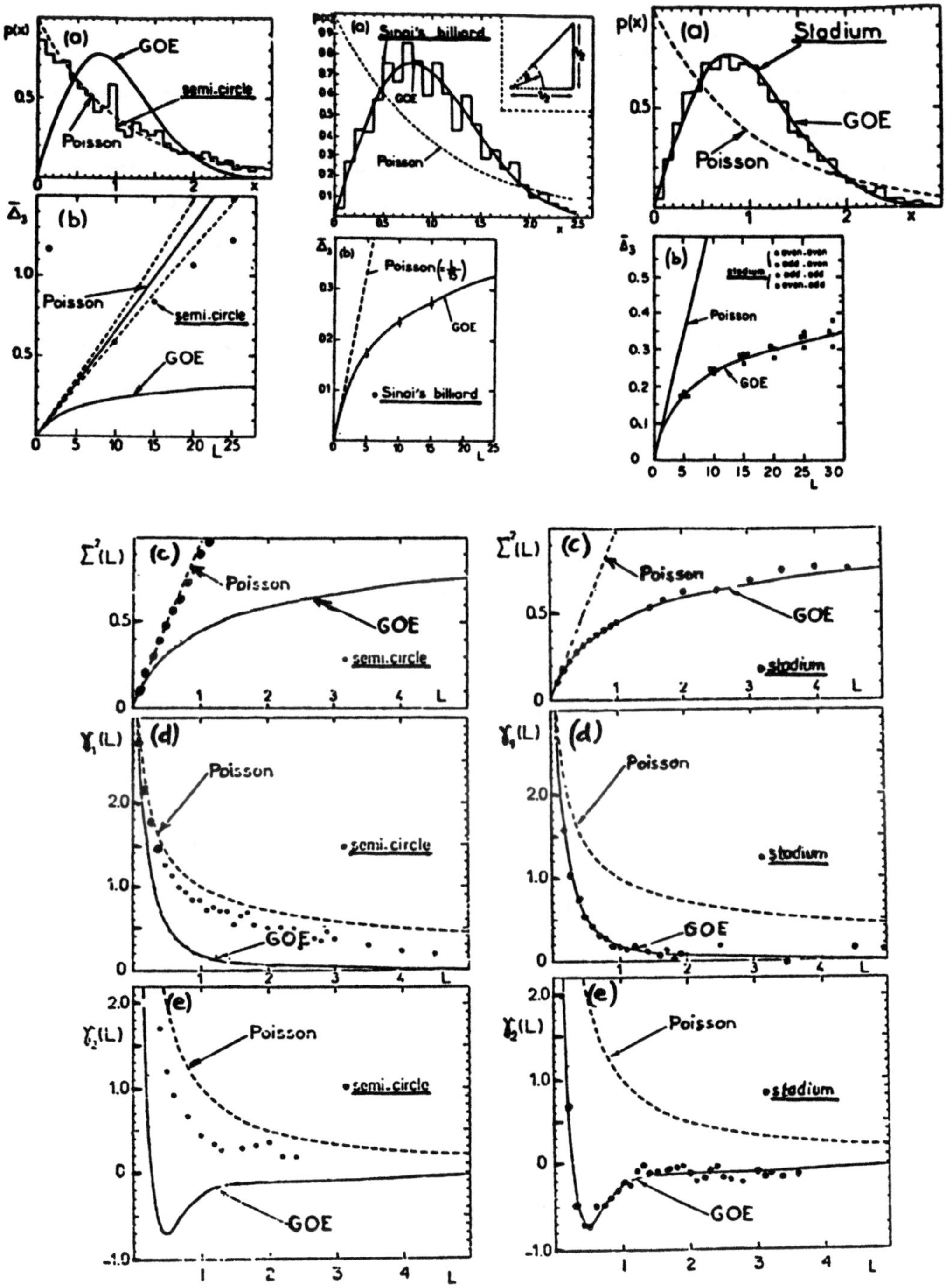

Fig.5 - Spectral fluctuations of the systems shown on Fig.4 : (a) spacing distributions p(x) ; (b) $\overline{\Delta}_3(L)$ statistic ; (c) number variance $\Sigma^2(L)$; (d) number skewness $\gamma_1(L)$; (e) number excess $\gamma_2(L)$ (taken from [BGS-84a,BGS-84b]).

with a circular boundary but fully chaotic (K-system) with Sinai's and Bunimovich's stadium boundaries. For the quantum problem we are interested in the spectrum of the Laplacian with Dirichlet boundary conditions (vanishing of the wave function on the boundary). And as explained before, we know how to separate out the average behaviour (see Eq.(4)). Results of the spectral fluctuations are reproduced in Fig.5. The message seems fairly clear : The system whose classical analogue is integrable shows Poisson fluctuations whereas the systems whose classical analogues are fully chaotic show GOE fluctuation patterns.

What happens if one considers not time-reversal systems but rather time non-invariant systems ? From RMT predictions one expects to have GUE-fluctuations and it is indeed what is found as shown by Seligman and Verbaarschot [SV-85b].These authors studied the motion of a particle in a combination of inhomogeneous magnetic and scalar force fields

$$H(\underset{\sim}{q},\underset{\sim}{p}) = \frac{1}{2}(p_1+a\,q_2^3)^2 + \frac{1}{2}(p_2 + a\,q_1^3)^2 + \\ + \alpha_1 q_1^6 + \alpha_2\, q_2^6 - \alpha_{12}(q_1 - q_2)^6 \tag{12}$$

By an adequate choice of the values of the parameters the classical motion is completely chaotic, and the spectral fluctuations of the quantum results are displayed in Fig.6 which, as expected, agree with GUE predictions. Berry and Robnik [BR-85] have studied the spectral fluctuations of a very interesting and subtle family of systems, which they call Aharonov-Bohm chaotic billiards. It consists of chaotic billiards with a single line of magnetic flux. The geometry of classical trajectories are unaffected by the flux, whereas the quantum mechanics -in particular the energy levels- are altered by the presence of the flux line. This system breaks time reversal invariance for the quantum particle and its spectral fluctuations are generically (i.e. except for special values of the quantum flux parameter) of the GUE type.

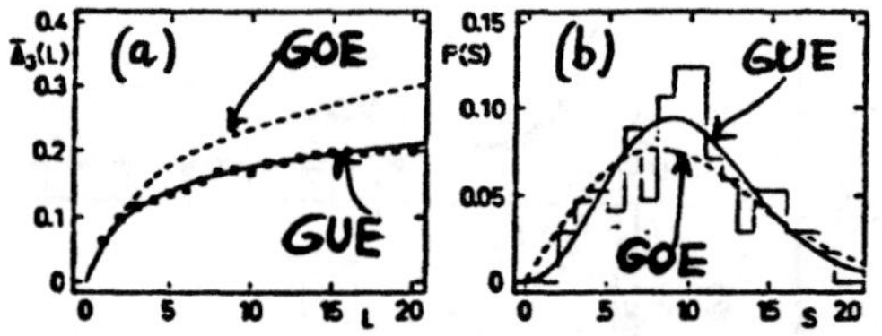

Fig.6 - Spectral fluctuations of the system governed by Eq.(12) : (a) $\overline{\Delta}_3(L)$; (b) spacing distribution p(x) (taken from [SV-85b]).

To summarize the numerical experience accumulated so far, one can say that classically integrable quantum systems show Poissonian spectral fluctuations whereas classically chaotic quantum systems show GOE(GUE) fluctuations if they are time reversal invariant (time reversal non invariant).

What is important now is to understand theoretically these findings. Recently Berry [Be-85] has given a semiclassical derivation of the spectral rigidity. He employs the semiclassical techniques introduced by Gutzwiller, Balian and Bloch, which consists in an asymptotic representation of the spectral density as a sum over all the periodic orbits of the classical system. He predicts universality in the regime $1 << L << L_{max}$, with L_{max} determined by the shortest classical closed orbit ; for $L >> L_{max}$ there is a non-universal saturation value of $\overline{\Delta}_3(L)$. In the universal regime, the long range predictions coincide with Poisson-, GOE- and GUE- predictions for classically integrable-, chaotic and time-reversal-, chaotic and time-reversal breaking- systems, in agreement with what is found numerically. It is important to notice that, in deriving the results, random matrix theory is not used and no statistical assumptions are made.

Some words concerning symmetries are in order. The random matrix predictions described in the previous section apply to levels having the same set of quantum numbers because the matrix representing the Hamiltonian is splitted into disconnected blocks, each block characterized by exact quantum numbers. If one considers a spectrum containing several sets of quantum numbers, the spectral fluctuations are obtained by an independent superposition of correlated spectra, with proper weighting factors. In order to derive the resulting fluctuations, consider more generally a spectrum obtained from the superposition of ℓ independent sequences with relative weights μ_i ($\Sigma\, \mu_i = 1$) and k-level cluster functions, number variance and $\overline{\Delta}_3$ statistic, $Y_{k,i}$, $\Sigma^2{}_i$, $\overline{\Delta}_{3,i}$ respectively. The corresponding quantities for the combined system are [Pa-79]

$$Y_k\ (x_1,\ldots,x_k) = \Sigma_i\ (f_i)^k\ Y_{k,i}(\mu_1 x_1,\ldots,\mu_k x_k) \qquad (13)$$

resulting in the additivity of Σ^2 and Δ_3

$$\Sigma^2(L) = \Sigma_i\, \Sigma^2{}_i(\mu_i L) \qquad (14)$$

$$\overline{\Delta}_3(L) = \Sigma_i\ \overline{\Delta}_{3,i}\ (\mu_i L) \qquad (15)$$

The spacing distribution of the combined spectrum is (see Appendix 22 of [Me-67])

$$p(x) = E(x) \left\{ \Sigma_i \mu_i^2\ \frac{p_i(\mu_i x)}{E_i(\mu_i x)} + \left[\ \Sigma_i \mu_i\ \frac{1-F_i(\mu_i x)}{E_i(\mu_i x)}\right]^2 - \Sigma_i\ \left(\mu_i\ \frac{1-F_i(\mu_i x)}{E_i(\mu_i x)}\right)^2 \right\} \qquad (16)$$

where

$$F_i(y) = \int_0^y p_i(x)dx , \qquad E_i(z) = \int_z^\infty [1 - F_i(y)]dy \tag{17}$$

and

$$E(z) = \prod_i E_i(z) \tag{18}$$

For instance, for the stadium billiard, one has four different symmetry classes with equal weights $\mu_i = 1/4$, each class following GOE predictions (see Fig.5). If one considers the total spectrum without separating with respect to the symmetry classes, the results are given in Fig.7 and compared to the superposition of four independent GOE sequences with equal weights.

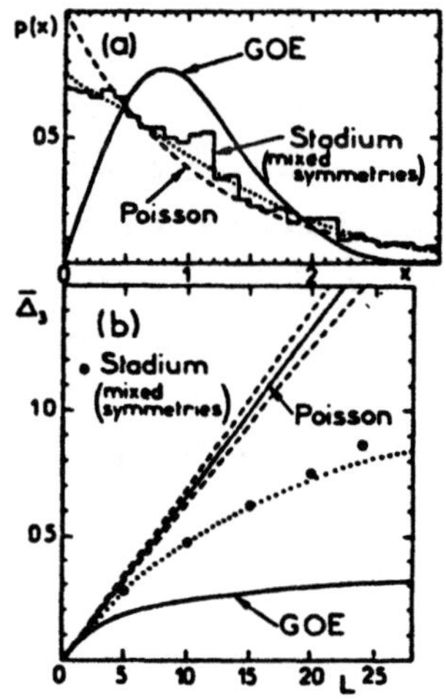

Fig.7 - Spectral fluctuations of the stadium compared to the predictions Eq.(16) and Eq.(15), dotted lines (taken from [BGS-84b]).

Robnik and Berry [RB-86] have discussed another problem of symmetries (which they call false T-violation) : there exist chaotic systems violating T but possessing an invariance under a combination of T and some other symmetry (which may be geometric -for example a reflection- but may also be of dynamical nature) such that they have GOE (and not GUE) fluctuations. Some examples are the system governed by Eq.(12) with $\alpha_{12} = 0$ (see[SV-85b]), Aharonov-Bohm billiards when the quantum flux is half integer and atoms in strong uniform magnetic fields (see next Section).

Until now we have only mentioned systems which are classically integrable or fully chaotic. But for d-dimensional generic systems the phase space is mixed : some orbits are on d-dimensional tori and others explore (2d-1)-dimensional regions chaotically. What are the fluctuations of the corresponding quantum systems ? To investigate this question several authors have studied systems that classically show a transition between the two limiting cases of totally regular and chaotic behaviour as a parameter is varied [HKC-84,SVZ-84,IY-85,CG-86]. An example is given in Fig.8 where, in correspondence with the classical order to chaos transition one can observe the transition in the spectra from Poisson to GOE fluctuations. Based on

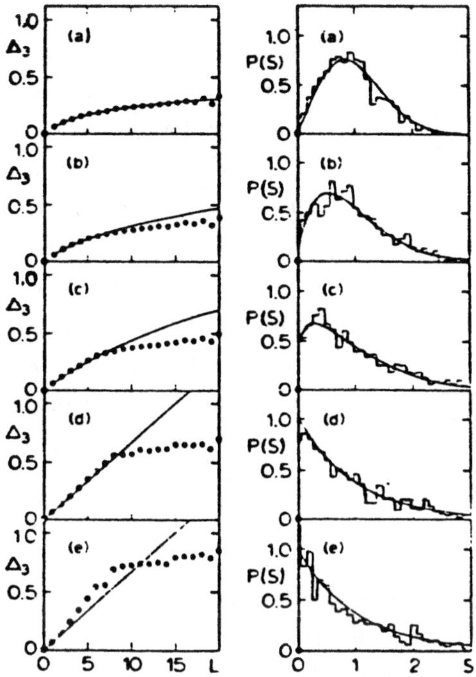

Fig.8 - Fluctuations for a polynomial Hamiltonian: (e),(d),(c),(b),(a), corresponding to regular to chaotic transition (taken from [SVZ-84]).

semiclassical arguments Berry and Robnik [BR-84] suggest that the spectral fluctuations should result from independently superposing a Poisson spectrum with relative weight μ_1, given by the sum of Liouville measures of all classical regular regions, and (ℓ-1) GOE spectra with weights μ_i i =2,...,ℓ given by the Liouville measure of each (ℓ-1) chaotic regions. For the case of only one chaotic region (generic case for d $\geq$3, due to Arnold diffusion) of measure $\bar{\mu}$, Eq.(16) gives [BR-84]

$$p(x) = \mu^2 e^{-\mu x} \operatorname{erfc}\left(\frac{\sqrt{\pi}}{2}\bar{\mu}x\right) + \left(2\mu\bar{\mu} + \frac{\pi}{2}\bar{\mu}^3 x\right) \exp\left(-\mu x - \frac{\pi}{4}\bar{\mu}^2 x^2\right) \qquad (19)$$

where μ denotes the sum of the measures of regular regions ($\mu + \bar{\mu} = 1$). In Fig.9 is illustrated the quality of this description when applied to different systems, including some polynomial Hamiltonians [ZMK-86,MHK-84] and billiard systems [IY-85]. One compares the value of $\bar{\mu}_{c\ell}$ calculated by studying the classical motion, with the value $\bar{\mu}_{qm}$ obtained by making a best fit of the spacing distribution, obtained from the quantum spectrum, with Eq.(19). One can see that the agreement is fair (see [SV-85a,ZMK-86] for a detailed discussion).

Let us now turn to the case of a GOE to GUE transition in the fluctuations. From the random matrix theory side, this problem has been completely solved by Pandey and Mehta [PM-83], who have derived closed analytic expressions for the k-level correlation and cluster functions. In Ref. [PM-83] one considers an ensemble of N × N random matrices

$$H_\alpha = H_{GOE}(v^2) + i\alpha A(v^2) , \qquad \alpha > 0 \qquad (20)$$

where H_{GOE} is a GOE matrix (v^2 is the variance of the non diagonal matrix elements) and A is an antisymmetric matrix with matrix elements distributed

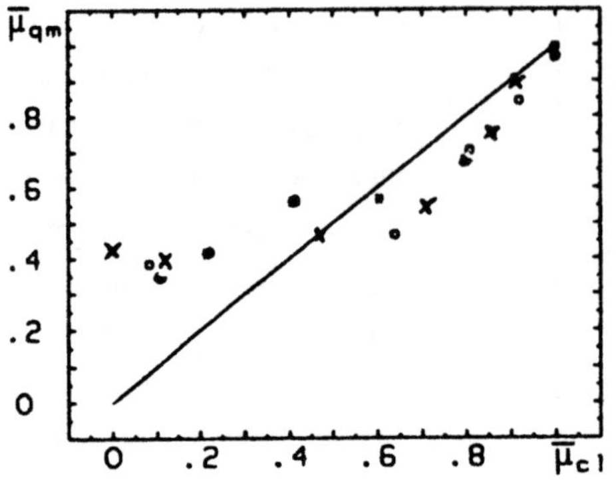

Fig.9 - Comparison of $\bar{\mu}_{c\ell}$ and $\bar{\mu}_{qm}$; see text for further explanation (taken from [ZMK-86])

normally (around zero with variance v^2) and independently. For $\alpha = 1$ one recovers the GOE and for $\alpha = 1$ the GUE. If v and α are finite the k-level correlation functions for $k > 1$ are discontinuous functions of α in the asymptotic-N limit : for $\alpha = 0$ one has GOE-results and for $\alpha \neq 0$ one has GUE-results. However, as a function of the parameter

$$\lambda = \frac{\alpha v}{D} \qquad (21)$$

where D is the mean spacing, the transition is continuous.

Can one exhibit systems showing this intermediate fluctuations between GOE and GUE ? We have studied a billiard problem with a uniform magnetic field perpendicular to the plane of the billiard [SGB-86]. The presence of the magnetic field changes the classical trajectories from sequences of straight line segments to sequences of circular arcs. This system is obviously not invariant under time reversal. The wave function satisfies the eigenvalue equation ($E = k^2$)

$$(\vec{p} - q\vec{A})^2 \, \vec{\Psi}(\vec{r}) = k^2 \, \Psi(\vec{r}) \qquad (22)$$

with Dirichlet boundary conditions. In (22) $\vec{A}$ is the vector potential which may be taken $A_x = -(B/2)y$, $A_y = (B/2)x$ and $A_z = 0$, where B is the strength of the magnetic field. With this choice Eq.(22) reads

$$[\,\Delta - iqB(x\frac{\partial}{\partial y} - y\frac{\partial}{\partial x}) - (\frac{qB}{2})^2\,(x^2+y^2)+k^2]\;\Psi(x,y) = 0 \qquad (23)$$

This system has the same scaling property as usual billiards, namely the spectrum depends trivially on the size of the billiard provided that the magnetic flux through the billiard is preserved (if one scales the size so that the surface S is multiplied by η , $S' = \eta S$, the spectrum is scaled by $E'_n = E_n/\eta$ provided that the magnetic field is scaled by the same factor $B' = B/\eta$). One can therefore keep constant the size of the billiard and just study the influence of the field strength. Now, if the billiard is chaotic without field, does it remain chaotic when the field is applied

[RB-85,Ro-86] ? The answer will depend on the energy of the particle. Indeed, for a given billiard, the classical motion depends strongly on the Larmor radius $R = k/qB$ of the trajectory. If it is large compared to the linear dimension ℓ of the billiard, the trajectory will be close to the one in the absence of the field. One expects that in this regime ($k \gg qB\ell$) the system remains chaotic, a fact which is confirmed by the study of the Poincaré section. On the opposite, when the Larmor radius is small compared to ℓ, there exist circular trajectories which do not hit the boundary and the system is now integrable in a large domain of the phase space. And in this Landau regime ($k \ll qB\ell$) the quantum spectrum will reflect the integrability of the system. Thus for a fixed magnetic field the spectrum will be divided in three parts : A low energy part ($k \ll qB\ell$) where it is quasi-Landau, a high energy part ($k \gg qB\ell$) where it shows rigidity and an intermediate energy part ($k \simeq qB\ell$) with transition properties. We are interested in the high energy part. One can easily estimate by using Eq.(4) that if $qBS \lesssim 2\pi$ the entire spectrum belongs to the high energy part. Attention must be payed to the symmetries : Eq.(23) is invariant under PT, where P is a symmetry with respect to the x or y axis. And if the billiard has the same symmetry axis, the spectrum will have GOE-fluctuations as mentioned before [RB-86].

We have studied an asymmetric stadium (see Fig.10) and solved Eq.(23) with Dirichlet boundary conditions, in the region $0 < qBS \lesssim 3$, where the entire spectrum is expected to show spectral rigidity. Let us discuss some results. In Fig.11 is shown the influence of the presence of the magnetic field on the fluctuation δ_n of the n-th level as a function of n (see Eq.(6)). One can note that both spectra are rigid (the fluctuations are small) and also that there are no missing levels (the fluctuations are around zero). The amplitude of the oscillations are slightly larger without than with magnetic field, as happens when going from GOE to GUE fluctuations. In Fig.12 is shown the cumulative level spacing distribution $F(y) = \int_0^y p(x)dx$. It can be seen that without field it follows closely GOE but when the magnetic field is turned on it departs from GOE towards GUE. The influence of the magnetic field is at first more important for small spacings than for large spacings and, by keeping increasing the field strength, one completes the transition to GUE fluctuations. This behaviour

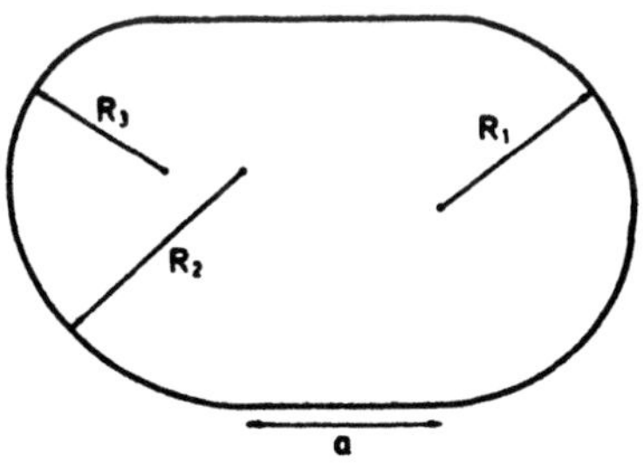

Fig.10 - Shape of the billiard : $R_1/a=1$, $R_2/a = 5/4$.

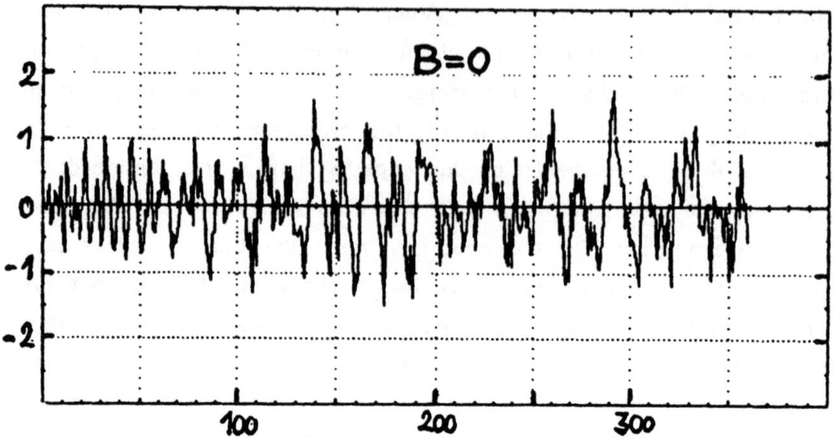

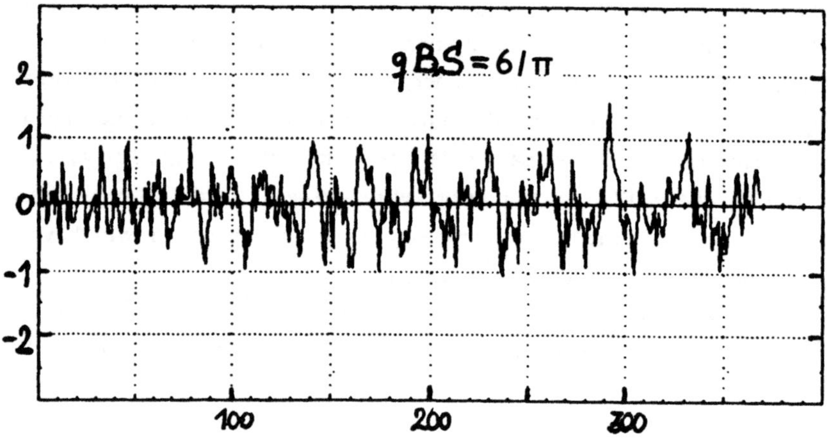

Fig.11 - Fluctuation δ_n of the n-th level plotted as a function of n for zero magnetic field and for qBS = 6/π (taken from [SGB-86]).

is very similar to the transition from GOE to GUE as a function of λ (see Eq.(21)). That the transition is faster at small than at larger scales can be seen by considering 2 × 2 matrices. One can easily derive the spacing distribution p(x) for this case [Fr-84]

$$p(x) = \frac{x}{4v^2\sqrt{1-\alpha^2}} \exp\left(-\frac{x^2}{8v^2}\right) \operatorname{erf}\left[\left(\frac{1-\alpha^2}{8\alpha^2v^2}\right)^{1/2} x\right] \tag{24}$$

which shows that, as soon as α is different from zero, the small x behaviour is quadratic and not linear in x (the same thing will be true in the large-N asymptotic case as a function of λ).

We have studied in detail other fluctuation measures and the results indicate that for a large enough magnetic field (qBS ~ 3) the spectrum of the billiard exhibits GUE properties while for small enough values of the field it is close to GOE fluctuations. But the existence of a transition is not really well understood.

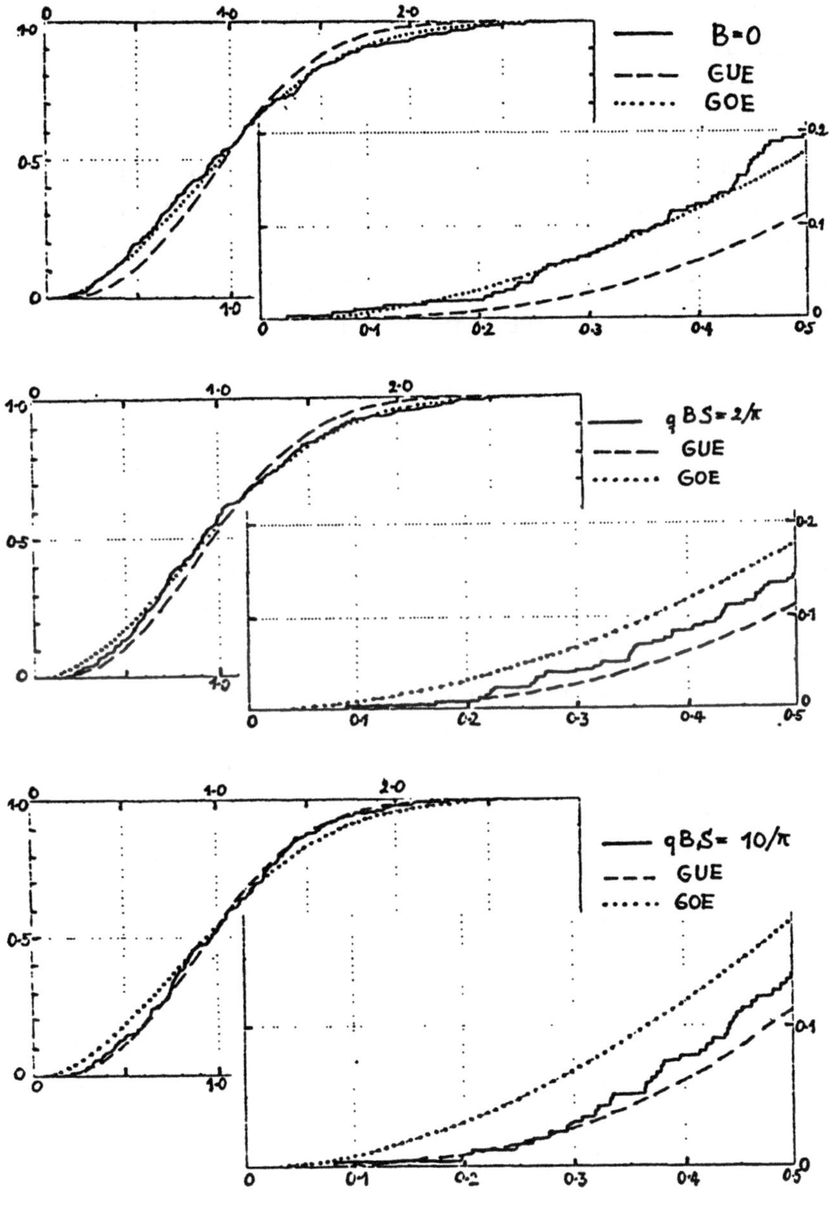

Fig.12 - Cumulative spacing distribution for qBS = 0, 2/π and 10/π . The behaviour between 0 and 0.5 is enlarged on the right (taken from[SGB-86]).

4. Some recent results

Hydrogen atom in a strong magnetic field

An extremely interesting system, for its fundamental 'simplicity' and for its astrophysical relevance as well, is the hydrogen atom in a strong uniform magnetic field (see, for instance, [Ga-85] and references therin). The conceptual importance of the problem follows from the fact that the two limiting cases of zero and infinite field strength, namely the Coulomb and the Landau (oscillator like) problems respectively, which have very different symmetries, are the only three dimensional problems which are exactly soluble, as well classically than quantum mechanically. The motion of the electron submitted to the action of the Coulomb and Lorentz forces is governed, in the symmetrical gauge with vector potential $\vec{A}=(1/2)(\vec{r}\times\vec{B})$ (the magnetic field is in the z-direction), by the Hamiltonian (in a.u.)

$$H = \frac{p^2}{2} - \frac{1}{r} + \frac{\gamma}{2} L_z + \frac{\gamma^2}{8}(x^2+y^2) \tag{25}$$

In (25) γ is the reduced magnetic field strength B/B_c, $B_c = 2.35 \times 10^5$ T. The two last terms in (25) are the paramagnetic interaction associated with the normal Zeeman effect which is trivial in the present context and which will be droped from now on, and the diamagnetic interaction. L_z (and parity) are the only constants of the motion besides the energy. The importance of the diamagnetic effects are characterized by the ratio of the diamagnetic to Coulomb energy, which is proportional to $\gamma n^3 \propto \gamma |E|^{-3/2}$, where n is the principal quantum number and E is the energy. To magnify diamagnetic effects one can : i) increase the magnetic field (values of γ up to 10^{-4} can be achieved under laboratory conditions) ii) work with Rydberg atoms (one can achieve typically values of $n \simeq 50$).

The phase space structure of this system (diamagnetic Kepler problem) has been recently studied in detail by Delande and Gay [DG-86]. By increasing the value of $\gamma/|E|^{3/2}$, which is the significant parameter, one evolves from a motion which is fully regular, then a connected chaotic region appears and keeps increasing until it occupies the whole surface of energy, for a critical value of the parameter. How is this behaviour reflected in the spectral fluctuations of the corresponding quantum system? One expects a transition from the Poisson to the GOE regime (as mentioned before, due to the particular spatial symmetry of the problem, although the system is not T-invariant one expects GOE-fluctuations). And this is indeed what is found by Delande and Gay when analyzing the computed spectrum. The results are illustrated in Figs. 13 and 14 . In Fig. 13 one can see the evolution of a stretch of the spectrum. The level repulsion or avoided crossings are

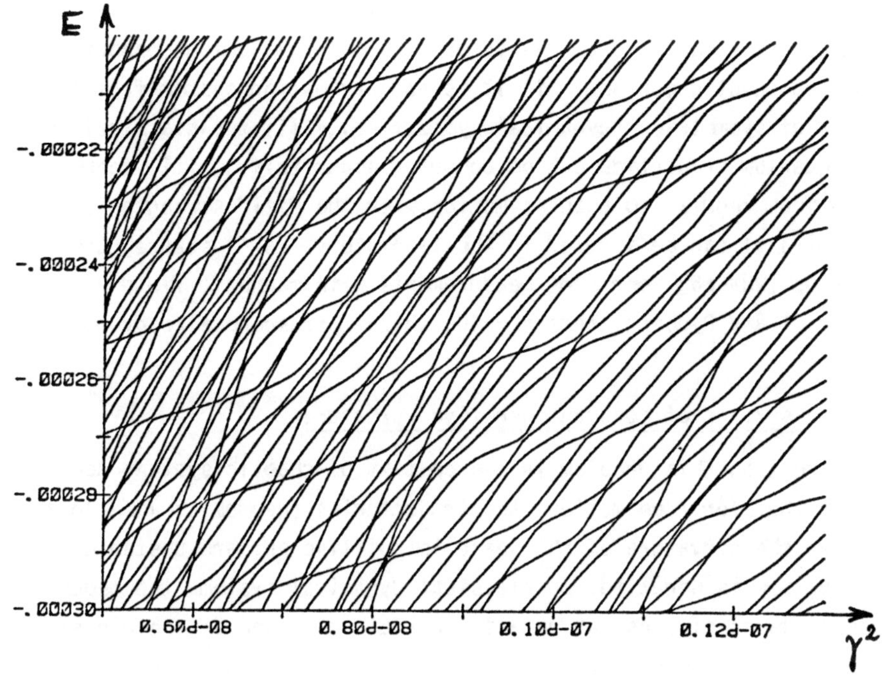

Fig.13 - Evolution of a segment of the spectrum as a function of the square of the reduced magnetic field strength γ, in the chaotic region (taken from [DG-86]).

clearly exhibited. And results for the spacing distribution and the spectral rigidity are reproduced in Fig.14 (similar results have been obtained by Wintzen and Friedrich [WF-86]).

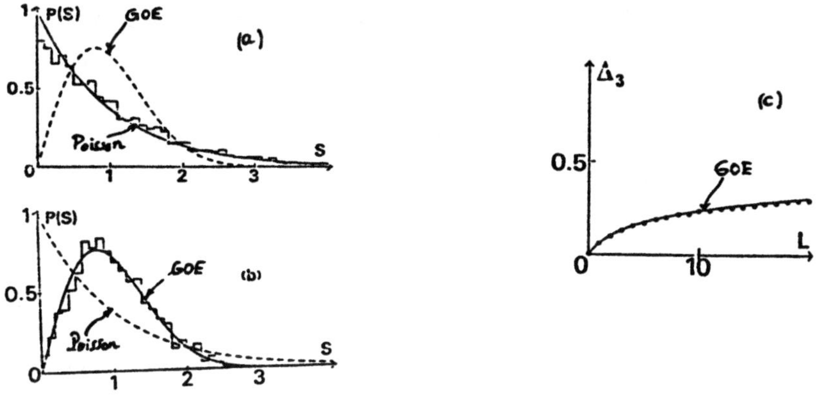

Fig.14 - (a) spacing distribution in the regular region ; (b) spacing distribution in the fully chaotic region ; (c) spectral rigidity Δ_3 in the fully chaotic region (taken from [DG-86]).

In summary, the Kepler diamagnetic problem offers a unique opportunity for studying chaotic motion : i) it is an interesting system from the point of view of symmetries (dynamical symmetries), ii) the classical as well as the quantum properties can be computed accurately ; iii) there is a transition from regular to chaotic motion when the field strength is increased ; iv) it can be studied in the laboratory. For instance, the Bielefeld group [WHW-86] performs measurements with magnetic field strengths $B \simeq 6T$ and Rydberg atoms $n \simeq 40$, which correspond already to the chaotic regime.

Billiards on a surface of constant negative curvature

The interest on this problem in the context of chaotic systems comes from the fact that, for free motion, the trajectories (geodesics) have local stability when the curvature of the space is positive and are unstable when the curvature is negative. In classical dynamical systems the geodesic flow on surfaces of constant negative curvature is one of the oldest problems and best understood examples of chaotic motion. It has provided the basis for the development of the theory of Anosòv systems and the proof of ergodicity of a gas with hard core interactions. In spite of the chaotic nature of the problem, there exists in this case a direct (and exact) connection between classical and quantum mechanics if the surface is compact (Selberg trace formula) : The trace of the propagator consists of a sum of terms characterizing the lengths of closed geodesics on the surface. One of the main issues is to see if, computing closed orbits and via a Selberg trace formula determine the quantum mechanical spectrum, constitutes in principle a general method which can be applied in many other cases, chaotic as well as integrable (Gutzwiller, see [Gu-84] for a recent review).

An example of surface of constant negative curvature is the Poincaré disk model of hyperbolic space : the open unit disk in the complex plane, endowed with the metric (in polar coordinates r, ϑ)

$$ds^2 = \frac{4}{(1-r^2)^2}\,[dr^2 + r^2\,d\vartheta^2] \qquad (26)$$

(see Balazs and Voros [BV-86] for a recent review on chaos on surfaces of negative curvature). In this model the geodesics are circular curves perpendicular to the unit circle and diameters of the circle. The Laplace-Beltrami operator

$$\Delta = \frac{1}{\sqrt{g}}\,\frac{\partial}{\partial q_k}\,\sqrt{g}\;g^{kj}\,\frac{\partial}{\partial q_j} \qquad (27)$$

where g_{kj} is the metric tensor, $g = \det(g_{kj})$, $g^{kj} = (g_{kj})^{-1}$ reads in this case

$$\Delta = \frac{(1-r^2)^2}{4} \; \frac{\partial^2}{\partial r^2} + \frac{1}{r}\,\frac{\partial}{\partial r} + \frac{1}{r^2}\,\frac{\partial^2}{\partial \theta^2} \tag{28}$$

We are now interested in the fluctuation properties of the spectrum of Δ on the triangle OLM (see Fig.15). Results are shown on Fig.16a. The spacing distribution is close to the Poisson case and presumably that by going higher in energy the agreement with the Poisson distribution would improve. This result came as a surprise and does not fit with the general view discussed until now. However, the triangle OLM is not generic because it has the very special property of tiling the space. If one modifies it slightly in such a way that the resulting triangle does not tile any more, one drastically changes the fluctuation properties as shown in Fig.16b.

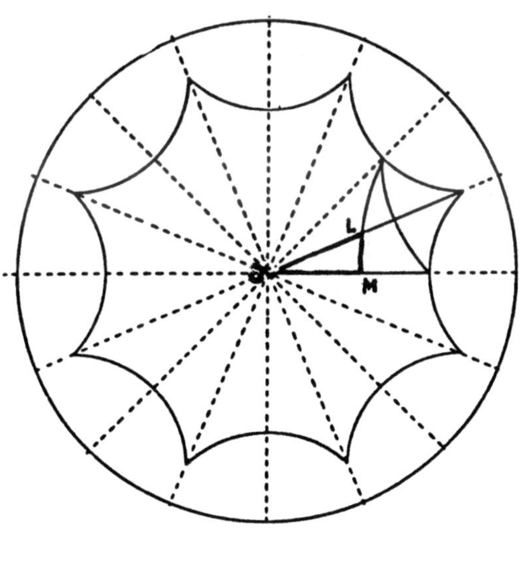

Fig.15 - From the initial problem, octogon with periodic boundary conditions, one is lead, by "désymmetrization", to consider the triangle OLM with Dirichlet boundary conditions. One has $M\hat{O}L = \pi/8$, $L\hat{M}O = \pi/2$, $O\hat{L}M = \pi/3$ and the surface is $S = \pi/24$ (see [BV-86], [Sc-86]).

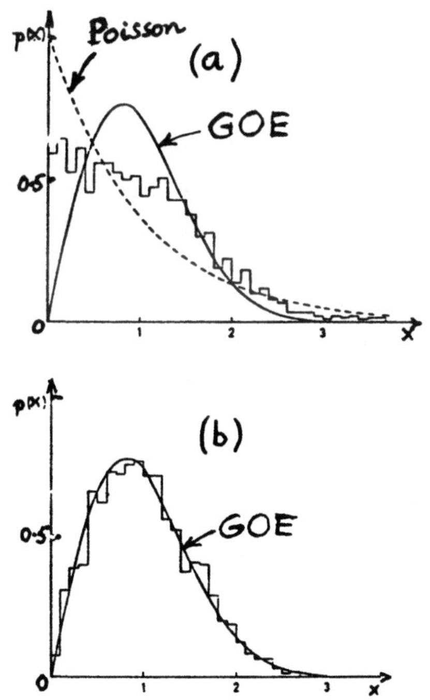

Fig.16 - Spacing distribution for the case (a) triangle OLM shown in Fig.15 ; (b) a triangle with angles $\pi/8$, $\pi/2$ and $(67/200)\pi$ with surface $S = \pi/28$. Each distribution contains 1500 spacings (taken from [Sc-86]).

Therefore it seems that for generic boundaries one has GOE-fluctuations, but that for boundaries like polygons that pave the space, even if the classical system is chaotic, one has Poisson-like fluctuations (the results in Fig.16 are shown as examples of a detailed analysis supporting these conclusions [Sc-86]). It is worth mentioning that Selberg trace formulae apply for tiling polygons but it is not known if they are still exact for cases like the one discussed here. With the possibility of performing extensive numerical results a better understanding of these problems is to be expected in the near future.

5. Remarks and conclusions

In the last two years, the gap between random matrix theories (RMT) and spectral fluctuations of classically chaotic quantum systems has been considerably filled. We know now that the fluctuations exhibited by the nuclear resonances, who gave rise to a large extent to the development of RMT, are not specific of nuclei. On the contrary, they are generic of fully chaotic (time reversal invariant) systems as demonstrated by the study of many low-dimensional model systems. It has also been possible to exhibit low-dimensional models showing : i) GUE-fluctuations for time reversal non-invariant systems, ii) transitions from Poisson –to GOE-fluctuations corresponding to the classical regular to chaotic regime transition. A transition, not fully understood, from GOE to GUE has also been found.

From a theoretical point of view, a semiclassical derivation of the spectral rigidity has been given and results agree with RMT predictions. But a more direct connection between RMT and properties of Hamiltonian systems, as well as a better understanding of 'Selberg trace formulae' are still missing. In this respect, the study of surfaces of negative curvature is promising.

From the experimental point of view, the search of the fluctuation properties of systems other than nuclei is challenging. For molecular systems new problems arise, namely the analysis of spectra with typically thousands of unresolved levels. Work in these directions is in progress.

The hydrogen atom in a strong uniform magnetic field (diamagnetic Kepler problem) deserves special mention : for its reachness concerning symmetries, for its experimental feasability, and because it can be theoretically treated in detail, it will probably become the chaotic system 'par excellence'.

At the risk of repetition, we insist with some warnings especially addressed to nuclear physicists. It is common belief that fluctuation patterns of the compound nucleus resonances are characteristic of the compound nucleus concept. This view is incorrect. One can hardly imagine a system as far from the compound nucleus as the hydrogen atom (in a magnetic field). And still, the level fluctuations are the same.

If the level fluctuations laws are universal, where does the specificity of the dynamics manifest itself ? If one restricts to spectral properties, it shows up, for instance, in the average behaviour of the spectrum. But it will mainly reveal itself in the structure of some selected eigenstates, a subject that we have not touched here, a subject that in nuclear physics is called theory of collective motion.

Acknowledgments

We are indebted to D. Delande and J.C. Gay for discussions and for permission to present results prior to publication.

References

[Be-85] M.V. Berry, Proc. Roc. Sco. Lond. **A400** (1985) 229

[BFF-81] T.A. Brody, J. Flores, J.B. French, P.A. Mello, A. Pandey and S.S.M. Wong, Rev. Mod. Phys. **53** (1981) 385

[BG-84] O. Bohigas and M.J. Giannoni, in 'Mathematical and Computational Methods in Nuclear Physics', J.S. Dehesa et al. (eds.), Lecture Notes in Physics **209**, Springer Verlag 1984, p.1.

[BGS-84a] O. Bohigas, M.J. Giannoni and C. Schmit, Phys. Rev. Lett. **52** (1984) 1.

[BGS-84b] O. Bohigas, M.J. Giannoni and C. Schmit, J. Physique Lett. **45** (1984) L-1015

[BHP-83] O. Bohigas, R.U. Haq and A. Pandey, in 'Nuclear Data for Science and Technology', K.H. Böckhoff (ed.), Reidel, Dordrecht, 1983, p.809

[BHP-85] O. Bohigas, R.U. Haq and A. Pandey, Phys. Rev. Lett. **54** (1985) 1645

[BR-84] M.V. Berry and M. Robnik, J. Phys. **A17** (1984) 2413

[BR-85] M.V. Berry and M. Robnik, J. Phys. **A19** (1985) 649

[BV-86] N.L. Balazs and A. Voros, preprint Saclay 1986

[Ca-85] G. Casati (ed.), 'Chaotic Behavior in Quantum Systems', Plenum, 1985

[CG-83] H.S. Camarda and P.D. Georgopulos, Phys. Rev. Lett. **50** (1983) 492

[CG-86] E. Caurier and B. Grammaticos, to appear in Europhys. Lett.

[DG-86] D. Delande and J.C. Gay, preprint 1986

[Fr-84] J.B. French, Lecture Notes, Rochester 1984, unpublished

[Ga-85] J.C. Gay in 'Photophysics and Photochemistry in the Vacuum Ultraviolet', S.P. Mc Glynn et al. (eds.), Reidel, 1985, p.631

[Gu-84] M. Gutzwiller, to appear in the Proceedings of the American Mathematical Society, '1984 Summer Research Conference on the Selberg Trace Formula'

[HKC-84] E. Haller, H. Köppel and L.S. Cederbaum, Phys. Rev. Lett. **52** (1984) 1665

[HMY-81] F.T. Hioe, E.W. Montroll and M. Yamawaki, in'Perspectives in Statistical Physics,' H.J. Raveché (ed.), North-Holland 1981, p.295

[HPB-82] R.U. Haq, A. Pandey and O. Bohigas, Phys. Rev. Lett. **48** (1982) 1086

[IY-85] T. Ishikawa and T. Yukawa, Phys. Rev. Lett. **54** (1985) 1617

[LLJ-86] L. Leviandier, M. Lombardi, R. Jost and J.P. Pique, Phys. Rev. Lett. **A56** (1986) 2449

[Me-67] M.L. Mehta, 'Random Matrices and the Statistical Theory of Energy Levels', Academic Press, 1967

[MHK-84] H.D. Meyer, E. Haller, H. Köppel and L.S. Cederbaum, J. Phys. **A17** (1984) L831

[MSP-84] S. Mukamel, J. Sue and A. Pandey, Chem. Phys. Lett. **105** (1984) 134

[Pa-79] A. Pandey, Ann. Phys. **119** (1979) 170

[PM-83] A. Pandey and M.L. Mehta, Comm. Math. Phys. **87** (1983) 449

[Po-65] C.E. Porter (ed.),'Statistical theories of spectra : fluctuations,' Academic Press, 1965

[RB-85] M. Robnik and M.B. Berry, J. Phys. **A18** (1985) 1361

[RB-86] M. Robnik and M.V. Berry, J. Phys. **A19** (1986) 669

[Ro-86] M. Robnik in 'Nonlinear Phenomena and Chaos', S. Sarkar (ed.), Adam Hilger, Bristol 1986

[Sc-86] C. Schmit, to be published

[SGB-86] C. Schmit, M.J. Giannoni and O. Bohigas, in 'Phase Space Approach to Nuclear Dynamics', M. Di Toro et al. (eds.), World Scientific, 1986 ; and to be published

[SV-85a] T.H. Seligman and J.J.M. Verbaarschot, J. Phys. **A18** (1985) 2227

[SV-85b] T.H. Seligman and J.J.M. Verbaarschot, Phys. Lett. **108A** (1985) 183

[SVZ-84] T.H. Seligman, J.J.M. Verbaarschot and M.R. Zirnbauer, Phys. Rev. Lett. **53** (1984) 215

[WF-86] D. Wintgen and H. Friedrich, preprint 1986

[WHW-86] D. Wintgen, A. Holle, G. Wiebusch, J. Main, H. Friedrich and K.H. Welge, preprint 1986

[ZMK-86] Th. Zimmermann, H.D. Meyer, H. Köppel and L.S. Cederbaum, Phys. Rev. **A33** (1986) 4334.

CHAOTIC MOTION AND STATISTICAL NUCLEAR THEORY

Hans A. Weidenmüller*
Institut de Physique Nucléaire, Université Paris Sud, (91) Orsay, France

1. Aim.

In this talk, I discuss average properties of finite quantum systems versus fluctuation properties of such systems. Average properties depend on the underlying dynamics as embodied in the Hamiltonian and are quite insensitive to whether these dynamics are regular, or stochastic. In contrast, the fluctuation properties do depend on this distinction and are very different for regular, and for chaotic motion. In the case of chaotic motion, the fluctuation properties are generic, and quite independent of the specific dynamics of the system. They can then be simulated and evaluated in a universal fashion. This is demonstrated for one particular example, the S-matrix for compound-nucleus scattering. Some remarks concerning entropy production in systems with chaotic dynamics conclude the presentation.

2. Average versus Fluctuation Properties in Finite Quantum Systems.

Average properties are defined as mean values over a large number of levels carrying the same conserved quantum numbers. They may be subject to secular variations : Their values depend on where in the spectrum the average has been taken. Fluctuations are the local variations about the averages. These concepts are applicable to physical systems (atoms, molecules, nuclei) if the mean level spacing d is small compared to the energy interval ΔE over which the average properties change significantly.

I consider three examples for averages and fluctuations. All three are taken from the context of nuclear physics.

(i) The mean level spacing d, and the level spacing fluctuations about this mean value.

(ii) The average value (over energy) $\langle S_{ab} \rangle$ of the nuclear scattering matrix, and the fluctuations about this value.

(iii) The strength function (or the mean decay width Γ) of a group of levels, and the fluctuations about this value.

It is characteristic of average values that they can usually be well approximated in terms of simple models. In the cases given above, the average level density can be well approximated in terms of the Weyl formula (which expresses the total number of levels with energy less than E in terms of the volume of phase space with energy less than E of the corresponding classical system); the average S-matrix is given in terms of an optical-model or coupled-channels calculation (the energy-average of S_{ab} describes fast processes involving only few degrees of freedom); and the strength function Γ/d can be extracted from the knowledge of the average of S_{ab}. These statements illustrate some of the claims regarding average properties made in section 1. There is another feature of central importance for this paper : For the examples given (and many others), average properties are defined in terms of energy-averages over the Green function of the system (the "one-point function"). For the level spacing, this follows from $d^{-1} \propto \mathrm{Im} \langle \mathrm{trace}\, (E+i\delta-H)^{-1} \rangle$ where E is the energy, H the Hamiltonian, and δ positive infinitesimal. The bracket denotes an average over energy. For the S-matrix (and, thereby, also for the strength function) this fact is demonstrated in section 3 below. Information on the fluctuation properties is contained in the higher moments of the same quantities. For the level spacings, we consider, for instance, the "two-point function" $\langle \mathrm{trace}\, [(E_1+i\delta-H)^{-1}]\ \mathrm{trace}\, [(E_2-i\delta-H)^{-1}] \rangle$. Here, the average is defined by keeping the distance E_2-E_1 fixed. The sign of the term $i\delta$ must be different in the two traces as otherwise we do not gain new information which would not be contained already in the one-point function. Higher moments must also be considered if the complete distribution function for level spacings is required. For the S-matrix it is indicated below that $\langle |S_{ab}(E)|^2 \rangle$ is needed to calculate average cross sections, the fourth moment of S_{ab} is needed to calculate the cross section autocorrelation function, etc...

In summary : Average properties relate to the one-point function and can be calculated within the framework of simple approximations. Fluctuation properties relate to the n-point functions with $n \geq 2$.

The point of central interest to this talk, extensively discussed by the two previous speakers, is this : In chaotic quantum systems, the fluctuation properties can be simulated by the random-matrix ensemble introduced by Wigner, the Gaussian Orthogonal Ensemble (GOE). It is defined as follows. We consider an ensemble of Hamiltonian matrices $H_{\mu\nu}$ of dimension $N \gg 1$. Time-reversal invariance allows us to choose the matrices in the ensemble real and symmetric, $H_{\mu\nu} = H_{\nu\mu} = H^*_{\mu\nu}$. The independent elements $H_{\mu\nu}$ with $\mu \geq \nu$ are uncorrelated random variables with a Gaussian probability distribution, zero mean values, and second moments given by

$$\overline{H_{\mu\nu}H_{\mu'\nu'}} = \frac{\lambda^2}{N}\ (\delta_{\mu\mu'}\delta_{\nu\nu'} + \delta_{\mu\nu'}\delta_{\nu\mu'}) \quad . \tag{1}$$

The bar denotes the ensemble average. The parameter λ with dimension energy determines the average level spacing d. A word of caution : The average properties of the GOE do not reproduce average properties of real physical systems, but the fluctuation properties do model the fluctuation properties of chaotic quantum systems. This must be borne in mind when one models properties of actual systems in terms of the GOE. To establish the claim just made, I refer to the numerical evi-

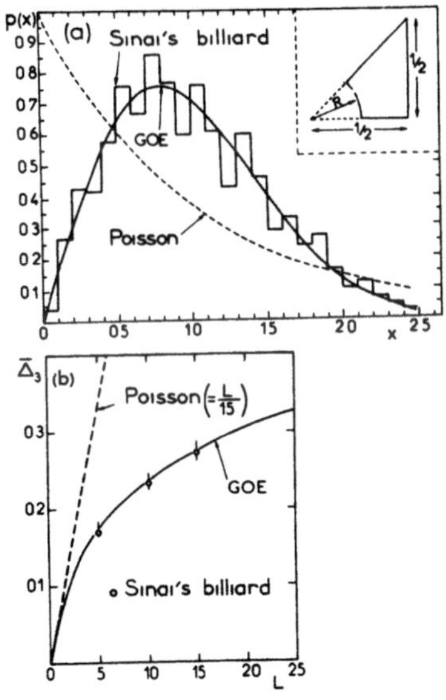

Fig. 1 Nearest-neighbour spacing distribution p(x) (upper part) and average Δ_3 statistic (lower part) for the quantized Sinai billiard. The solid curves result from the GOE. (Taken from ref. [1]).

dence and the analytical arguments presented by the two previous speakers, and confine myself to showing three examples. Figure 1, taken from reference [1], shows in the upper part as histogram the nearest-neighbour spacing distribution p(x), i.e. the distribution of spacings of neighbouring eigenvalues, of the quantized Sinai billiard, where x = s/d is the spacing s in units of d. The solid line is the GOE prediction for $N \to \infty$. In classical mechanics, the Sinai billiard is the prototype of a chaotic system. We see that the nearest-neighbour spacing distribution of its quantum version is perfectly reproduced by the GOE. The same is true of the average Δ_3 statistic, a measure of the stiffness of the spectrum extensively discussed by M. Berry in this conference, shown in the lower part of figure 1. Figure 2, taken from ref. [2], shows the same kind of plots, sampled from a totally different system : Sequences of nuclear energy levels of the same spin and parity yield the histogram (upper part) and the dots (lower part); the dashed line gives the errors expected because of the finite size (1726 spacings) of the sample. The data suggest that at least several MeV

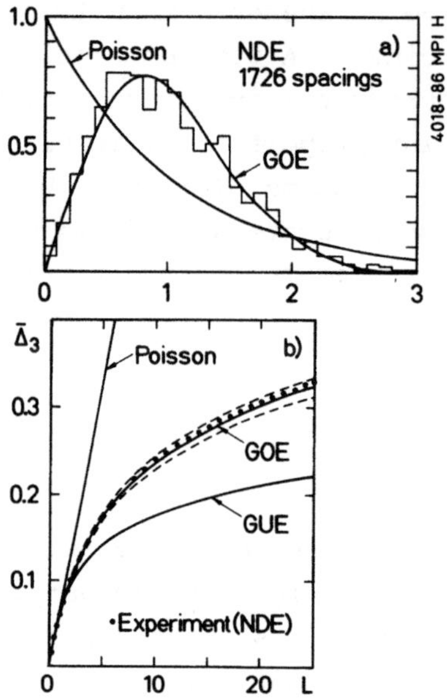

Fig. 2 Same as figure 1, except that the data are taken from level spacings measured in atomic nuclei. (Taken from ref. [2]).

above the ground state, nuclei behave like chaotic quantum systems : The fluctuation properties of energy levels coincide with those of the quantized Sinai billiard, and can be simulated by the GOE. The nuclear data also yield information about the fluctuations of partial widths. Figure 3, taken from ref. [3], shows in the histogram the data for ^{167}Er, plotted for a suitably chosen measure of the probability distribution versus $x = \Gamma_n^o/\overline{\Gamma_n^o}$ where Γ_n^o is the reduced neutron width. The solid line is the GOE prediction (Porter-Thomas distribution), again for $N \to \infty$. Unfortunately, the statistics on wave-function distributions for quantized classically chaotic systems is not good enough to be shown here, although no disagreement with GOE predictions has been recorded.

Evidence similar to that in figure 1 (figure 2) is also available from other quantized chaotic systems (from atoms and molecules); in either case, the statistics is not as impressive as in figures 1 and 2. For the theoretical treatment and understanding of the quantum-mechanical many-body problem, at least at sufficiently high excitation energies, the results presented above lead to the following conjecture [1].

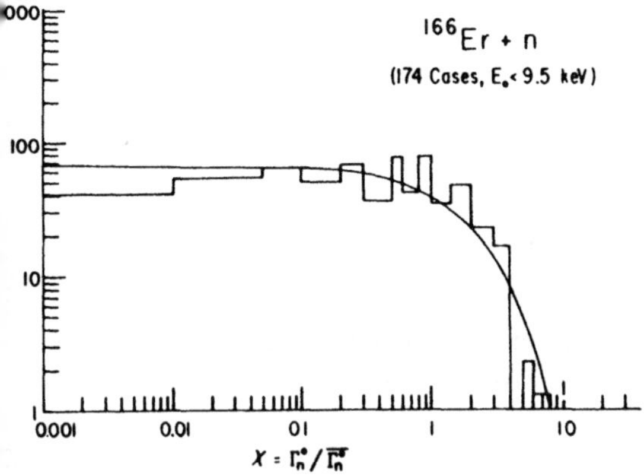

Fig. 3 Distribution of reduced neutron widths Γ_n^o (in units of the mean value $\overline{\Gamma_n^o}$) for the compound nucleus ^{167}Er. Solid line : GOE prediction. (Taken from ref. [3]).

The many-body problem is generically chaotic. Average properties, given by the underlying dynamics, can be calculated within the usual approximation schemes (mean-field theories and their extensions). Fluctuation properties are generic for chaotic quantum systems. They are independent of the specific dynamical properties of the system under consideration. In that sense, they carry no information content. They can be simulated by the GOE.

In the simplistic fashion in which it was just formulated, the conjecture certainly does not hold universally : Over sufficiently long level sequences (never attained in practice in nuclei or in atoms and molecules) the GOE fluctuation properties are not expected to coincide with those of chaotic quantum systems. Another qualification is necessary in the nuclear context because of the presence of slowly relaxing (collective) degrees of freedom even at high excitation energy. Disregarding such worries, I address in the remainder of this talk another question : Given the conjecture, can we implement the scheme ? More precisely, given the average values, can we calculate the fluctuation properties of physical observables in terms of a GOE simulation ? In the case of the nuclear scattering matrix and of compound-nucleus scattering, this question will be answered in the affirmative.

3. A model for the statistical nuclear S-matrix.

As early as 1936, N. Bohr [4] recognized stochasticity as an essential element of nuclear reactions. Data on neutron capture reactions led him to introduce the concept of the compound nucleus, modelled in classical terms in figure 4 for a neutron-induced reaction. The close similarity between this picture and Boltzmann's hard-sphere gas on the one hand, and the Sinai billiard on the other, bears witness to the depth of Bohr's insight.

Fig. 4 Bohr's model of the compound nucleus. (Taken from the second of refs. [4]).

Typical modern data on compound-nucleus scattering are shown in figures 5 and 6. Figure 5, taken from ref. [5], shows neutron scattering data on the target nucleus ^{232}Th. The isolated resonances with average spacing $d \approx 10$ eV and average width $\Gamma \approx 1$ eV are clearly visible. Their fluctuation properties are described by the GOE. While at and near neu-

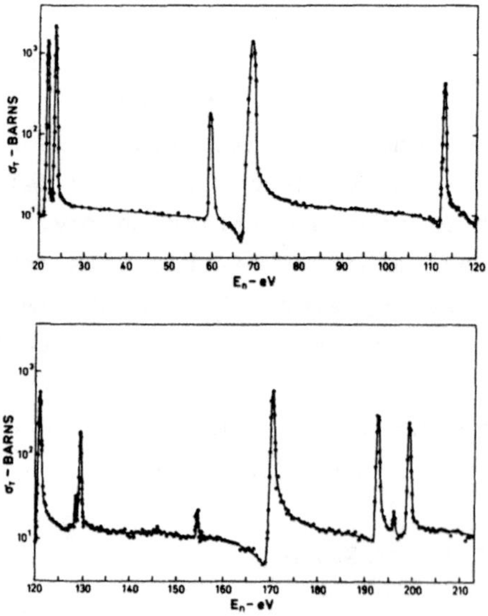

Fig. 5 Elastic cross section for the reaction n + ^{232}Th versus neutron energy in eV. (Taken from ref. [5]).

tron threshold we have $\Gamma << d$ for most nuclei, the situation changes drastically with increasing energy (Γ and d^{-1} both grow roughly exponentially), so that $\Gamma >> d$ a few MeV above neutron threshold. A cross section typical for this domain is shown in figure 6 taken

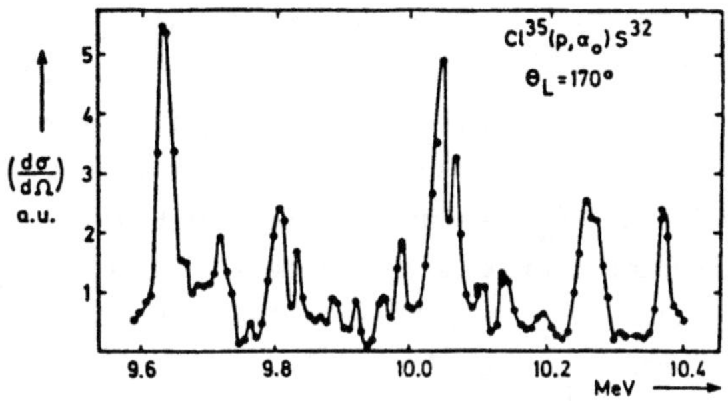

Fig. 6 Fluctuations of the compound-nucleus cross section versus energy of the incident particle. (Taken from ref. [6]).

from ref. [6]. The strong dependence on energy is caused by the fact that the S-matrix contains a sum over many overlapping resonances with stochastic properties. While at one energy, the sum happens to consist largely of coherent contributions, at another, the contributions accidentally cancel : The cross section is a random process.

For nuclear reaction theory as well as for nuclear physics applications in astrophysics and reactor technology, a knowledge of average cross sections and of cross-section fluctuations (as measured, for instance, by the variances of the cross sections) is required. In the case of figure 5, the correlation width Γ of the cross-section fluctuations yields information about the average lifetime $\hbar/\Gamma$ of the compound nucleus and is likewise of interest. By modelling the fluctuations in terms of a GOE, we expect to be able to predict these quantities in terms of the average S-matrix. In contradistinction, we expect that it is impossible for <u>any</u> theory - dynamical or statistical - to calculate correctly the values of the cross sections shown in figures 5 and 6 for any <u>fixed</u> energy, because of the chaoticity of the dynamics.

I define the GOE model for the nuclear S-matrix for the simplest case: Absence of direct reactions (so that S is diagonal, $\overline{S_{ab}} = \delta_{ab}\,\overline{S_{aa}}$), and absence of precompound reactions (so that only one class of compound-nucleus resonances need be considered). As shown in refs. [7,8], both these simplifications can easily be removed. With a,b,... denoting the Λ open channels, and $\mu,\nu,\ldots$ denoting the N compound nuclear resonances, the element $S_{ab}(E)$ of the nuclear S-matrix pertaining to states of fixed spin and parity is written as

$$S_{ab} = \delta_{ab} - 2i\pi \sum_{\mu,\nu=1}^{N} W_{\mu a}(D^{-1})_{\mu\nu}W_{\nu b} \quad ; \tag{2}$$

$$D_{\mu\nu} = E\delta_{\mu\nu} - H_{\mu\nu} + i\pi \sum_{a=1}^{\Lambda} W_{\mu a}\, W_{\nu a} \quad . \tag{3}$$

The real, non-stochastic matrix elements $W_{\mu a}$, independent of energy E, describe the coupling of levels μ and channels a. For N = 1, eqs. (2,3) yield simply a single-level Breit-Wigner formula; for $N \neq 1$ they represent the extension of this formula to N resonances which may or may not overlap depending on the strength of the $W^2_{\mu a}$ versus d. The S-matrix of eq. (2) becomes an ensemble of random matrices if we take for $H_{\mu\nu}$ the GOE defined in Eq. (1). In the limit $W_{\mu\alpha} \to 0$, this model for S_{ab} reduces to the GOE model for N bound states used to produce, in the limit $N \to \infty$, the solid curves in figures 1, 2, and 3. Put differently, the model of eqs. (2,3) with $H_{\mu\nu}$ a GOE is a straightforward extension of the GOE model for bound states to the case of compound-nucleus reactions. It is expected to apply for any value of the ratio Γ/d. The form of S_{ab} in eqs. (2,3) is obtained in any general theory of nuclear reactions; it guarantees unitarity.

The parameters $W_{\mu a}$ and λ of the model (with λ occurring in eq. (1)) must be deduced from the knowledge of average properties like $\overline{S_{ab}} = \delta_{ab}\, \overline{S_{aa}}$. (In the spirit of section 2, these quantities are supposed to be known). As mentioned above, λ is given in terms of the average level spacing d. As for the $W_{\mu a}$, the invariance of eqs. (1) under orthogonal transformations of the basis implies that the moments of S_{ab} depend only on the orthogonal invariants $\sum_\mu W_{\mu a}W_{\mu b}$. The diagonality of $\overline{S_{ab}}$ requires $\sum_\mu W_{\mu a}W_{\mu b} = \delta_{ab} \sum_\mu W^2_{\mu a}$. The number of parameters $\sum_\mu W^2_{\mu a}$ is equal to the number of diagonal S-matrix elements $\overline{S_{aa}}$. We see that the model of eqs. (2,3) is completely determined in terms of the average properties $\overline{S_{ab}}$ and d.

The model of eqs. (2,3) is largely equivalent to the model of compound-nucleus reactions studied by Moldauer and others over the last 25 years [9]. It differs in one essential detail : It emphasises and utilises the orthogonal invariance of the GOE, in avoiding the diagonalisation of $H_{\mu\nu}$ and the introduction of the associated eigenvalues and eigenfunctions. Keeping an orthogonally invariant formulation is important for the exact solution, i.e. for the explicit calculation of the second moment of $S_{ab}(E)$.

4. Solution of the model.

The space available for this presentation prevents me from giving even an outline of the method of solution of the problem formulated in eqs. (2,3). In lieu of this, I discuss why the problem is difficult, and how it relates to other problems in theoretical physics, thereby attempting to make it plausible that it can be solved.

The difficulty is this. The random variables $H_{\mu\nu}$ occur in the denominator of S_{ab}. Averaging a product of several elements of S is not possible in a straightforward fashion, although the weight factors are simple Gaussians. [For the one-point function, this problem can be circumvented. All poles of S lie on the same side of the real energy axis. The calculation of the energy average is completed by contour integration, yielding $\langle S_{ab}(E)\rangle = S_{ab}(E+iI)$ where I is the width of the averaging interval. This method fails for higher moments which have poles on either side of the real axis]. The curves labelled GOE in figures 1 and 2 were obtained analytically by Dyson and Mehta [10] in the limit $N \to \infty$. This method of calculation is tailored specially to spectral fluctuation properties, and apparently cannot be extended to handle problems such as the one formulated in eqs. (2,3). The moments method of Mon and French [11], extremely useful for the calculation of the one-point function of matrix ensembles different from the GOE, is of very limited use for the calculation of the n-point functions with $n \geq 2$. Inasmuch as the problem formulated in eqs. (2,3) is a generic one - embodying the hopes and difficulties of describing fluctuation properties of any many-body system - we look for a general method capable of handling GOE fluctuations for a wide class of problems. Here it is.

Random Hamiltonians, proposed by Wigner in the context of nuclear physics, are widely used in the statistical mechanics of disordered systems. I recall such diverse problems as spin glasses, the integer quantum Hall effect, and Anderson localisation. In all these cases, random interactions are used to model systems which are too complex to allow for any other treatment. The randomness is supposed to model a very complex interaction, a disordered geometrical arrangement, or a complex chemical composition. What is the connection between such models, and the random-matrix problem formulated in eqs. (1) to (3) ? To be specific, I consider the case of the Anderson model depicted in figure 7. We see a regular lattice. The lattice sites are supposed to be occupied by atoms of a very different chemical composition. The motion of a single electron on the lattice is described in the tight-binding approximation : The electron can hop (arrows) from one site to the next neighbours. To simulate disorder in the chemical composition, the single particle energy at each site as well as the hopping matrix elements are not considered as fixed, but are taken as random variables drawn from some suitable probability distribution. An extension of this model

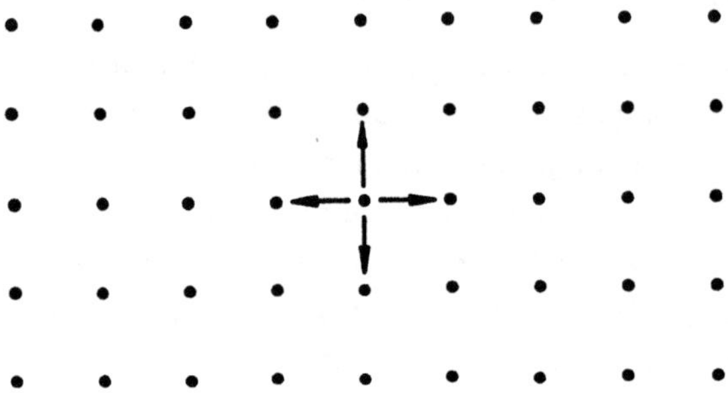

Fig. 7 Schematic version of the Anderson localisation problem in two dimensions as explained in the text.

consists in considering N orbitals at each site, all of them with random properties as given by a GOE Hamiltonian. Restricting the model of figure 7 to a single site, we obtain a GOE problem, similar to the one formulated in eqs. (1) to (3).

We note that the problems in disordered solids are more general (more complex) than the nuclear physics problem : They involve geometrical and topological aspects which are absent in eqs. (1) to (3). More general problems usually require more general methods for their solution. It is perhaps not surprising that these methods, in modified form applied to the (simpler) nuclear physics problem, turn out to be very powerful, allowing for an exact calculation of the two-point function of the S-matrix [12].

The problem is handled as follows. Any element of the S-matrix, and the product of any two elements, are written as derivatives with respect to some suitably chosen auxiliary variables J at J = 0 of a generating function. This function gives a functional representation of the problem, and is normalized to unity at J = 0 by virtue of a judicious use of both commuting and anticommuting integration variables. Taking the ensemble average of the generating function, one can map the problem onto a non-linear sigma model of the type studied in field theory, especially as "Skyrmion model" in the context of the QCD description of the nucleon. The method is quite general and applies similarly to localisation theory, to the integer quantum Hall effect, etc. In the special case of the problem formulated in eqs. (1) to (3), the non-linear sigma model is sufficiently simple that it can be solved exactly.

The availability of an exact solution in the context of this general method is of interest beyond the realm of nuclear physics : The exact solution can be used to test other approaches, and this has been done [13].

In the realm of nuclear physics, faith in the exact solution of the compound-nucleus problem is supported by the following observations. The second moment of $S_{ab}(E)$ is, in the framework of eqs. (1) to (3), found to be universal over the entire GOE spectrum [12]. The exact solution agrees [14] with series expansions obtained earlier for $\Gamma \ll d$ and for $\Gamma \gg d$; in particular, it yields for $\Gamma \gg d$ the Hauser-Feshbach formula as the leading term in an asymptotic expansion in powers of d/Γ. For intermediate values of Γ/d, table 1 (taken from ref. [14]) shows that the exact solution agrees with earlier numerical results based on Monte-Carlo-simulations of the model of eqs. (1) to (3). It also agrees with numerical results obtained in the framework of a seemingly very different approach invoking maximum entropy as presented by P.A. Mello at this Conference.

5.7(1)E-2	5.66E-2	8.7(1)E-2	8.55E-2
6.9(2)E-3	6.93E-3	8.5(1)E-2	8.55E-2
2.1(1)E-3	2.19E-3	8.4(1)E-2	8.56E-2
3.42(7)E-2	3.38E-2	8.4(1)E-2	8.57E-2
2.8(1)E-3	2.83E-3	8.5(1)E-2	8.52E-2
9.3(5)E-4	8.98E-4	8.6(1)E-2	8.53E-2
1.31(4)E-2	1.35E-2	8.4(1)E-2	8.54E-2
1.2(1)E-4	1.20E-4	8.7(1)E-2	8.53E-2
1.9(1)E-3	1.74E-3	8.8(1)E-2	8.55E-2
5.8(3)E-4	5.53E-4	8.6(1)E-2	8.56E-2

Table 1 This is part of a table contained in ref. [14], showing a comparison between Monte-Carlo calculations (first and third column) and the corresponding exact result (second and fourth column) for $\overline{|S_{ab}|^2}$ and for some few-channel problems. The figures in brackets estimate the statistical error of the last digit given in the Monte-Carlo calculations.

The result has meanwhile found several applications. I only mention the calculation of the time-delay function needed for the interpretation of experiments using the crystal blocking technique [15], and a study of the $^{26}Al(p,n)^{26}Mg$ compound cross section relevant for nuclear astrophysics [16]. The method is capable of being expanded. It has been used to obtain novel insights into precompound cross sections [8], and into the study of time-reversal invariance in nuclear reactions [17].

5. Comments on Entropy Production.

Entropy production has been much discussed recently, both within and outside the domain of nuclear reaction theory [18]. In the sequel, some comments on this problem are made from the point of view of ensemble averaging.

I consider the density matrix in channel space ρ_{ab}, and define the entropy $\mathcal{S}$ by

$$\mathcal{S} = - \text{trace } (\rho \ln \rho). \tag{4}$$

With the initial density (prior to any reaction) given by $\rho_{ab}^{(i)} = \delta_{ab}\delta_{a_0 a}$ where a_0 is the incident channel we have $\mathcal{S}^{(i)} = 0$ since the state is fully specified. The density matrix $\rho_{ab}^{(f)}$ after scattering is obtained by transforming $\rho^{(i)}$ with the unitary S-matrix,

$$\rho_{ab}^{(f)} = (S\, \rho^{(i)} S^{+})_{ab} \quad . \tag{5}$$

Because of the unitarity of S, we have $\mathcal{S}^{(f)} = -\text{trace } (\rho^{(f)} \ln \rho^{(f)}) = \mathcal{S}^{(i)} = 0$: There is no entropy produced in the reaction [18].

The result is very different, however, if we consider instead of $\rho^{(f)}$ the ensemble average $\overline{\rho^{(f)}}$ for the stochastic problem formulated in eqs. (1) to (3). The difference arises because $\overline{|S_{ab}|^2}$ cannot be written as the product of an element of a unitary matrix with its complex conjugate, although flux conservation as expressed by $\sum_b \overline{|S_{ab}|^2} = 1$ still holds. In the special case where many channels are open ($\Lambda >> 1$), and where the absorption in each channel is big ($\overline{S_{aa}} \to 0$ for all a), we can use the Hauser-Feshbach formula

$$\overline{|S_{ab}|^2} = (1 + \delta_{ab}) \frac{T_a T_b}{\sum_c T_c} \tag{6}$$

with $T_a = 1 - \overline{|S_{aa}|^2} \simeq 1$. We then find, omitting terms of order Λ^{-1}, that

$$\mathcal{S}^{(f)} = - \text{trace } (\overline{\rho^{(f)}} \ \ln \overline{\rho^{(f)}}\ ) \simeq \ln \Lambda \quad . \tag{7}$$

This shows that in this case the entropy (defined in terms of $\overline{\rho^{(f)}}$) attains its maximum possible value. If we use the same definition (7) involving $\overline{\rho^{(f)}} = <\rho^{(f)}>$ for a very different physical situation (direct reactions without compound nucleus formation), the S-matrix is a smooth function of energy, therefore $<\rho^{(f)}> \simeq \rho^{(f)}$, and the entropy increase is negligible.

This simple discussion suggests that a physically useful definition of entropy in a quantum system with stochasticity should involve the following two features. (i) A coarse graining in the form of an average over energy and over an interval which is large compared to the mean level spacing. (ii) The object one considers should involve the two-point function, and not the one-point function. It was emphasised throughout this talk that the difference between regular and chaotic motion is visible only in the two-point function. If entropy production is to be a measure of quantum chaoticity, the entropy definition must surely involve this function.

6. Conclusions

In this talk, I have argued that average properties of finite quantum systems, related to the one-point function, carry dynamical information and can often reliably be calculated in the framework of semiclassical or other approximations. Fluctuation properties, on the other hand, depend on whether the system is regular or chaotic. In the generic case of chaotic motion, fluctuations are universal, carry no information content, and can be simulated by the GOE. This opens a new approach to the quantal many-body problem. I have discussed a general method of handling GOE fluctuations in the specific case of compound-nucleus scattering, and have related it to methods used in the theory of disordered systems. Much of the material presented is not specific to nuclei and equally applicable to small metallic particles, and large atoms and molecules. In that sense, I have addressed a generic problem in the statistical mechanics of finite quantum systems.

References

[1] O. Bohigas, M.J. Giannoni, and C. Schmit, Phys. Rev. Lett. 52 (1984) 1

[2] R.U. Haq, A. Pandey, and O. Bohigas, Phys. Rev. Lett. 48 (1982) 1086

[3] T.A. Brody, J. Flores, J. Bruce French, P.A. Mello, A. Pandey, S.S.M. Wong, Rev. Mod. Phys. 53 (1981) 385

[4] N.Bohr, Nature 137 (1936) 344; see also Nature 137 (1936) 351

[5] Aa. Bohr and B.R. Mottelson, Nuclear Structure, W.A. Benjamin, New York-Amsterdam 1969, Vol. 1 p. 178

[6] T.Ericson and T.Mayer-Kuckuk, Ann. Rev. Nucl. Sci 16 (1966) 183

[7] H. Nishioka and H.A. Weidenmüller, Phys. Lett. 157 B (1985) 101

[8] H. Nishioka, J.J.M. Verbaarschot, H.A. Weidenmüller and S. Yoshida, submitted to Ann. Phys. (N.Y.).

[9] C. Mahaux and H.A. Weidenmüller, Ann. Rev. Nucl. Part. Sci. 29 (1979) 1

[10] F.J. Dyson and M.L. Mehta, J. Math. Phys. 4 (1963) 701

[11] K.K. Mon and J.B. French, Ann. Phys. (N.Y.) 95 (1975) 90

[12] J.J.M. Verbaarschot, H.A. Weidenmüller and M.R. Zirnbauer, Phys. Rep. 129 (1985) 367

[13] J.J.M. Verbaarschot and M.R. Zirnbauer, J. Phys. A 17 (1985) 1093

[14] J.J.M. Verbaarschot, Ann. Phys. (N.Y.), in press

[15] J.J.M. Verbaarschot and S. Yoshida, Z.Phys. A 322 (1985) 621

[16] J. Thomas, M.R. Zirnbauer and K. Langanke, Phys. Rev. C (in press)

[17] D. Boosé, H.L. Harney, and H.A. Weidenmüller, submitted to Phys. Rev. Lett.

[18] R. Balian, M. Veneroni, and N. Balasz, Europhysics Letters 1 (1986) 1, and references therein.

TESTS OF SPECTRUM AND STRENGTH FLUCTUATIONS WITH PROTON RESONANCES

G.E. Mitchell
North Carolina State University, Raleigh, North Carolina, USA and
Triangle Universities Nuclear Laboratory, Durham, North Carolina, USA

1. INTRODUCTION

The measurement of nuclear resonances has been closely associated with statistical theories of nuclear reactions and spectra since the discovery of neutron resonances in 1936 by Moon and Tillman [Mo 36] and Rasetti et al. [Ra 36] Such theories can be tested most directly with a complete, pure set of states, and this ideal is most closely approximated in resonance studies. Lynn [Ly 68] provides an excellent description of neutron resonance reactions. Neutron resonances provided the basis for a phenomenological description of global properties such as average level density and strength function, as well as the first evidence on spectrum fluctuations (the nearest neighbor spacing distribution) and strength fluctuations (the reduced width distribution). The situation a generation ago is presented in the compilation by Porter [Po 65]. Statistical theories of spectra provided explicit predictions about the spectrum fluctuations. Sufficiently complete and pure level sequences were first obtained by Rainwater and collaborators for neutron resonances in heavy nuclei [Li 72a, 72b]; the data agreed with the predictions of the GOE version of random matrix theory [Br 81]. Subsequently our group obtained proton resonance data of suitable quality to test these predictions in intermediate nuclei [Wi 75].

For the strengths the basic result is the Porter-Thomas (PT) distribution [Po 56], which implies that the amplitudes are Gaussian random variables. There are many sets of data which appear to agree with the PT distribution. However, the PT distribution is intrinsically single channel. Channel-channel correlations are important both to test statistical theory and to consider what extensions to statistical theory are required [Br 81]. Correlations have been measured in neutron resonance capture. However, all of these measurements were for width correlations [Ch 72, Mu 79].

We developed a method to measure both width and amplitude correlations for proton transition amplitudes [Di 75]. This new method permits more sensitive and direct tests of statistical theory and of doorway state theory. These unique results will be emphasized here. First the experimental method is briefly described — the cornerstone is very good beam energy resolution [Bi 76]. The resonance analysis utilizes a multilevel, multichannel R-matrix approach to provide most of the resonance parameters. Angular distributions of the reaction products are then measured to obtain the amplitude and width correlations. Results for the spacings [Wi 75, Wa 81] are described in section 3. The amplitude correlation results are presented in section 4. Most of these results have been presented in a comprehensive review by Mitchell et al. [Mi 85] Results for isolated doorway states [La 78, Mi 79] are discussed, as well as tests of the width amplitude distribution. The separate determination of width and amplitude correlations permits the most direct test yet of the Gaussian assumption [Mi 85]. The average value of the measured correlations is surprisingly large. The existence of these correlations is ascribed to direct reactions [La 83], confirming a prediction of direct reaction theory. However, the origin of the large magnitude of the correlations is not yet explained. The final section provides a summary and outlook.

2. EXPERIMENTAL METHOD AND RESONANCE ANALYSIS

For an isolated compound nuclear resonance the shape of the elastic cross section as a function of energy depends upon the orbital angular momentum ℓ, the total angular momentum J, the channel spin s, and the scattering angle θ. With the Coulomb amplitude larger than (or comparable to) the resonance amplitude, the resonance shapes for different ℓ differ strongly, while the shapes for the same ℓ and different J are not as distinct. For protons of a few MeV on nuclei with $A \gtrsim 30$, the Coulomb term dominates. For zero spin targets the J uncertainty is reduced to two values ($\ell \pm 1/2$), but can be removed in general only by a separate experiment. We use the (p,p'γ) reaction to determine J. Of course in special cases there is little or no ambiguity, such as for $\ell = 0$ resonances, or for resonances which are enhanced locally by a doorway state. (The restriction to zero spin targets has been removed and we have interesting new results [Ne 85]; however, these will not be discussed here.) Thus for an isolated resonance, one can make a catalog of resonance shapes as a function of J^π and θ, and determine many of the resonance parameters by inspection. Examples are given in figure 2.1.

The practical problems are in two general categories. First, the experimental resolution is finite. In order to have sufficient levels for statistical analysis, one wants high level density. Since the Porter-Thomas distribution favors small widths, for a given experimental resolution the number of levels missed increases rapidly with level density. Thus extremely good beam resolution is necessary to achieve a complete (or nearly complete) set. The neutron measurements used to test GOE theory have been described by Liou et al. [Li 72a, 72b] Our method [Bi 76] for obtaining very good energy resolution is shown schematically in figure 2.2. This method involves two beams from the same ion source — H^+ to perform the experiment and HH^+ to serve as a control beam. The HH^+ beam is bent through an electrostatic analyzer and is held fixed by adjusting the voltage on the plates of the electrostatic analyzer. This same correction (suitably amplified) is applied to the target. Thus the target potential changes to accelerate (or decelerate) the protons and correct for time dependent energy fluctuations. With solid targets the overall resolution is 300 eV; with a gas target the best resolution was about 100 eV. With this resolution it is possible to study nuclei up to $A \simeq 70$. The data are recorded on a VAX 11/750 computer for subsequent analysis. Other key experimental factors are good absolute energy stability and uniform, thin targets (about 1 $\mu g/cm^2$).

The other practical difficulty arises in the analysis. Instead of isolated levels, one observes many levels (over 400 in a few cases) of several different spins and parities. Sample data for three isotopes of calcium are shown in figure 2.3. The practical problems with fitting several hundred resonances with multilevel, multichannel R-matrix analysis are enormous. Very high quality, good resolution data are essential for correct spin assignments, and thus for pure, complete sets. The procedure for the elastic scattering analysis for zero spin targets is described by Bilpuch et al. [Bi 76] and for non-zero spin targets by Nelson et al. [Ne 85]

Our group performed an extensive set of experiments in the nuclear 1f-2p shell [Bi 76], which emphasized the study of the fine structure of analogue states. From these data two $1/2^+$ sequences (^{44}Ca, ^{48}Ti), and one $1/2^-$ sequence (^{44}Ca) in the vicinity of a $1/2^-$ analogue state, were considered sufficiently pure and complete for level fluctuation analysis [Wi 75]. Later we obtained another $1/2^+$ sequence (^{56}Fe) which is the most pure and complete sequence for proton resonances [Wa 81]. These data are discussed in section 3.

Our new and unique results (amplitude correlations) are obtained with the (p,p'γ) reaction. Angular distributions are measured for protons inelastically scattered to the 2^+ first excited state and for the deexcitation γ-rays emitted when that state decays to the ground state. Coincidence measurements are also employed. The analysis requires expressions which relate the experimental coefficients of the angular correlations to the reduced width amplitudes in the various channels. The distributions are expressed in terms of ratios of the reduced width amplitudes, and a least-squares search is performed to determine the

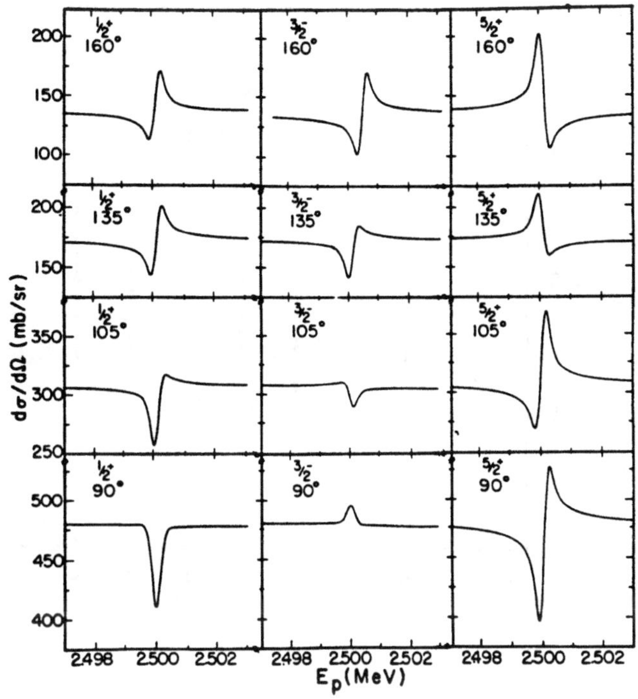

2.1 Differential cross section versus energy for ℓ = 0, 1, and 2 resonances as a function of scattering angle.

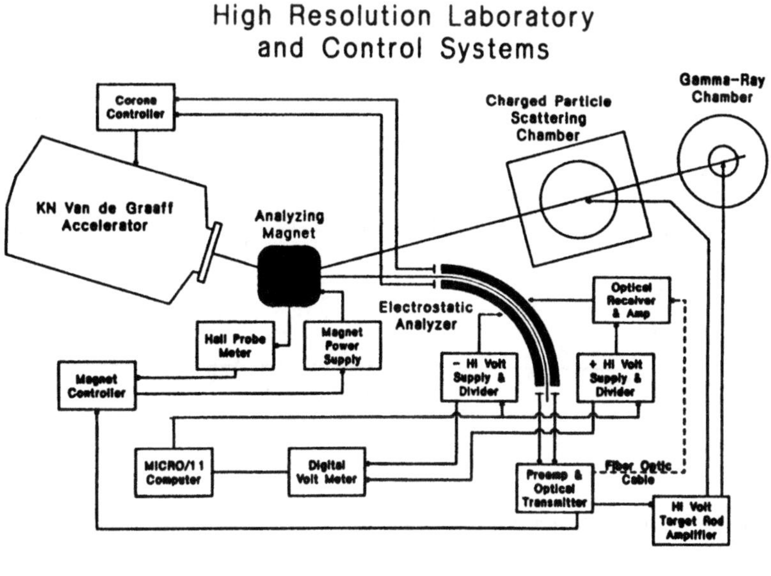

2.2 Schematic of the high resolution laboratory.

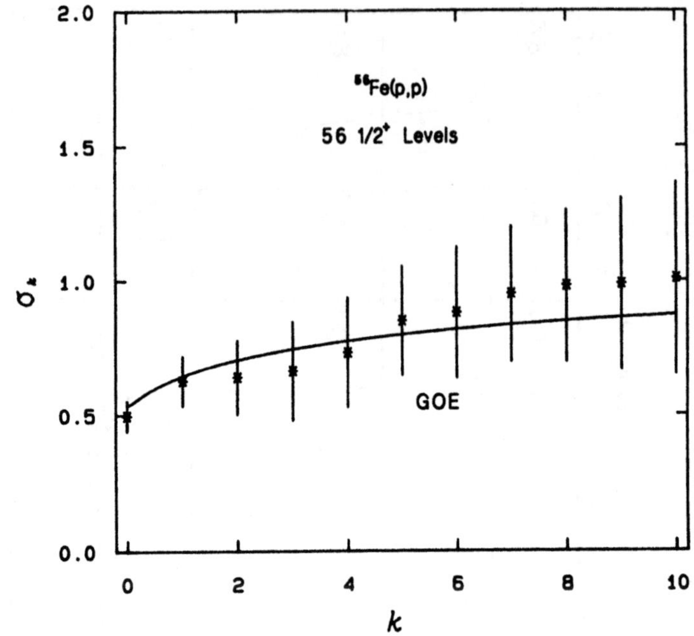

3.2 Widths of the k-th nearest neighbor spacing distributions as a function of k. The solid curve is the GOE prediction.

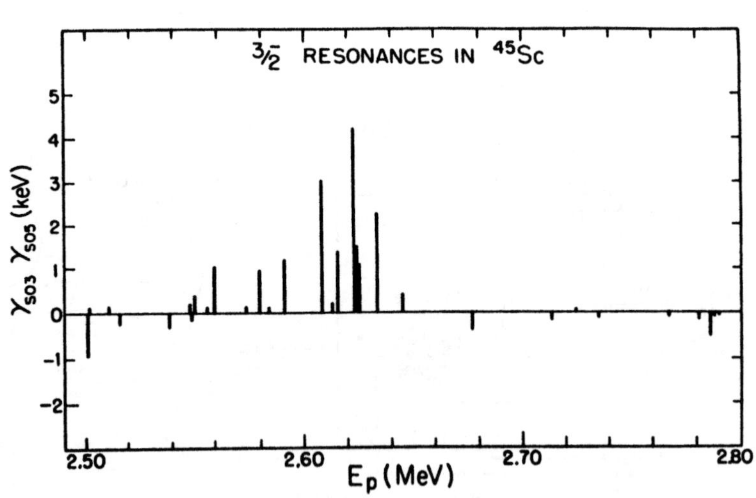

4.1 Products of reduced width amplitudes versus energy in the vicinity of the $3/2^-$ analogue in ^{45}Sc.

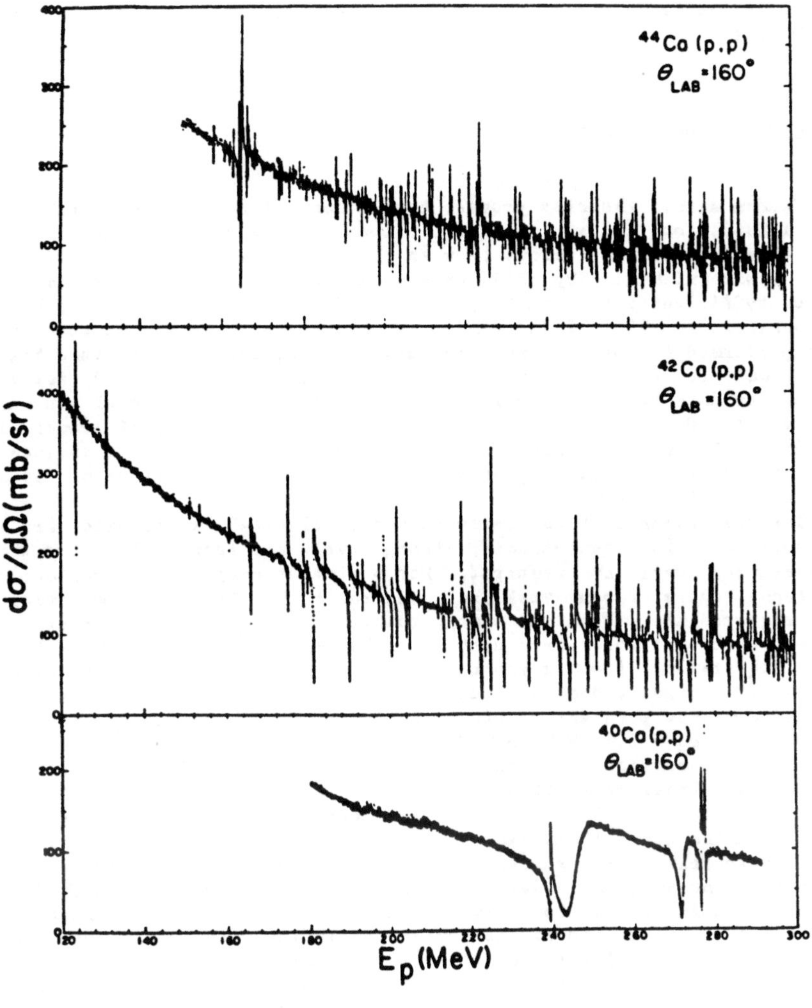

2.3 Measured differential cross section versus energy for three isotopes of calcium. The solid line is a multilevel R matrix fit to the data. For ^{44}Ca the fit is to 429 resonances.

Details of the experimental procedure and the analysis are presented by Mitchell et al. [Mi 85]

3. SPECTRUM FLUCTUATIONS

Random matrix theories provide both a convenient framework for the study of level fluctuations and explicit predictions for a variety of experimental properties. Early comparisons between theory and experiment were inconclusive (e.g., Dyson and Mehta [Dy 63]), because there were no data of suitable quality. Data of sufficiently high quality were first obtained for erbium by Rainwater and collaborators [Li 72a, 72b]. Similar slow neutron resonance measurements were later performed for other heavy elements. Subsequently we obtained high quality spacing data for proton resonances on several nuclei in the 1f–2p shell. Since most of our data have been included in general analyses of level fluctuations by Brody et al. [Br 81] and Haq et al. [Ha 82], here only a brief summary is provided. The overall conclusion for both neutron and proton resonance data is that the experimental data agree very well with the predictions of the GOE version of random matrix theory.

The requirements of completeness and statistical significance are almost contradictory. For completeness, strong levels are ideal, while for statistical purposes many levels are required. For a given strength function, one can have a few strong levels or many weak ones, but not both. One must compromise between these requirements. With our energy resolution (about 300 eV overall) two nuclei from our earlier studies were most suitable — ^{44}Ca and ^{48}Ti. To satisfy the purity requirement, $\ell = 0$ resonances are considered, since in this case the $J = \ell \pm 1/2$ ambiguity does not occur. There was one exceptional case for which we obtained a rather pure set of $1/2^-$ resonances. In the vicinity of the strong $1/2^-$ analogue near $E_p = 2.04$ MeV in ^{44}Ca(p,p) the $1/2^-$ resonances were enhanced relative to the $3/2^-$ resonances. In addition, the $1/2^-$ strength function is larger than the $3/2^-$ strength function in this mass region. For these three sets of data the results agreed very well with the predictions of GOE [Wi 75]. These data have been included in the general analyses.

To illustrate our results, consider an experiment by Watson et al. [Wa 81] The nucleus chosen (^{56}Fe) lies near the peak of the s-wave strength function and the level density is not too large. Since the average width is about 50 times the observation limit imposed by the finite resolution, one should miss few $1/2^+$ levels. The purity of the sequence depends upon the ability to distinguish the ℓ-values of the resonances. This is easy except for weak s- and p-wave resonances. The detailed interference shapes are such that one is unlikely to misassign a p-wave resonance as an s-wave resonance; thus the purity of the sequence is very good. Some weak s-wave resonances may be misassigned, causing a few missing levels.

In the energy range $E_p = 3.12 - 4.00$ MeV, 56 $1/2^+$ resonances were observed. For level fluctuation analysis a problem with these data is the relatively large energy range covered. Thus a "correction" is necessary to unfold the changing level density before comparing the observed fluctuations with theory. Fortunately the results appear insensitive to the details of the unfolding process. The cumulative number of levels N(E) versus energy is shown for the uncorrected and corrected sequences in figure 3.1. The Dyson–Mehta Δ_3 statistic is the mean squared difference between a staircase function N(E) and a straight line

$$\Delta_3 = \min_{A,B} \frac{1}{2L} \int_{-L}^{L} [N(E) - AE - B]^2 \, dE,$$

where A and B are parameters of the straight line and $N(E)<0$ for $E<0$. The energy origin is the center of the interval. The expectation value of Δ_3 is

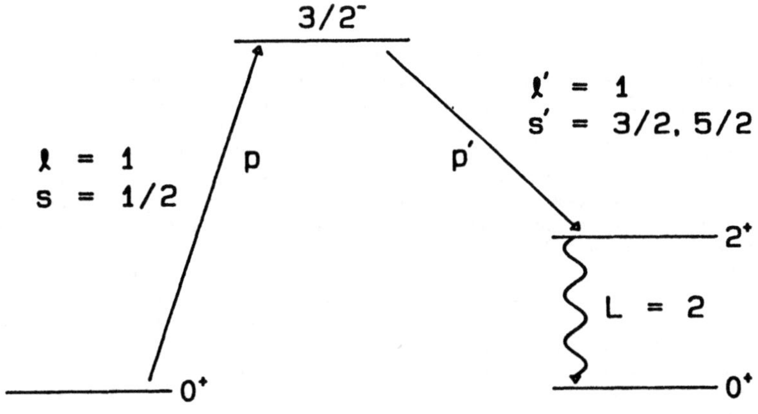

2.4 Schematic for the inelastic scattering experiments.

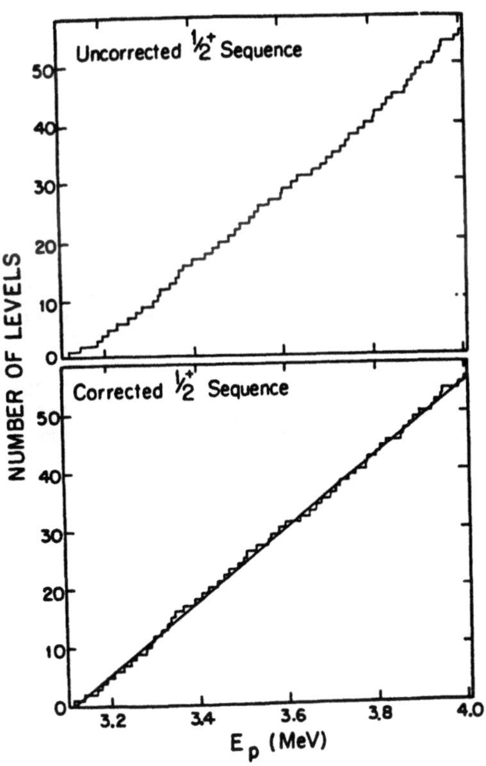

3.1 Number plot for $1/2^+$ resonances in ^{57}Co. The corrected sequence has the energy dependence of the level density removed. The solid curve is the best fit straight line to the data.

values of these mixing parameters which best satisfy all measured correlations simultaneously.

A proton ($i_p = 1/2$) is incident on a target ($I^\pi = 0^+$) with relative orbital angular momentum ℓ. The two particles combine to form a compound state with angular momentum $J = \ell \pm 1/2$ and parity $(-1)^\ell$. This compound state then decays by emission of a proton ($i_p' = 1/2$) with relative orbital angular momentum ℓ', leaving the residual nucleus in its 2^+ first excited state. This state decays by emission of a γ-ray of multipolarity L = 2, leaving the nucleus in its 0^+ ground state.

For simplicity, we present results for only one representation for the angular momentum coupling — the channel spin representation $\vec{s} = \vec{i} + \vec{I}$, and similarly in the exit channel). Because I = 0, in the entrance channel there is only a single value of s for a state of given J. This is not the case for inelastic decay where there is more than one value of both ℓ' and s'. The measured angular correlations of inelastically scattered protons and of deexcitation γ-rays depend strongly on the interference between amplitudes in different channels.

The situation for $3/2^-$ compound nuclear states is illustrated in figure 2.4. Both proton and γ-ray angular distributions are proportional to $1 + a_2P_2(\theta)$. Neglecting the $\ell' = 3$ channels, there are two exit channels (s' = 3/2 and 5/2). We adopt the notation $\gamma_{rep,\ell',2s'}$ for the amplitudes. Because one can measure only the relative sign of these amplitudes and not the absolute signs, the two amplitudes are combined into a single variable which contains the relative sign by defining the mixing ratio

$$\delta_s \equiv \frac{\gamma_{s15}}{\gamma_{s13}}.$$

It is usually more convenient to work with the mixing angle

$$\varphi_s \equiv \tan^{-1} \delta_s.$$

The coefficients of the angular distribution equations can be expressed in terms of this mixing angle

$$a_{2p} = -\frac{4}{5}\cos^2\varphi_s + \frac{1}{5}\sin^2\varphi_s$$

$$a_{2\gamma} = \frac{1}{10}\cos^2\varphi_s - \frac{2}{5}\cos\varphi_s\sin\varphi_s + \frac{2}{5}\sin^2\varphi_s.$$

One can show that a_{2p} and $a_{2\gamma}$ cannot both be zero for a given resonance. This provides a simple determination of J for $\ell = 1$ resonances, since both coefficients must be zero for a $J^\pi = 1/2^-$ resonance. In addition, the interference term in $a_{2\gamma}$ permits the determination of the sign of δ.

Thus one obtains a definitive measurement of J (1/2 or 3/2) and both the magnitudes and relative signs of the amplitudes (γ_c^2, $\gamma_{c'}^2$, $\gamma_c\gamma_{c'}$). This information for a set of resonances provides both width and amplitude correlations, where the linear correlation coefficient is defined by

$$\rho(x,y) = \frac{\Sigma_i(x_i-\langle x\rangle)(y_i-\langle y\rangle)}{[\,\Sigma_i(x_i-\langle x\rangle)^2\,\Sigma_i(y_i-\langle y\rangle)^2\,]^{1/2}}.$$

For the amplitude correlations one assumes $\langle\gamma_c\rangle = \langle\gamma_{c'}\rangle = 0$. The determination of amplitude correlations is the unique feature of our measurements, which make possible new tests of statistical theory.

This approach also has been applied to $\ell = 2$ and $\ell = 3$ resonances. The $\ell = 2$ case is very important, since we observe many $\ell = 2$ resonances, but essentially no $\ell = 2$ analogues. These measurements distinguish between $3/2^+$ and $5/2^+$ resonances. There are three significant exit channels for $3/2^+$ and $5/2^+$ resonances (one $\ell' = 0$ and two $\ell' = 2$ amplitudes). For the $5/2^+$ case the two singles measurements provide a unique solution, while for the $3/2^+$ case there are two solutions from the singles data alone. In this case, we also measure the angular distribution of the γ-rays in coincidence with inelastically scattered protons to obtain a unique solution.

$$\langle \Delta \rangle_3 = [\ln(n) - 0.0687]/\pi^2.$$

For this sequence the experimental value of Δ_3 is 0.42, which compares well with the theoretical prediction of 0.40. It is clear from figure 3.1 that the experimental data have very few large fluctuations.

The sequence also was tested with the Dyson F statistic. The agreement was again excellent, consistent with no missing or spurious levels. The linear correlation coefficient for adjacent spacings has a predicted value of −0.27. The experimental value of ρ is −0.23, with a statistical uncertainty of about 0.1.

The nearest neighbor spacing distribution agrees with the Wigner distribution. It is also of interest to compare the higher order spacing distributions with theoretical predictions. The spacings are defined as

$$s_{k,i} = E_{i+k+1} - E_i$$

$$x_{k,i} = s_{k,i}/\langle s_0 \rangle$$

The distributions for P(k,x) become approximately Gaussian as k increases, but the width of the distribution grows very slowly. French et al. [Fr 78] provide an approximate expression for the variance

$$\sigma_k^2 = (\ln[2\pi(k+1)] + \gamma + 1)/\pi^2 - 5/12,$$

where $\gamma = 0.577$ (Euler's constant) and k is the number of levels between the two levels in question. (Note that if the levels were uncorrelated σ_k^2 would increase as k+1.) The experimental results for the variances of the higher order spacing distributions are shown in figure 3.2, along with the GOE prediction.

Overall, the proton spacing data agree very well with GOE, confirming the existence of both short and long range order in the level fluctuations.

4. STRENGTH FLUCTUATIONS

4.1 General

According to the Bohr compound nucleus model, resonance amplitudes are random in magnitude and phase. In random matrix theory the width amplitude distribution is assumed multivariate Gaussian, the corresponding width distribution in each channel is a chi-squared of one degree of freedom (the Porter-Thomas distribution) and there are no channel correlations. The simplest statistical model may be extended to include direct reactions. In that case the randomness in phase is removed and channel correlations occur. For an isobaric analogue state the strength fluctuations are modified by a strong energy variation due to the symmetry mixing. A strong channel correlation is predicted.

It seems very surprising that relatively few measurements of channel correlations have been performed. The bulk of the existing correlation data is from neutron capture experiments, which determined width correlations. Before the present work, no amplitude correlations had been measured in nuclear physics.

Our primary interest was in statistical properties. However, to test our approach, we first considered an isolated doorway state. For this case, as in subsequent tests of statistical theory, the amplitude correlations provide the most sensitive test yet.

4.2 Isobaric Analogue State

For an isolated common doorway state, Lane [La 71] showed that the reduced width amplitudes should be maximally correlated, i.e., that $\rho(\gamma_{\lambda c},\gamma_{\lambda c'}) = 1$, and that $\{\gamma_{\lambda c}\gamma_{\lambda c'}\}$ should have the same phase for all levels λ (that of the doorway state). As Mahaux and Weidenmüller comment [Ma 79]: "Isobaric analogue resonances provide the best understood example of isolated doorway states." The theory of the fine structure of analogue states was summarized by Lane [La 69], while the experimental status was reviewed by Bilpuch et al. [Bi 76] Measurements by Graw et al. [Gr 74] and Davis et al. [Da 75] indirectly demonstrated that the relative phase was constant. We measure the effect directly.

The fragmented $3/2^-$ analogue near $E_p = 2.62$ MeV on ^{44}Ca is common to the elastic and inelastic channels; for each individual $3/2^-$ state λ we determined the elastic width, two inelastic widths, and the relative sign between the two inelastic amplitudes. The product $\gamma_{\lambda c}\gamma_{\lambda c'}$ is plotted in figure 4.1. By inspection the relative sign is the same for 15 resonances comprising the analogue.

One can consider these multichannel results as generalizing the usual strength function. Thus

$$s_{cc'} \equiv \langle \gamma_{\lambda c'}\gamma_{\lambda c'} \rangle / D$$

may be considered the "off-diagonal strength function", where D is the average level spacing. Lane et al. [La 78] showed that one can fit the analogue fine structure for the two inelastic channels and predict the fine structure distribution for the product $\gamma_{\lambda c}\gamma_{\lambda c'}$ (the off-diagonal strength function). They obtained excellent agreement between theory and experiment, lending additional support to the present theory of analogue fine structure.

With this new information on amplitudes, one can test directly the theoretical predictions for the amplitude correlation. The measured correlation is +0.93, in agreement with the expected value of one. The mixing angle distribution is also interesting, since in this case all resonances are weighted equally. For a doorway state with a picket fence background the mixing angle should be the same for all fragments λ. The measured mixing angles are approximately constant, as shown in figure 4.2 for the analogue fragments.

4.3 Overview of data

We have studied $3/2^-$ resonances in ^{45}Sc, ^{49}V, and ^{51}Mn, $5/2^+$ resonances in ^{45}Sc, ^{49}V, ^{51}Mn, ^{57}Co, and $3/2^+$ resonances in ^{49}V and ^{57}Co. All except the ^{57}Co data are included in the review by Mitchell et al. [Mi 85] For ^{57}Co the $5/2^+$ data have been published [Ra 84], and the $3/2^+$ data [Ra 86] will be published shortly.

For each data set γ^2 was plotted as a function of energy, and except for a few $3/2^-$ analogues, little energy variation was observed in either the elastic or inelastic channels. The partial inelastic widths for $5/2^+$ resonances in ^{49}V are shown in figure 4.3. For each channel the width distribution was compared with the Porter-Thomas distribution. The agreement was usually excellent. The experimental width cutoff (in dimensionless units $y_i = \gamma_i^2 / \langle \gamma^2 \rangle$) was usually less than 0.01 (i.e., no more than a few percent of the levels were missed).

Since the spacing and width distributions appear to behave according to statistical predictions, and there is no apparent "intermediate structure" in the summed strengths, one expects the products of amplitudes $\gamma_{\lambda c}\gamma_{\lambda c'}$ to follow the predictions of the statistical model: the products should have random signs and the summed product should fluctuate about zero.

Experimentally, this is not true. For example, for $5/2^+$ resonances in ^{49}V (figure 4.4), the product is essentially always positive for one pair of channels. The amplitude correlation for this pair is +0.88! Large correlations (both width and amplitude) are the rule in these data. The average amplitude correlation and

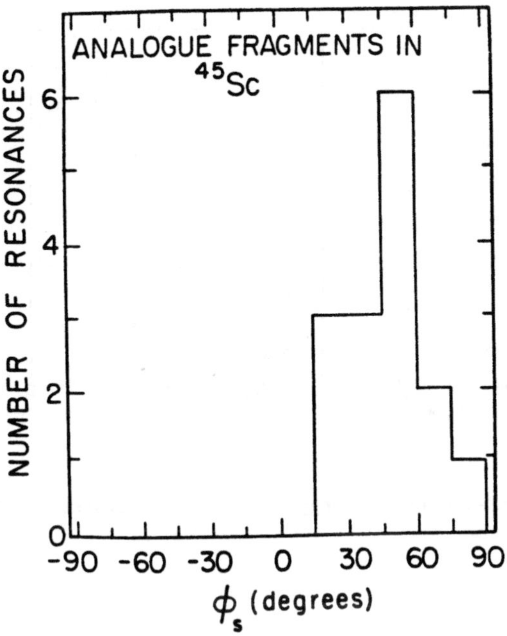

4.2 Distribution of the mixing angle for the analogue fragments in ^{45}Sc.

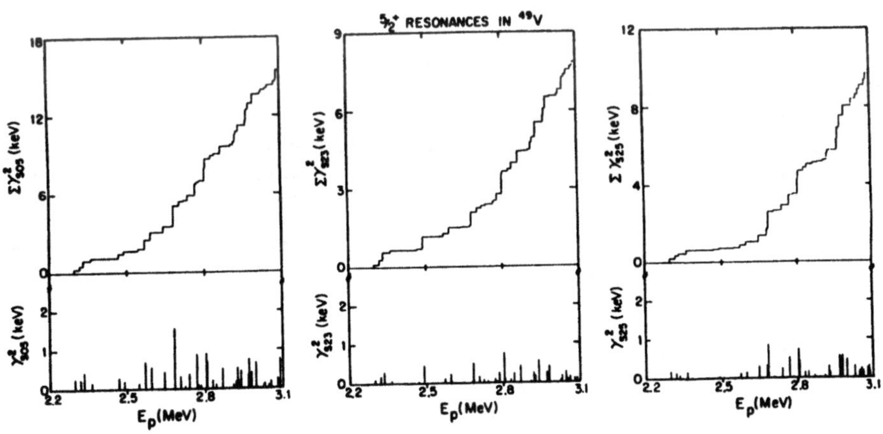

4.3 Plots of partial inelastic reduced widths versus energy for $5/2^+$ resonances in ^{49}V.

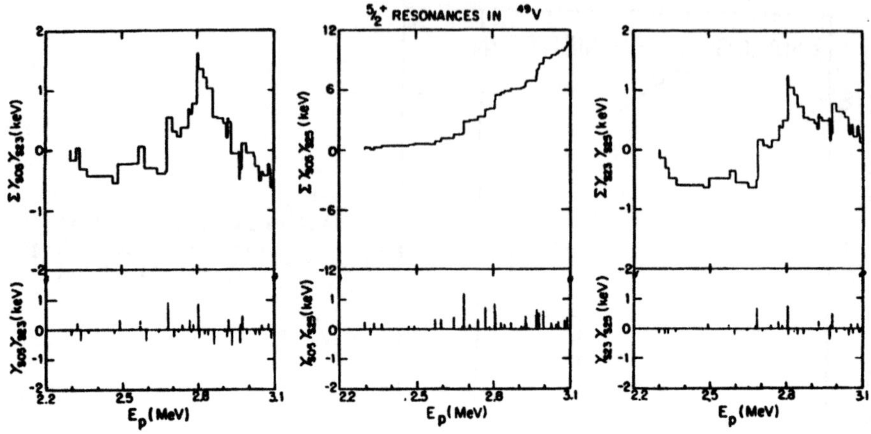

4.4 Products of reduced width amplitudes versus energy for $5/2^+$ resonances in ^{49}V.

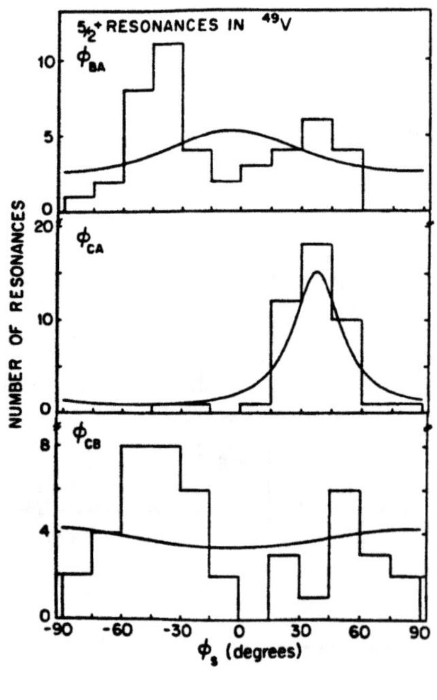

4.5 Distributions of the mixing angles for $5/2^+$ resonances in ^{49}V. The smooth curves are the multivariate Gaussian (Kreiger-Porter) distributions.

width correlation for 15 sets of independent correlations (1 for each set of $\ell = 1$ resonances, 2 for each set of $\ell = 2$ resonances) is > 0.5.

These correlations are very large. However, since the sample size is limited (30–70), the question of statistical significance is crucial. The confidence intervals were determined using the bootstrap method of Efron [Ef 79]: a set of data is generated by sampling with replacement from the original data set N times (where N is the number of members of the original set). The statistic of interest (here, the linear correlation coefficient) is then calculated from the generated set. The procedure is repeated many times and the resulting distribution of the statistic is used as an estimate of the parent distribution. Although the confidence intervals are frequently rather broad, presumably due to the limited sample sizes, a large fraction of the coefficients are significantly different from zero at the 99% level of confidence.

4.4 Width Amplitude Distribution

A multivariate reduced width amplitude distribution, P, was obtained by Krieger and Porter [Kr 63]. Their fundamental assumptions are: (i) P is independent with respect to levels, i.e., P is factorizable into functions corresponding to the various levels λ, and (ii) P is independent with respect to a proper orthogonal transformation. For m channels the result is

$$P_{cc'} = \prod_{\lambda} \frac{|M|^{1/2}}{(2\pi)^{m/2}} \exp[-\tfrac{1}{2}(\gamma_\lambda, M\gamma_\lambda)]$$

where M is the inverse of the covariance matrix $\Sigma \equiv (\gamma_\lambda \times \gamma_\lambda)$, $|M|$ is the determinant of M, and γ_λ is an m component vector. Unless M is diagonal, $P_{cc'}$ is not factorizable with respect to channels, and correlations may exist between amplitudes in different channels. It is sufficient to consider the channels pairwise. For the two channel case the explicit form of P is

$$P(\gamma_1,\gamma_2) = \frac{|M|^{1/2}}{\pi} \exp[-\tfrac{1}{2}(M_{11}\gamma_1^2 + M_{22}\gamma_2^2 + 2M_{12}\gamma_1\gamma_2)],$$

where a factor of two arises from the restriction $\gamma_1 \geq 0$, and

$$|M|^{-1} = \langle\gamma_1^2\rangle\langle\gamma_2^2\rangle - \langle\gamma_1\gamma_2\rangle^2 \qquad M_{11} = \langle\gamma_2^2\rangle\,|M|$$

$$M_{22} = \langle\gamma_1^2\rangle\,|M| \qquad M_{12} = -\langle\gamma_1\gamma_2\rangle\,|M|.$$

Integrating over all but one channel c results in the usual Gaussian expression for γ_c. The corresponding distribution for $\gamma_c^2/\langle\gamma_c^2\rangle$ is the Porter–Thomas distribution [Po 56]. Note that the Porter–Thomas distribution is a singlet distribution and therefore says nothing about channel correlations.

It is convenient [Ch 80] to transform to a polar system (r,θ), where $r=(\gamma_1^2+\gamma_2^2)^{1/2}$ and $\theta = \tan^{-1}(\gamma_2/\gamma_1)$. After integration over r one obtains the marginal probability density function

$$P(\theta) = \frac{|M|^{1/2}}{\pi}(M_{11}\cos^2\theta + M_{22}\sin^2\theta + M_{12}\sin 2\theta)^{-1}.$$

If there is no correlation between channel c and c', then $M_{12} = 0$ and $P(\theta)$ is symmetric about $\theta = 0$. If in addition $M_{11} = M_{22}$, then the distribution is uniform.

The Krieger–Porter distribution has no adjustable parameters and thus can be compared directly with the experimental distributions of the mixing angles. The mixing angle distributions for $5/2^+$ resonances in ^{49}V are shown in figure 4.5. Considering the limited sample size, the agreement between experiment and theory seems reasonable. These results constitute the first direct proof of the reduced width amplitude distribution.

4.5 Representation Dependence of Correlations

All of the data presented here are in the channel spin representation. One could equally well analyze the data in the particle angular momentum representation ($\vec{j} = \vec{l} + \vec{i}_p$). However, the numerical values of the correlations are a function of the representation. This issue did not arise in previous correlation measurements. For uniformity we normally present data in the channel spin representation. However, to determine the maximum value of a correlation, we consider all possible representations obtained by an orthogonal transformation. These maximum values are used to determine whether statistically significant non-zero correlations are observed. (As an example, consider the two channel case. There is always a representation where $\rho_A = 0$. However, a large value of ρ_A is of interest, since this is not ensured by the transformation itself.)

For a total of 15 independent correlations the results are $\langle \rho_A \rangle = 0.54$ and $\langle \rho_W \rangle = 0.49$. From the bootstrap analysis 11 of the 15 amplitude correlations are statistically different from zero at the 99% confidence level.

4.6 Interpretation of Large Correlations

The existence of direct reactions implies correlations, while the Bohr compound nuclear theory implies zero correlations. Our results indicate very large correlations between the inelastic channels. We ascribe these correlations to direct reactions. The observed amplitude correlations constitute the first explicit confirmation of the prediction of direct reaction theory [Bl 57, Hü 67]. In fact the correlations are so large that DWBA is not a valid approximation. The details are given by Lane et al. [La 83] and Mitchell et al. [Mi 85]

The physical origin of these large correlations is not known. Some explanations have been advanced, but in the absence of nuclear structure calculations these explanations remain speculative. Since there is no experimental information about amplitude correlations in other mass regions, the generality of these results is also unknown.

4.7 Test of Gaussian Assumption

It is assumed in statistical theory that each set of amplitudes $\{\gamma_{\lambda c}\}$ are distributed according to a Gaussian distribution with mean zero and variance $\langle \gamma_c^2 \rangle$. Because the absolute signs of the reduced width amplitudes cannot be determined, a direct test of this aspect of the theory is impossible. Width distributions do not provide a sufficiently sensitive test of the amplitude distribution. However, the present data in principle provide a rather direct experimental test. If amplitudes are Gaussian, the amplitude correlation and width correlation are simply related [Kr 63, La 71] by

$$\rho^2(\gamma_1,\gamma_2) = \rho(\gamma_1^2,\gamma_2^2).$$

Thus a nonzero value of $\rho^2(\gamma_1,\gamma_2) - \rho(\gamma_1^2,\gamma_2^2)$ implies that the underlying amplitude distribution is non-Gaussian. Our data provide the first possibility of testing this relationship. The unexpectedly large values for the correlations make this test feasible.

The experimental results are mixed. Most of the data sets agree with the relation between ρ_A and ρ_W (and thus with the Gaussian assumption), but some individual data sets do not. This has led to extensive analysis on the effects of experimental errors by Hofmann et al. [Ho 83] and on the effects of the finite range of the data by Harney [Ha 84a]. Harney derived analytic formulae for the finite range of data error in the test of the bivariate Gaussian distribution of a pair of amplitudes by the width correlation-amplitude correlation relation. These errors impose severe constraints upon tests of the equality. While such considerations (errors and finite sample size) are not incorporated directly into the calculations presented here, the use of the bootstrap method to determine confidence interval provides an indirect measure of their importance.

Although individual sets of data appear to disagree with the amplitude correlation-width correlation relation, the overall data are in agreement. For all of the data combined, Harney [Ha 84b] obtains a result ($\rho_A^2/\rho_W = 0.93 \pm 0.11$), in excellent agreement with a joint Gaussian distribution for the amplitudes. However, a calculation by Shriner [Sh 85] following Harney's approach, but considering the difference $\rho_A^2 - \rho_W^2$, led to a non-zero result for the combined data. These results do not disprove the Gaussian assumption, but they do indicate that this basic assumption is not as well established empirically as one would prefer.

5. SUMMARY

Measurements of proton resonances provide information on global and local properties of spacings. This information is complementary to neutron resonance measurements. The proton measurements have advantages and disadvantages relative to the neutron measurements. The major disadvantage is the limitation to $A \lesssim 70$. The major advantage is the ability to study resonances of different ℓ and J. For four level sequences which seem pure and complete, the proton data agree very well with GOE. Thus these data confirm the neutron resonance results for spectrum fluctuations, with a different projectile in a different mass region. and with a completely different set of experimental errors and analysis ambiguities.

However, the special value of the proton resonance measurements lies in the study of strength fluctuations. The first experimental determination of amplitude correlations has permitted more sensitive tests of several theoretical properties:

(1) These results confirm the prediction that $\rho(\gamma_{\lambda c},\gamma_{\lambda c'}) = 1$ and that $\{\gamma_{\lambda c}\gamma_{\lambda c'}\}$ have the same sign for the fragments of an analogue.

(2) The observation of amplitude correlations confirm the prediction of direct reaction theory. When direct reactions are present, amplitude correlations are required.

(3) The average values of the width and amplitude correlations are very large. The generality of these effects is unknown.

(4) The comparison of the distribution of the ratio of reduced width amplitudes with the Krieger-Porter (multivariate Gaussian) distribution provides a valuable qualitative check. However, the most direct test of the Gaussian assumption is given by a separate determination of ρ_A and ρ_W. Although the average overall results agree with the Gaussian assumption, the issue is not fully resolved.

For the future one needs amplitude correlations determined with larger sample sizes and for different reactions and different mass regions. These amplitude correlation experiments may be considered the second generation of experiments which test statistical theories: these special tests of strength fluctuations require pure sets of resonances as a starting point. Extension of the measurement of amplitude correlations to other reactions and nuclei should be both difficult and rewarding.

ACKNOWLEDGEMENTS

This work was performed in collaboration with E.G. Bilpuch. The author would like to thank our former students: T.R. Dittrich, J.R. Chandler, W.K. Wells, W.A. Watson III, B.H. Chou, K.M. Whatley, J.F. Shriner Jr., and P. Ramakrishnan.

Special appreciation is extended to C.R. Westerfeldt for assistance in the experiments, and to A.M. Lane for aid with the theoretical interpretation. Informative discussions and communications with many colleagues are gratefully acknowledged, including O. Bohigas, J.B. French, H.L. Harney, P.A. Mello, A. Pandey, T.H. Seligman and H.A. Weidenmüller. Work supported by the U.S. Department of Energy, Office of High Energy and Nuclear Physics, under contract No. DE-AS05-76ER03624.

REFERENCES

[Bi 76] E.G. Bilpuch, A.M. Lane, G.E. Mitchell, and J.D. Moses, Phys. Rep. 28, 145 (1976)
[Bl 57] C. Bloch, Nucl. Phys. 4, 503 (1957)
[Br 81] T.A. Brody, J. Flores, J.B. French, P.A. Mello, A. Pandey, and S.S.M. Wong, Rev. Mod. Phys. 53, 385 (1981)
[Ch 72] R.E. Chrien, Statistical Properties of Nuclei, ed. J.B. Garg (Plenum, New York 1972), p. 233
[Ch 80] B.H. Chou, G.E. Mitchell, E.G. Bilpuch, and C.R. Westerfeldt, Phys. Rev. Lett. 45, 1235 (1980)
[Da 75] S. Davis, C. Glashausser, A.B. Robbins, G. Bissinger, R. Albrecht, and J.P. Wurm, Phys. Rev. Lett. 34, 215 (1975)
[Di 75] T.R. Dittrich, C.R. Gould, G.E. Mitchell, E.G. Bilpuch, and K. Stelzer, Phys. Lett. 59B, 230 (1975)
[Dy 63] F.J. Dyson and M.L. Mehta, J. Math. Phys. 4, 70 (1963)
[Ef 79] B. Efron, SIAM Review 21, 460 (1979)
[Fr 78] J.B. French, P.A. Mello, and A. Pandey, Ann. Phy. 11B, 277 (1978)
[Gr 74] G. Graw, H. Clement, J.H. Feist, W. Kretschmer, and P. Pröschel, Phys. Rev. C 10, 2340 (1974)
[Ha 82] R.U. Haq, A. Pandey, and O. Bohigas, Phys. Rev. Lett 48, 16 (1982)
[Ha 84a] H.L. Harney, Z. Phys. A 316, 177 (1984)
[Ha 84b] H.L. Harney, Phys. Rev. Lett. 53, 537 (1984)
[Ho 83] H.M. Hofmann, T. Mertelmeier, and H.A. Weidenmüller, Z. Phys. A 311, 289 (1983)
[Hü 67] J. Hüffner, C. Mahaux, and H.A. Weidenmüller, Nucl. Phys. A 105, 489 (1967)
[Kr 63] T.J. Krieger and C.E. Porter, J. Math. Phys. 4, 1272 (1963)
[La 69] A.M. Lane, Isospin in Nuclear Physics, ed. D.H. Wilkinson (North Holland, Amsterdam 1969), p. 609
[La 71] A.M. Lane, Ann. Phys. (N.Y.) 63, 171 (1971)
[La 78] A.M. Lane, T.R. Dittrich, G.E. Mitchell, and E.G. Bilpuch, Phys. Rev. Lett. 41, 454 (1978)
[La 83] A.M. Lane, G.E. Mitchell, E.G. Bilpuch, and J.F. Shriner,Jr., Phys. Rev. Lett. 50, 321 (1983)
[Li 72a] H.I. Liou, H.S. Camarda, and F. Rahn, Phys. Rev. C 5, 1002 (1972)
[Li 72b] H.I. Liou, H.S. Camarda, S. Wynchank, M. Slagowitz, G. Hacken, F. Rahn, and J.Rainwater, Phys. Rev. C 5, 974 (1972)
[Ly 68] J.E. Lynn, The Theory of Neutron Resonance Reactions (Clarendon, Oxford 1968)
[Ma 79] C. Mahaux and H.A. Weidenmüller, Ann. Rev. Nucl. Part. Sci. 29, 1 (1979)
[Mi 79] G.E. Mitchell, T.R. Dittrich, and E.G. Bilpuch, Z. Phys. A 289, 211 (1979)
[Mi 85] G.E. Mitchell, E.G. Bilpuch, J.F. Shriner,Jr., and A.M. Lane, Phys. Rep. 117, 1 (1985)
[Mo 36] P.B. Moon and R. Tillman, Proc. R. Soc. London 153, 421 (1936)
[Mu 79] S.F. Mughabghab and R.E. Chrien, Neutron Capture Gamma-Ray Spectroscopy, ed. R.E. Chrien and W.R. Kane (Plenum, New York 1979), p. 265
[Ne 85] R.O. Nelson, E.G. Bilpuch, and G.E. Mitchell, Nucl. Inst. Methods, A 236, 128 (1985)
[Po 56] C.E. Porter and R.G. Thomas, Phys. Rev. 104, 483 (1956)

[Po 65] C.E. Porter, Statistical Theories of Spectra: Fluctuations (Academic, New York 1965)

[Ra 36] F. Rasetti, E. Segre, G. Fink, J.R. Dunning, and G.B. Pegram, Phys. Rev. 49, 104 (1936)

[Ra 84] P. Ramakrishnan, G.E. Mitchell, E.G. Bilpuch, J.F. Shriner,Jr., and C.R. Westerfeldt, Z. Phys. A 319, 315 (1984)

[Ra 86] P. Ramakrishnan, G.E. Mitchell, E.G. Bilpuch, C.R. Westerfeldt, and J.F. Shriner,Jr., to be published

[Sh 85] J.F. Shriner,Jr., Phys. Rev. C 32, 694 (1985)

[Wa 81] W.A. Watson III, E.G. Bilpuch, and G.E. Mitchell, Z. Phys. A 300, 89 (1981)

[Wi 75] W.M. Wilson, E.G. Bilpuch, and G.E. Mitchell, Nucl. Phys. A 245, 285 (1975)

SURVEY OF CORRELATION PROPERTIES OF POLYATOMIC MOLECULES VIBRATIONAL ENERGY LEVELS USING F T. ANALYSIS

R. JOST and M. LOMBARDI
Service National des Champs Intenses (C.N.R.S.)
BP 166 X, 38042 Grenoble Cédex, FRANCE.
and
Laboratoire de Spectrométrie Physique (U.S.T.M.G.)
BP 87, 38402 Saint Martin d'Hères Cédex, FRANCE.

I. Introduction

In the last few years molecular spectroscopists have begun to study the highly excited vibrational levels of polyatomic molecules. In this high energy regime vibrational quantum numbers can no longer be intrinsically assigned (in contrast with vibrational levels at low energy). One can only characterize these levels by their correlation properties[1], we shall consider :

- Short range correlations which are characterized by the Next Neighbor Distribution, (N.N.D.). These correlations range from a POISSON (random or uncorrelated spectra) to a WIGNER distribution (which shows "level repulsion").
- Long range correlations are characterized by the $\Sigma^2(L)$ and $\Delta_3(L)$ function. They describe the behavior which ranges from an uncorrelated spectra (POISSON statistic) to a spectra with "spectral rigidity".

In nuclear physics, spectra obtained many years ago by low energy neutron scattering show the phenomena of "level repulsion" and "spectral rigidity" [1]. These results have stimulated theoretical developments such as Random Matrix Theory (R.M.T.) [2] and the Gaussian Orthogonal Ensemble (G.O.E.) [2]. In the field of molecular physics the theoretical predictions have preceeded experimental results. Many classical trajectory calculations of two-dimensional systems show features of "irregular" behaviour[3]. These classical trajectories have also been calculated with the potential energy surfaces of the electronic ground state of triatomic molecules. All these calculations suggest a smooth change of the dynamics from regular to irregular behavior as the energy exceeds a certain threshold. At an intermediate energy regime, the phase space is embedded with both regular and irregular regions. Quantum calculations with models of (usually) two degrees of freedom, produce eigenvalues which display strong correlations properties[4]. Very few quantum calculations have employed the realistic potential surfaces of polyatomics, due to the prohibitive size of the matrix to be diagonalized. Even calculations concerning the simplest polyatomic, triatomic molecules involve three vibrational degrees of freedom. Comparison of the classical and the quantum calculations for two dimensional systems, show a qualitative agreement, i.e., the correlation properties of eigenvalues evolve from POISSON to G.O.E. in the same energy range where the classical trajectories display a transition from "regular" to "irregular" behavior[5]. Up to now, there is no quantitative relationship between these classical and quantum descriptions.

The main interest in molecular physics for these problems arise from the need to understand the phenomenon of vibrational energy redistribution, which is the basis of the usual R.R.K.M. theory for unimolecular reactions. This theory supposes complete intramolecular thermalisation of vibrational energy before reaction (the isolated molecule acts as its own "thermal bath"). The basis of this theory is believed to be that at relevant excitation energy all classical trajectories are ergodic. A few classical trajectories addressing directly this problem for two-dimensional systems have been made recently[6].

In this paper we will summarize in chapter V the experimental results obtain to date concerning the statistical correlation properties of the vibrational energy levels of polyatomics. First we present, in chapter II, a general review of the properties of vibrational energy levels and, in chapter III, their relationship with experimental molecular spectra. In chapter IV the Fourier Transform method is presented as a tool to displays correlations in experimental spectra.

II. General considerations on the vibrations of polyatomic molecules.

In this section, we shall only consider molecules without rotation (see III). The number of vibrational degrees of freedom, i.e. the number of vibrational modes for non linear species is $N = 3n-6$ (for $n \geq 3$), where n is the number of atoms (diatomics have only one vibrational mode and they are not considered here). For each electronic state, the bottom of the N dimensional potential energy surface can be approximated by N harmonic oscillators. This means that at "low" energy, the vibrational levels are well characterized by the normal modes description, except for the occurence of an occasional Fermi resonance. When we consider levels at high energies, E, the density of states increases as $\rho(E) \propto [E/\langle\omega\rangle]^{N-1}$ where $\langle\omega\rangle$ is the geometric mean of the frequencies. Typically, the magnitude of $\langle\omega\rangle$ is of the order of 1000 cm^{-1} (0.12 eV) and the height of the well ranges from 10^4 cm^{-1} to few 10^4 cm^{-1} (a few eV) for the ground state of stable molecules. Consequently, there is a very rapid increase in the vibrational density of states with energy and with the number of atoms. The coupling terms in the Hamiltonian, like higher order cross terms in the potential or kinetic energy operators induce an increasing number of Fermi resonance as the density of states becomes larger. The absence of a complete set of spectroscopic constants, or accurate ab initio potential energy surfaces, precludes any <u>precise</u> calculation of eigenvalues at high vibrational energies. Nevertheless a model of oscillators coupled with cubic and quartic cross terms in the potential, with a harmonic basis set can be considered. This model yields a very sparse matrix. The validity of such a model stems from the spectroscopy of the lower excited vibrational levels which shows a very rapid decrease of matrix elements with increasing the order of the cross terms. The size of the cubic coupling matrix elements may be as large as a few hundred cm^{-1} in the particular case of stretch-bend C-H coupling but, most of the matrix elements are much smaller: 0.1 to 10 cm^{-1} for cubic and quartic coupling terms. We expect from rough calculations on random matrix[7] that the mechanism of overlapped Fermi resonances plays a crucial role in terms of statistical properties when the density of states becomes larger than 1 to 100 levels per cm^{-1}, depending on the molecule. The effect of these coupling on vibrational energy redistribution has been studied- by techniques which depend mainly on the phenomenon of mixing of wavefunctions (observed via probability of transitions), not on energy level positions. The various vibrational modes which are mixed at a given energy emit different spectra, which may be used to label them (at least qualitatively)[8]. Some more recent work use picosecond time resolved spectroscopy to study these couplings[9]. The conclusion of such recent studies is that one can find lack of vibrational redistribution at surprisingly high energy in large molecules, increasing the interest to find other measures of these coupling on relevant molecules and at relevant energies. On one hand, the fast convergence of the polynomial expansion in harmonic coordinates may be no longer valid for large amplitude vibrational motion. A contrario, for a large molecule, a great amount of vibrational energy, enough to dissociate it for example, can be attained by distributing the energy over the large number of modes available with only very few (0, 1, 2) quanta in each mode. This kind of vibrational level, named "combination level", which correspond to very harmonic motion at large energy, is in fact the vast majority of states present at energies interesting for chemical studies like photochemistry, intermolecular vibrational relaxation, and multiphoton excitation. But the minority (in terms of numbers) of very anharmonic states which concentrate the energy on one given mode (for example a local mode which leads to dissociation) play probably an important role in chemical process like dissociation, isomerization, fluorescence decay, etc...

III. Experimental technique on molecular spectra.

According to usual ideas coming from previous work of nuclear physicists, in order to study the statistical correlations between levels, it is necessary to consider only those levels which have the same good quantum numbers (spin, parity and total angular momentum). Thus the spectroscopic method must be able to sort these levels by their quantum numbers in order to avoid the superposition of sets (see chapter IV). In addition, the spectrum must contain nearly all the levels in a given energy range (no "missing" levels). The experimental signal to noise ratio must be large enough to let all the levels appear with sufficient intensity. The spectrum should yield a complete set of levels, therefore the resolution should be sufficient to avoid overlapping lines (no blended lines). Only gas phase laser spectroscopy is able to produce molecular data at high vibrational excitation with a typical resolution of MHz (for c.w. laser) to GHz (for pulsed laser) required for statistical analysis.

Nevertheless, most of experimental molecular spectra do not satisfy the requirement cited above. It is why, a new technique of analysis, the Fourier Transform described in chapter IV has been evolved to deal with experimental spectra which suffers imperfection.

This new technique, which takes full advantage of the fact that large (but imperfect) stretches of levels can be obtain, clearly opens new avenues to analyze the large amount of data which can be obtained in molecular physics in chemically interesting situations.

The classical spectroscopic techniques relevant to study the vibrations of polyatomic molecules are summarized as follows :

A. <u>Infrared spectroscopy</u> (vibrational excitation).

This technique has produced a great deal of information about the fundamental vibrational frequencies and the overtones of CH and OH stretches in the ground state. However infrared studies at high vibrational excitation are restricted to mainly these types of overtones and then do not allow to observe the numerous "combination" levels since the transitions occur between levels on the same potential surface. Furthermore, rotational congestion occurs and precludes to observed dense sets of levels (see B, below).

B. <u>Excitation spectroscopy</u> (electronic excitation).

In this technique, transitions from the vibrationless electronic ground state (S_0) to the (first) excited state (S_1) allow to study the rovibronic levels in the excited state. But Franck Condom factors do not permit the observation of every vibrational level because many of them are too small.

Furthermore, as in I.R. spectra, rotational congestion occurs : at room temperature, there is a large number of rotational levels populated in the molecule as the rotational constants range from 0.1 cm^{-1} to a few cm^{-1} (10^{-4} - 10^{-3} eV). Even though the selection rules ($\Delta J = 0, \pm 1$) limit the number of rotational transitions, the number of rotational lines appearing in the spectra is pretty large (typically 10^2 to 10^3 per band). These lines are spread over a region of about 100 cm^{-1} ($\approx 10^{-2}$ eV). This spread is much larger than the mean spacing of vibrational levels we want to study (see above) and consequently precludes the observation of every vibrational level. The use of a supersonic jet (a free expansion of a mixture of a carrier gas (He,Ar...) with the molecule of interest) cools the rotational temperature to about 1K for a few μsec. The corresponding rotational spectrum is much less congested as few rotational levels are populated in the vibrationless ground electronic state. But, even with a supersonic jet, the possibility of the overlap of rovibronic bands remains (see NO_2 results below).

A further spectral selectivity is possible with an additional spectroscopic step : double resonance technique, such as Stimulated Emission Pumping (S.E.P.) or Microwave-Optical Double Resonance (MODR) and also Anticrossing Spectroscopy (A.S.) are possible techniques for producing spectra corresponding to a simple set of good quantum numbers. For example, in S.E.P., a transition from the vibrationless ground state populate a single rovibronic level (J = 0 for example), in the excited state. Then a second transition, by stimulated emission, transfer population, according to well defined selection rules, down to one (or very few) rotational levels of high vibrational levels of the ground state. The spectral simplicity achieved by this method allows to consider each line (or set of very few lines) of the spectrum as a vibrational eigenvalue with a known angular momentum. We shall present three examples of results mainly obtained by double resonance method in NO_2, Acetylene (C_2H_2) and Methylglyoxal (CH_3-O-C-C-H-O).

A very related problem is the coupling between rotational and vibrational degrees of freedom. Two examples concerning NO_2 and H_2CO will be discussed at the end of the NO_2 chapter. Beforehand, in chapter IV, we present a new method of analysis of the correlation properties of spectra : the Fourier transform.

IV. The Fourier Transform : a new method to analyse the correlation properties of spectra.

The statistical method like N.N.D. and Δ_3 have been developped in order to analyse the available data in nuclear physics, i.e. a relatively small set of levels (typically 50 per nucleus). These levels constitute a high quality data set with respect to signal to noise, resolution and spectral purity (single J^{π} assignment). Then, these data have been analysed with the N.N.D. and Δ_3 in terms of stick spectrum. In a recent paper, L. Leviandier et al[10] introduce the Fourier Transform (F.T.) as a tool to measure statistical correlation properties able to treat a noisy, poorly resolved and spectrally impure spectrum. In this method, the raw spectrum is Fourier transformed, without extracting a stick spectrum, to obtain a function C(t). The correlation properties can be determined from the smoothed, ensemble averaged, square of the modulus of C(t), i.e. $\overline{|C(t)|}^2$. Consider a spectrum composed of lines with the line shape L(f). The amplitude of lines is assumed to be the product of a deterministic envelope A_E by a a stochastic function A_S of the position. Then, $\overline{|C(t)|}^2$ contains two components as sketched in fig. 1.

a) The "fast component", which is the square of the F.T. of the envelope A_E, is proportional to N^2, i.e. the square of the nomber of levels. This "fast component" gives a very large peak at the origin. For example, it is a $(\sin Nt/t)^2$ function when A_E is a rectangle function.

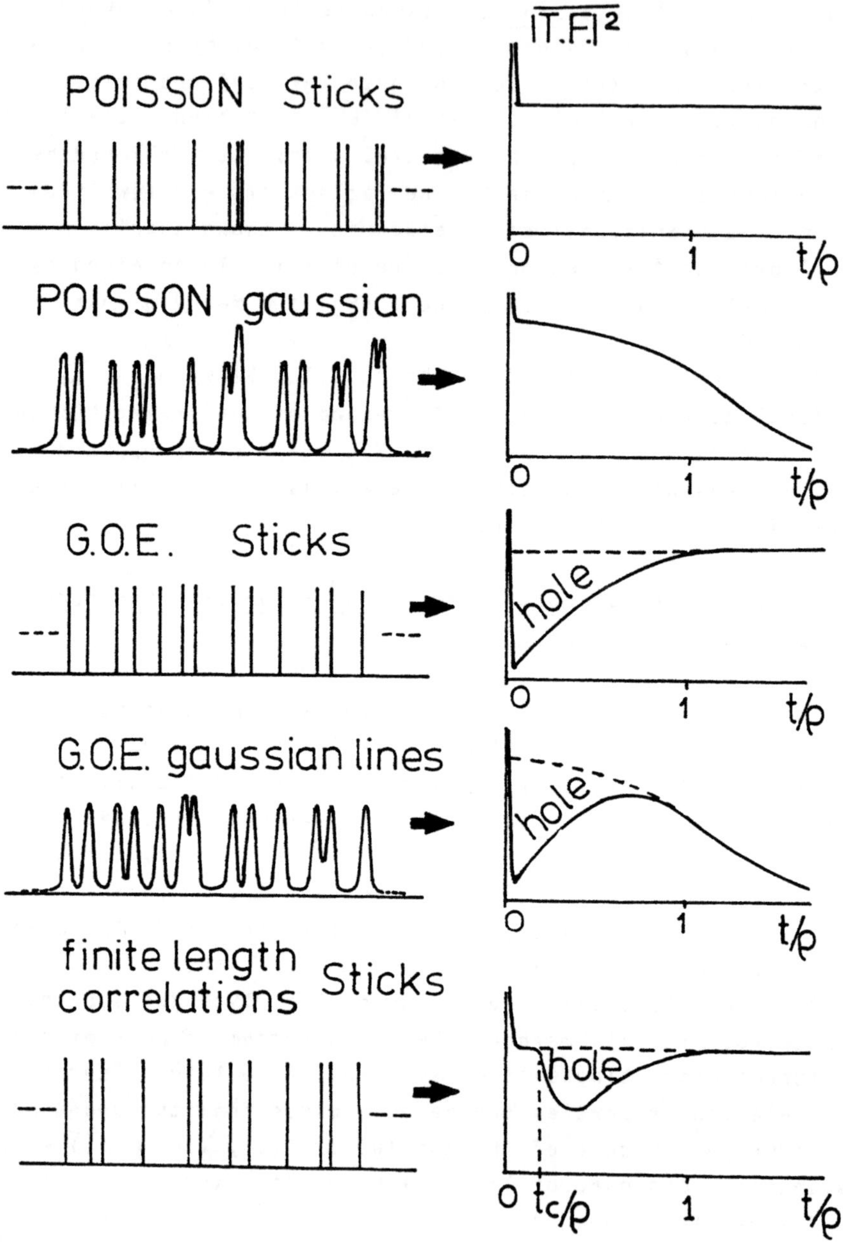

Fig. 1 : Typical spectra and their corresponding smoothed (or ensemble averaged) square of Fourier Transform $|F.T|^2$. The "fast component" appears near the origin. Unsmoothed $|F.T|^2$ displays 100 % fluctions as shown in figure 4 (down). t/ρ is a dimensionless variable, where ρ is the density of lines.

b) The "slow" component is composed of the F.T. of the individual lineshapes L(f). Its amplitude is proportional to N.

Furthermore, if there are correlations in the spectrum, the shape of $\overline{|C(t)|}^2$ changes after the fast component : a "correlation hole" appears. This can be explained in the following way : the random position of lines, f_i, are described by one and two level correlation functions : $R_1(f) = d$ which is the level density, and $R_2(f_1,f_2)$ which is the joint probability that there is a level at f_1 and another level at f_2. The two levels correlations can be expressed as $R_2(f_1,f_2) = R_1(f_1).R(f_2).(1 - Y_2(d.f_1,d.f_2))$ where Y_2 is the two levels cluster function given by Metha[2]. If Y_2 is not zero, i.e. if correlations exist, the "slow component" is multiplied by $[1 - b_2(t)]$, where $b_2(t)$ is the F.T. of Y_2. The shape of $[1 - b_2(t)]$ is shown on figure 1 for two limiting cases : POISSON ($b_2(t) = 0$) and G.O.E. The decrease of the amplitude of $1 - b_2(t)$ near the origin (for $t/\rho < 1$) is called a "correlation hole". The shape of this "hole" depends on the particular system and experimental situation. Furthermore, for any random spectrum (Poisson or G.O.E), the F.T. of a single spectrum is 100 % randomly modulated with a frequency of the order of the reciprocal of the width of the envelope. This random modulation is mathematically analogous to the speckle phenomenon familiar to laser users and it can be reduced either by an ensemble averaging or by smoothing of the spectrum or both.

The relationship between spectra and its $\overline{|C(t)|}^2$ is shown for typical examples in figure 1.

Some comments are necessary in order to better understand the meaning of the F.T. of a spectrum and espacially why correlations properties can been seen even when the spectrum is "bad" (see above).

First this F.T. method, looks familiar for molecular physicists dealing wiht the theory of radiationless transitions in "intermediate" molecules. Consider phenomena like intramolecular transfer of energy between a single vibrational non stationary state which can be directly excited by a pulse of light (with a laser) and a dense manifold of "dark" (i.e. for which there are no optical transitions) vibrational states. This dense set may belongs to the same electronic state (pure vibrational energy transfer) or an other electronic states with the same electronic spin (internal conversion to the ground state) or with different spin multiplicity (intersystem crossing between singlet and triplet). The "doorway state" which carry oscillator strength is "diluted" by coupling between a lot of neighbouring states, giving a spectrum composed of a lot of lines whose amplitude are the product of a deterministic envelope A_E which represente the smoothed energy dependence of the amount of the doorway state contained in each the molecular eigenstate, and a stochastic component because the coupling can be considered as random. The time evolution of the physical system is then gouverned by two components. A sufficiently short burst of light (a picosecond laser pulse) excite the pure (non stationary) doorway state which is in fact a coherent superposition of a stretch of stationary molecular eigenstates. All these eigenstates start to radiate in phase, but, since they have differents frequencies, they dephase in a time of the order of the reciprocal of the width of A_E, giving the "fast component". After this time the molecular eigenstates radiate incoherently giving the slow component whose time constant is the reciprocal of the width of individual spectral line (lifetime). The interesting point is the relationship between the correlations properties of eigenstates and theirs corresponding time evolution. This question has been addressed by Delory and Tric as early as in 1974[11] but up to now, no example of correlation hole has ever been observed in time resolved experiments. Only biexponential decay corresponding to a POISSON statistics have been reported. The speckle "noise" on the slow component, which can also be called many levels quantum beats, seems to have been overlooked in time resolved experiments because the fluorescence decay is smoothed, due to poor time resolution and/or superposition of several decay corresponding to simultaneous observation of different rotational levels.

As second comment, it is intersting to notice the relationship between the correlation hole of the F.T. and the $\Sigma^2(L)$ function used as standard test of long range correlations.

The shape of the correlation hole can be numerically related with the shape of the $\Sigma^2(L)$ function. For example, the ordinate and the slope at the origin of the slow component is related with the POISSON and G.O.E. contribution of the $\Sigma^2(L)$ function for large L. We conclude that the correlation hole of the F.T. of a spectrum is mainly a measure of long range statistical correlations properties of this spectrum. This explains why the F.T. is relatively insensitive to lack of resolution in the spectrum : long range correlations properties are lost, in principle, only if resolution is worse than the considered range of these properties. By contrast N.N.D. informations are rapidly lost as soon as the resolution is worse than the average level spacing.

As third comment, the superposition of m independent G.O.E. spectra produces a correlation hole m times narrower. The same informations about correlations can be recovered if one F.T. a stretch containing m times more levels. This explain why this F.T. method allows to display correlations even in impure spectrum which can be considered as the superposition of independent G.O.E.'s. This point is to be contrasted with autocorrelation or N.N.D. techniques for which the correlations properties, like level repulsion, disappears with only two independent G.O.E.

At this point we should remark that the superposition of independent set of levels may be due to profound and interesting physical reasons like for example when there is a strong anharmonic coupling inside subsets of modes of a molecule but not coupling between these subsets (this mean that there are remaining good quantum numbers).

As a result the F.T. method enables to measure the number of independent subsets, by measuring the ratio of the time corresponding to the onset of the correlation hole to the time corresponding with the level density (mean spacing).

As fourth comment, for a given "pure" subset, "intermediate coupling", i.e., coupling much smaller than for G.O.E. leads to <u>finite range</u> of the correlations properties. This gives a very characteristic shape to F.T. (see fig. 1). The correlation hole exist only between $t = \varrho$ and t_c, $t_c < t_o$, where t_c is a characteristic time for vibrational redistribution.

We emphasize the difference with G.O.E. for which t_c is basically 0. This t_c can be measured even for "bad" spectra as soon as the stretch of levels is larger than a few times reciprocal of t_c.

The crucial point was to found an integral method to measure long range properties, avoiding the step of defining individual components (the extraction of a stick spectrum from the experimental spectrum) which cannot be done when the resolution is worse or comparable with the average spacing of lines.

As a conclusion, the "correlation hole" is the signature of the correlation properties.

Examples of F.T. of experimental spectrum are given below.

V. Experimental results.

A. Excitation Spectrum and Microwave Double Resonance of NO_2

In 1975 Smalley et al[12] have obtained the excitation spectrum of NO_2, rotationally cooled (T_R = 3 K) in a supersonic jet in the region of 6708 A - 5708 A, i.e. from 14900 cm^{-1} up to 17500 cm^{-1} above the vibrationless electronic ground state. They observed 140 vibrational bands. The strong non adiabatic interaction between $^2A_1$ and $^2B_2$ explain the main features of the observed spectrum[13]. We discuss here the statistical analysis done by Haller et al[14] on this data set for which they found a Wigner statistics for N.N.D. First, in the range of 14900-16580 cm^{-1}, there are 20 "hot bands" identified among the 83 bands observed ("hot bands" originate from excited vibrational level(s) of the electronic ground state). In the remaining region of the spectrum (16580-17500 cm^{-1}) there are probably numerous hot bands; thus, there is a total of the order of 35 "hot bands" among the 140 observed bands. Second, each band consists of 10 to 50 rotational lines which spread over 10 cm^{-1}. The mean spacing between the bands is 19 cm^{-1}, thus many bands overlap and give a high probability that two bands are considered as one. This effect would produce a spurious "level repulsion". For example, in reference (12), there are many more lines in bands number 113 and 22 than in other bands (see bands number 115 and 95 for instance). This is a strong indication that bands number 113 and 22 are composed of at least two overlapped vibrational bands. As a conclusion the validity of the statistical analysis done by Haller et al on NO_2 is doubtful because there are too many spurious and missing levels. Furthermore, this analysis demonstrates the necessity to obtain a pure data set and consequently to use double resonance techniques.

Recently, Lehmann and Coy[15] have obtained, by a microwave-optical double resonance technique, a very dense spectrum of NO_2. An optical transition to high excited vibrational level at 16800 cm^{-1} to 17100 cm^{-1} above the vibrationless ground state level was excited. Specific rotational levels (J = 7 to 10) are selected with a microwave rotational transition in the vibrationless $^2A_1$ ground state. The number of observed transitions is a factor of 8 greater than expected if only the allowed rotational transition to any B_2 symmetry vibronic level were observed. Hardwick[16] has suggested that the numerous

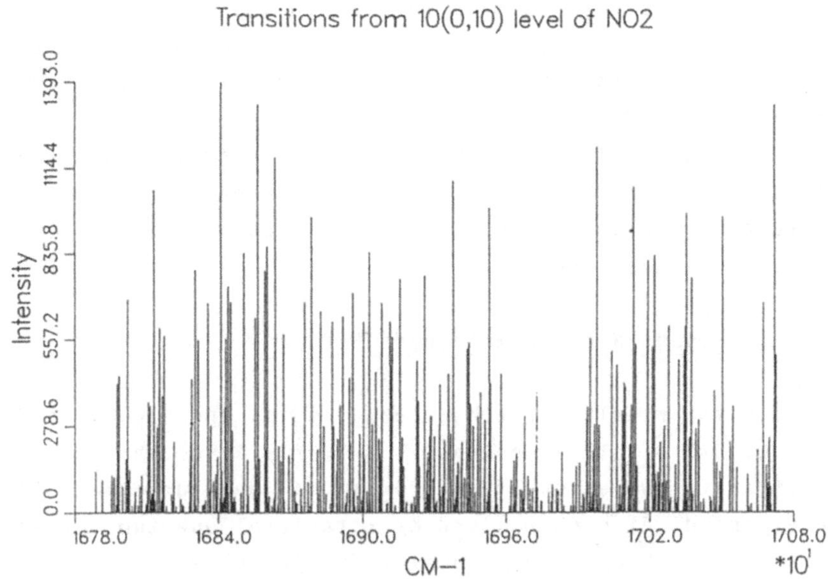

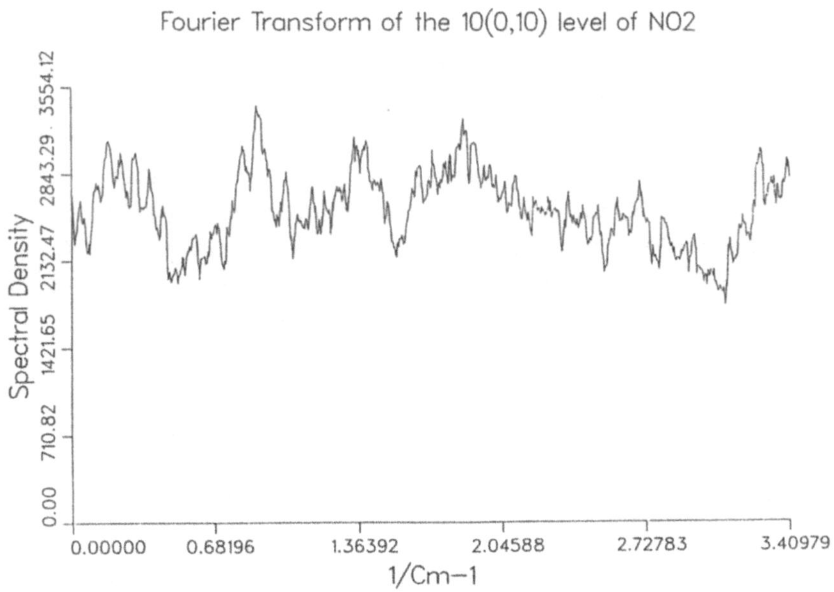

Fig. 2 : Stick experimental spectrum of M.O.D.R. of NO_2. and the corresponding, partly smoothed $|F.T|^2$. There is no correlation hole near the origin.

forbidden transitions observed are related to quantum ergodicity in these highly excited vibrational levels. The number of observed lines is a factor of 3 less than the number predicted if all selection rules wer broken as predicted by Hardwick. However, the analysis of these spectra by the F.T. method, as described above, displays no correlations[17] as shown of fig. 2. This breaking of rotational selection rules may also be explained by a molecular axis switching instead of vibrational coupling.

B. S.E.P. spectra on formaldehyde H_2CO

Very similar results have been obtained by the S.E.P. technique at MIT by H.L. DAI *et al*[18] on H_2CO : the only lines which appeared in the spectra at low values of the rotational quantum number ($J \leqslant 3$) are those expected. At higher J and K_a rotational quantum numbers, the spectra rapidly become more complex and the observed level densities at J = 10, K = 2 are several times larger than the known total density of vibrational levels. This increase in the density of accessible vibrational levels was the result of a rotation induced mixing of the anharmonic vibrational basis function (Coriolis coupling) which compromised the "goodness" of both vibrational and K_a quantum numbers. A further analysis of these results, with a criterion for chaos developped by Heller *et al*[19] shows that the increasing complexity in the spectrum with increasing angular momentum corresponds nonetheless with decreasingly chaotic behavior. This can be understood in terms of an available phase space volume that expands more rapidly than the occupied phase space volume as J increases.

In conclusion one can study the "pure" vibrational coupling by considering J = 0 levels and study rotational-vibrational coupling (Coriolis) by studying higher J (and K) rotational levels.

C. Stimulated Emission Pumping (S.E.P.) spectra of acetylene.

In 1983, Abramson *et al*[20] have observed, by S.E.P., very high vibrational energy levels (at 27900 cm^{-1}) of the ground electronic state of acetylene. A pulsed dye laser (the PUMP) excites (at 45300 cm^{-1}) a specific rovibronic level of the ^1A excited electronic state. A second pulsed dye laser (the PUMP) stimulates transitions down to high vibrational levels of the ^1Xg ground state. A decrease of the total fluorescence of the upper level is observed when the frequency of the PUMP laser is in resonance with a transition to a ^1Xg high vibration level. They observe a new type of vibrational behavior. The transitions are organized in clump of typically 50 lines with the same J values. The analysis of the spacings between adjacent lines (N.N.D.) from a stick representation of the spectrum shows that they approximatively follow a Wigner distribution in each clump. Further analysis by Mukamel et al[21] shows that there are missing levels in the stick spectrum, either due to the limited resolution which forbids the observation of too close levels or because the weakest lines are hidden in the noise. The analysis of the intensity distribution according to a modified Porter-Thomas distribution gives an estimation that 70 % of the levels are missing ! On the other hand, analysis with the Δ_3 statistics indicates that only about 15 % of the levels are missing. Further analysis by Sundberg et al[22] gives 75 % and 6 % of missing levels for intensity distribution analysis and Δ_3 statistics respectively. Clearly more experiments, especially at higher resolution, are required in order to avoid missing levels. But the Fourier Transform (see chapter IV) of the original spectrum as displayed in figure 3 shows that a strong correlation exists in the original spectrum . The strong correlation found in the stick representation of the acetylene spectrum, which might be due to a subjective analysis, is confirmed by the Fourier Transform technique.

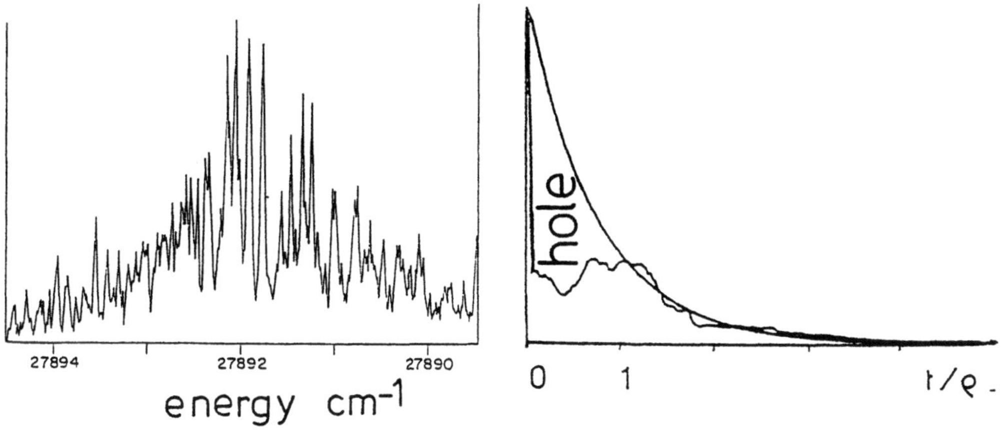

Fig. 3 : S.E.P. Acetylene spectrum near 27 900 cm^{-1} and the corresponding smoothed, Fourier Transform ($|F.T|^2$) which displays a correlation hole. A hole appears also on the $|F.T|^2$ of the corresponding stick spectrum.

D. Anticrossing spectra of glyoxal $C_2H_2O_2$ and methylglyoxal.

In the singlet-triplet anticrossing technique, a (strong) magnetic field is applied in order to Zeeman tune triplet levels into resonance with a single rovibrational level of a singlet electronic state. The anticrossing is a lorentzian, which is detected as a decrease in the fluorescence of the excited singlet. A strong magnetic field, provided by the Service National des Champs Intenses (S.N.C.I.) (Grenoble, France) allows one to obtain many anticrossings (and vibrational levels). Our Bitter coil gives a field up to 8 Tesla, which allows the observation of numerous triplet vibrational levels within $\pm$ 7 cm^{-1} of the selectively laser excited singlet level of S_1. The ground electronic state (S_0) of glyoxal (or methylglyoxal) is a singlet, and the first triplet state, T_1, is at lower energy than the first singlet excited state S_1. At the energy of the vibrationless level of S_1, the corresponding vibrational density of states of triplets, T_1, is $\approx$ 1 level per cm^{-1} for glyoxal, and $\approx$ 20 levels per cm^{-1} for methylglyoxal (see fig. 4). A supersonic jet of glyoxal is locates within the 100 mm bore diameter of the Bitter coil. A single rotational (N = 0) vibrational level of S_1 is selectively excited by c.w. ring dye laser. In contrast with an optical or microwave spectrum, all of the anticrossings have the same amplitude (a decrease of up to 50 % of the fluorescence intensity). The only "missing" anticrossings are due to magnetic field inhomogeneity broadening (1 Gauss 0.7 MHz), natural linewidth[23], and overlapping anticrossings.

Consequently, we expect only a few missing levels in the anticrossing spectrum.

For glyoxal, the width of the anticrossings varies over three orders of magnitude, from $\approx$ 1 gauss to $\approx$ 10^3 gauss. These widths are related to the magnitude of the singlet-triplet spin-orbit vibronic coupling. The excitation of the rotationless N = 0 singlet levels avoid the superposition of several anticrossing spectra which correspond to different Zeeman sublevels. According to the first order singlet-triplet anticrossing selection rules due to spin-orbit vibronic coupling[24], only triplet (S = 1) rotational level with N = 1, $M_N = \pm 1$, $M_S = \mp 1$ (and K = 0 or K = 1 according to the vibrational symmetries) can anticross the N = 0 excited singlet. The distribution of the widths (as contrasted with the distribution of line intensities in the spectra considered previously) results in a more complex shape for the "slow component" of the F.T. of the spectrum.

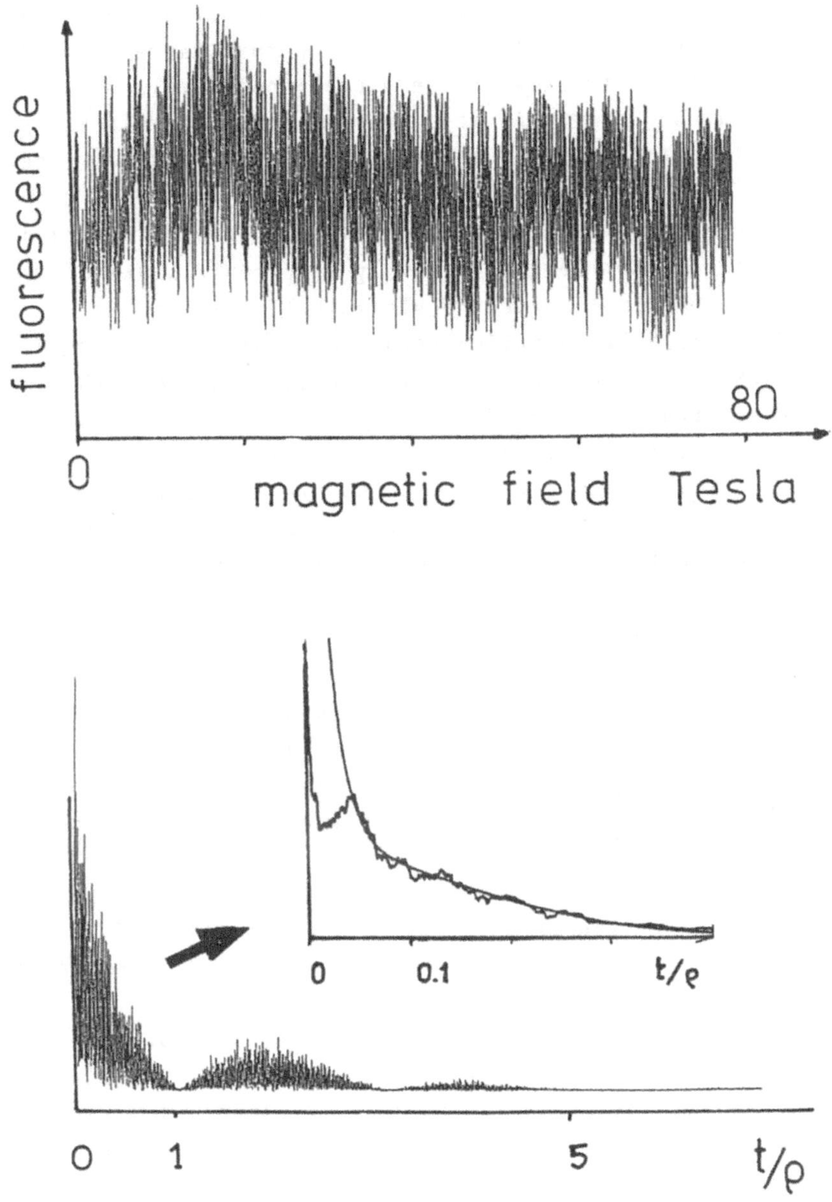

Fig. 4 : 0 to 8 Tesla anticrossing spectrum of Methylglyoxal and the unsmoothed |F.T| of this spectrum. The arches are due to hyperfine doublets of triplets levels which are separated by 28 Gauss. The smoothed |F.T| near the origin displays a correlation hole for $t/\rho \lesssim 0.05$.

The F.T. of the anticrossing spectrum of glyoxal does not display any "correlation hole", we conclude that there is no significant correlation property between the vibrational levels of the T_1 electronic state at 3500 cm^{-1}. At this energy, above the vibrationless level the vibrational density of states is 1.5 levels per cm^{-1} per symmetry.

In contrast, significant correlation properties are observed between vibrational levels of the T_1 electronic state of methylglyoxal at 3000 cm^{-1} above the vibrationless level.

The 0 to 8 Tesla anticrossing spectrum of methylglyoxal (CH_3-CO-CHO) of figure 4 looks like noise ! However all the features of this spectrum are reproducible. There are approximately 5000 anticrossings in this spectrum detailled below :

- The F.T. of this spectrum displays arches (fig. 4) which are due to hyperfine doublets (splitting of 28 gauss), induce by Fermi contact interaction. These arches correspond to deterministic properties of the spectrum.
- Furthermore, the $M_s = +1$ and $M_s = -1$ anticrossing spectra are superposed.
- We do not know all the anticrossing selection rules for methylglyoxal but, assuming the same kind of interaction (spin-orbit-vibronic) as in glyoxal, the observed anticrossing is composed of both N = 1, K = 0 and N = 1, K = 1 components. These component may be uncoupled or coupled.
- The two vibrational symmetry classes of the C_s point-group of methylglyoxal (or the classes of the corresponding G6 molecular symmetry group)[25] may also give two superposed anticrossing spectra.

As a result, between four and sixteen sets of independent symmetry are contained in our anticrossing spectrum.

A "correlation hole" appears in the smoothed F.T. of the anticrossing spectrum (figure 4) at $0 \lesssim t \lesssim 0.05$. The estimated shape of the "slow component" is also drawn. The width of the very narrow "correlation hole", 0.05, (instead of 1 for <u>one</u> G.O.E. spectrum, see above) is due to the superposition of several "pure" anticrossing spectra, as explained in chapter IV : the F.T. of a spectrum composed of m "correlated" spectra displays a "correlation hole" which is diminished in width by a factor of m.

Furthemore, a detailled inspection of the F.T. near the origin shows that the correlations disappear for large L ($L \gtrsim 100$) or conversely for small times $t \lesssim 0,01$ (in unit of reciprocal mean spacing).

Nevertheless, the existence of the "correlation hole" definitively shows significant correlations properties in triplet vibrational levels of methylglyoxal.

Summary and general conclusions.

The range of the number of degrees of freedom, the available range of density of states, and a variation in the coupling strength with energy make polyatomic molecules a fertile testing ground for the study of correlation properties in physical systems.

We are in the early age of the experimental study of the statistical properties of these vibrational levels. Only a few experimental molecular spectra display significant correlation properties. According to the discussion presented in chapters I to III, only two step state to state process, like optical-optical, or microwave-optical or optical-anticrossing, are able to produce spectra with the spectral purity required for the analysis of statistical correlation properties.

The understanding of these correlation properties may provide new insignts for molecular dynamics such as intramolecular vibrational redistribution and quantum chaos. The new technique of F.T. described in section IV is a crucial progress in analysis because it enables one to study large amounts of data at chemically relevant energies and complexities.

The existence of finite range of correlation properties (and the possibility of measuring very long streches of levels in molecular physics) is a new feature not contained in G.O.E., which, by construction, have an "infinite" range of correlation properties.

REFERENCES

(1) T.A. BRODY, J. FLORES, J.B. FRENCH, P.A. MELLO, A. PANDEY and S.S. WONG.
Rev. Mod. Phys. 53 385 (1981) and references cited within.

(2) M.L. MEHTA, Random Matrix (Academic, New-York 1967)

(3) G.E. POWELL and J.C. PERCIVAL
J. Phys. A : Math. Gen. 12 n°11 2053 (1979)

(4) a. O. BOHIGAS, M.J. GIANNONI and C. SCHMIT,
Phys. Rev. Lett. 52 1 (1984)
b. E. HALLER, H. KOPPEL and L.S. CEDERBAUM
Phys. Rev. Lett. 52 1665 (1984)
c. T.H. SELIGMAN, J.J.M. VERBAARSCHOT and M.R. ZIRNBAUER
Phys. Rev. Lett. 53 215 (1984)

(5) T.H. SELIGMAN, J.J.M. VERBAARSCHOT and M.R. ZIRNBAUER
J. Phys. A : Math. Gen. 18 2751 (1985)

(6) J. KATO, J. Chem. Phys. 82 3020 (1985)

(7) R. JOST, Second Conf. on Quantum Chaos, Mexico 1986.

(8) 1. C.S. PARMENTER, J. Phys. Chem. 86 1735 (1982)
2. B. FOURMANN, C. JOUVET, A. TRAMER, J.M. LE BARS and P. MILLIE
J. Chem. Phys. 92 25 (1985)

(9) P. FELKER and A. ZEWAIL
J. Chem. Phys. 82 2961, 2975, 2994, 3003 (1985)

(10) L. LEVIANDIER, M. LOMBARDI, R. JOST, J.P. PIQUE
Accepted in Phys. Rev. Lett. (1986)

(11) J.M. DELORY and C. TRIC, Chem. Phys. 3, 54 (1974)

(12) R.E. SMALLEY, L. WHARTON and D.H. LEVY
J. Chem. Phys. 63, n°11, 4977 (1975)

(13) E. HALLER, H. KOPPEL and L.S. CEDERBAUM
J. Mol. Spectros. 111, 377 (1985)

(14) E. HALLER, H. KOPPEL and L.S. CEDERBAUM,
Chem. Phys. Letters 101 215 (1983)

(15) K.K. LEHMANN and S.L. COY, J. Chem. Phys. 83 3290 (1985)

(16) J.L. HARDWICK, J. Mol. Spectrosc. 109 85 (1985)

(17) K.K. LEHMANN, S.L. COY, M. LOMBARDI and J.P. PIQUE
to be published

(18) H.L. DAI, R.W. FIELD and J.L. KINSEY,
J. Chem. Phys. 82, n°4, 2161 (1985)

(19) M.I. DAVIS and E.J. HELLER
J. Chem. Phys. 80 5036 (1984) and cited paper.

(20) E. ABRAMSON, R.W. FIELD, D. IMBRE, K.K. INNES and J.L. KINSEY
J. Chem. Phys. 80 2298 (1984)

(21) S. MUKAMEL, J. SUE and A. PANDEY
Chem. Phys. Lett. 105 134 (1984)

(22) R.L. SUNDBERG, E. ABRAMSON, J.L. KINSEY and R.W. FIELD
J. Chem. Phys. 83 466 (1985)

(23) J. DEROUARD, R. JOST and M. LOMBARDI
Journal de Physique Lettres 37, L 135 (1976)

(24) P. DUPRE, R. JOST and M. LOMBARDI
Chem. Phys. 91 355 (1984)

(25) P.R. BUNKER in Vibrational Spectra and Structure Vol. 3
J.R. DURIG ed. (Dekker, New-York), 1975
and in Molecular Symmetry and Spectroscopy,
Acad. Press, New-York 1979

NUCLEAR LEVEL DENSITIES AND LEVEL SPACING DISTRIBUTIONS

T.von Egidy, A.N. Behkami* and H.H. Schmidt
Technische Universität München, D 8046 Garching, Germany

Abstract

Extensive and complete level schemes and neutron resonance densities of 72 nuclides have been used to determine the spin cut-off parameter, parameters for level density formulae and level spacing distributions. The level spacing distributions indicate either very different nuclear structures or the existence of a further good quantum number at low excitation energies in addition to spin and parity.

1. Introduction

Statistical interpretation of nuclei yields important information on the nuclear structure[1,2]. The nuclear level densities play an essential role in all statistical theories, and their dependence on the atomic mass number A, the pairing energy and the shell structure has frequently been discussed[3-5]. The level spacing distribution of levels with the same quantum numbers (in particular spin and parity) is supposed to follow a Wigner distribution[6]; this means that close spacings are reduced due to level repulsion. Consequently, information on correlations between levels and on the number of good quantum numbers can be obtained from level spacing distributions.

The purpose of the present paper is to test statistical theories at low excitation energies (below 10 MeV). 72 nuclei from ^{20}F to ^{250}Cf have been selected including a total of 1672 levels. These nuclei have rather extensive and complete level schemes. In most cases at least the first twenty levels are known with spin and parity assignment. A large fraction of the selected nuclei has been carefully studied with (n,γ), (d,p) and other reactions. It can be assumed that these level schemes are about 95% complete in a given energy and spin

---

* Permanent Address: University of Shiraz, Shiraz, Iran

range. Neutron resonance densities[7] are available of most of these nuclei. These detailed experimental informations have been applied to determine A-dependent spin cut-off parameters, to fit the parameters in level density formulae and to test the formulae, and to obtain level spacing distributions. Details of the present contribution will be published elsewhere[8].

2. Statistical formulae

The nuclear temperature T can be defined by the level density $\rho(E)$[9]

$$1/T = d(\log \rho(E))/dE. \quad (1)$$

Integration yields the constant temperature Fermi gas formula

$$\rho(E) = (1/T)\exp((E-E_o)/T), \quad (2)$$

T and E_o (ground state back shift) being fit parameters to experimental data. The Bethe formula of the level density[3,4,10] is

$$\rho(E) = \exp(2\sqrt{a(E-E_1)})/(12\sqrt{2}\,\sigma\, a^{1/4}(E-E_1)^{5/4}), \quad (3)$$

with the level density parameter a and the back shift E_1. The spin cut-off parameter σ determines the spin distribution and the spin dependent level density[3,10]:

$$\rho(E,J) = f(J)\rho(E) \;;\; f(J) = \exp(-J^2/2\sigma^2)-\exp(-(J+1)^2/2\sigma^2). \quad (4)$$

σ is related to an effective moment of inertia I_{eff} and to T[4,9]:

$$\sigma^2 = I_{eff}\cdot T/\hbar^2 \text{ and for a rigid body } \sigma^2 = 0.0150\cdot A^{5/3}\cdot T. \quad (5)$$

A is the mass number. Gilbert and Cameron[3] calculated σ for the Bethe formula with reduced moment of inertia

$$\sigma^2 = 0.0888\cdot A^{2/3}\sqrt{a(E-E_1)}. \quad (6)$$

It is evident that the Bethe formula (3) is not meaningful at low excitation energies in particular if $E_1>0$.

The parameters a and E_1 or T and E_o, respectively, were fitted to the experimental level spacings $S_i = E_{i+1}-E_i$ and to the experimental neutron resonance densities ρ_{res} with

$$\chi^2 = \Sigma\{(S_i-D(E_i'))/D'(E_i')\}^2 + \{(\rho_{res}-\rho(B_n))/\Delta\rho_{res}\}^2; \quad (7)$$

$E_i' = (E_i+E_{i+1})/2$; B_n = neutron binding energy; and $D(E,J_1,J_2) = 1/\{(\exp(-J_1^2/2\sigma^2)-\exp(-(J_2+1)^2/2\sigma^2))\cdot\rho(E)\}$ in the spin range $J_1 \rightarrow J_2$. The weight D' is equal to D, but not varied in the fit. This weight corresponds to the average spacing variation.

The spacing P(S/D) of levels with all spins and parities

follows an exponential distribution[2]

$$P(S/D) = \exp(-S/D), \qquad (8)$$

while the spacing distribution of levels having the same good quantum numbers is expected to be described by a Wigner distribution[6]

$$P(S/D) = (\pi/2)(S/D)\exp\{-(\pi/4)(S/D)^2\}. \qquad (9)$$

Brody proposed an interpolation formula[11]

$$P(S/D) = (\omega+1)\alpha(S/D)^{\omega}\exp(-\alpha(S/D)^{\omega+1}) \text{ with} \qquad (10)$$

$$\alpha = (\Gamma\{(\omega+2)/(\omega+1)\})^{\omega+1}.$$

This correspondends to the exponential and Wigner distribution for $\omega = 0$ and $\omega = 1$, respectively.

3. Results and discussions

The spin cut-off parameter was determined by fitting the spin distribution (4) to the experimental spin distributions in each nucleus with an A-dependent ansatz $\sigma = x A^y$ resulting in

$$\sigma = (0.98 \pm 0.23) A^{(0.29 \pm 0.06)}, \quad \bar{\chi}^2 = 1.13. \qquad (11)$$

The errors are large because they are strongly correlated. The formula gives σ-values 3.1 and 3.9 for Fe and Sb, respectively, in agreement with previous experimental values for energies between 4 and 6 MeV in this region[12].

The parameters of the level density formula (2) with σ from (11) and of formula (3) with σ from (6) were determined with a least squares fit and χ^2 from (7). Both formulae reproduce the experimental level sequences equally well with $\bar{\chi}^2 \approx 1$. The values T and a are displayed in Fig. 1 and Fig. 2, respectively, as a function of A. A surprising result is that the temperature T is a rather smooth function of A and can be fitted with an ansatz $T = xA^y$, if Pt, Hg, Pb and Bi are excluded; the result is

$$T = (19 \pm 2) \cdot A^{-(0.68 \pm 0.02)}, \quad \bar{\chi}^2 = 3.3 \qquad (12)$$

The density parameter a shows much more variation with the shell structure of the nuclides. A similar fit as for T gives

$$a = (0.21 \pm 0.03) \cdot A^{(0.87 \pm 0.03)}, \quad \bar{\chi}^2 = 14.6.$$

The ground state back shift parameters E_o and E_1 are related to the pairing energy[3]. The newest experimental pairing

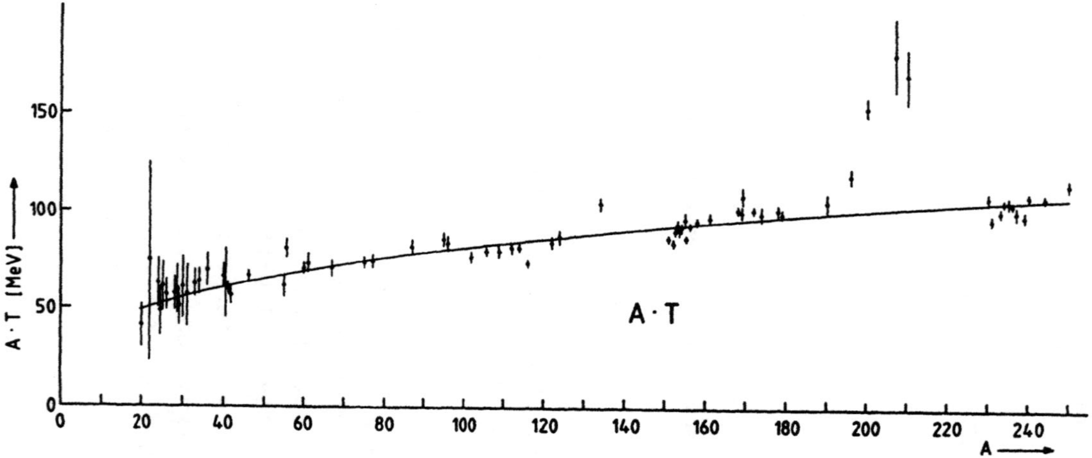

Fig. 1: Nuclear temperatures T times A determined with the constant temperature formula.

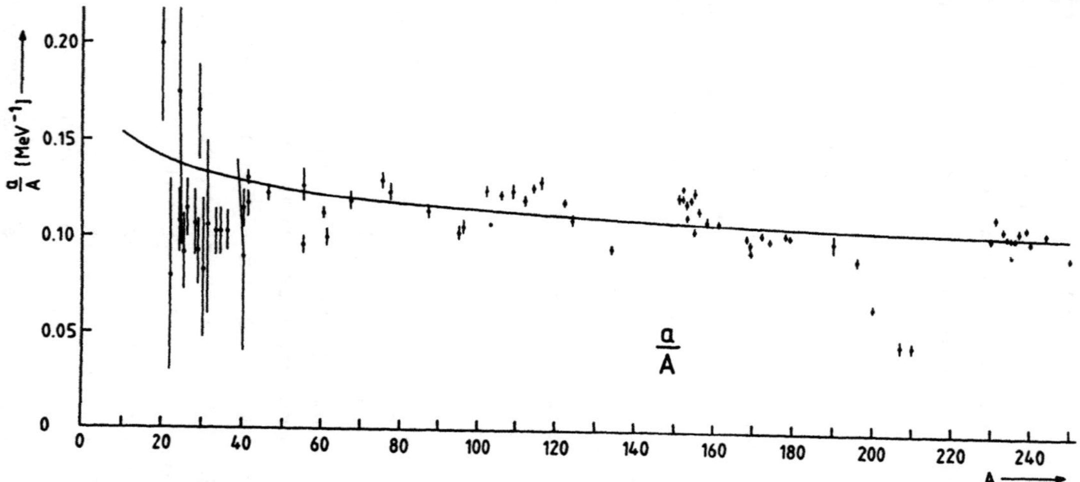

Fig. 2: Bethe parameter a divided by A.

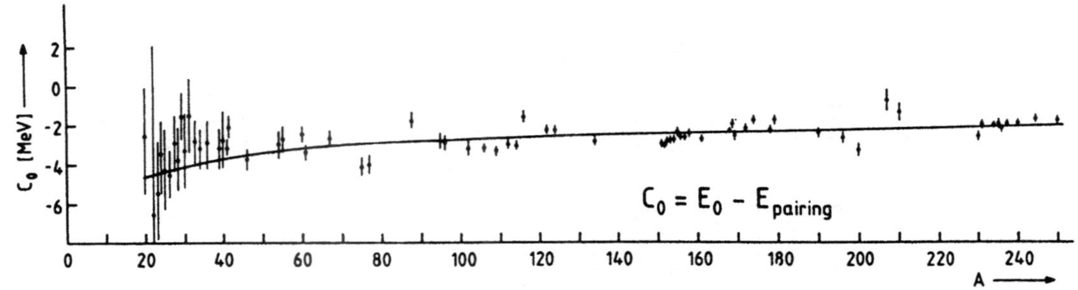

Fig. 3: Corrected backshift parameter C_o of the constant temperature formula.

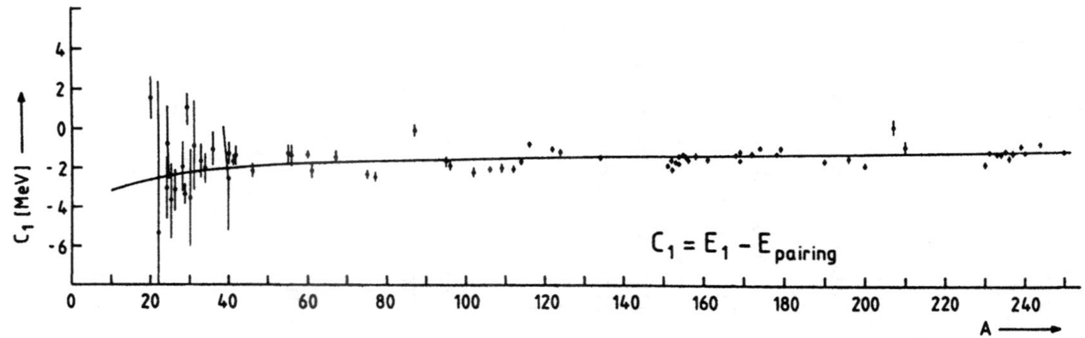

Fig. 4: Corrected back shift parameter C_1 of the Bethe formula.

energies E_{pair} were used which are calculated from mass differences by Wapstra and Audi[13]. The values $C_o = E_o - E_{pair}$ and $C_1 = E_1 - E_{pair}$ are shown in Fig. 3 and Fig. 4, respectively. It was tried to reproduce C_o and C_1 with a function of A:

$$C_o = E_o - E_{pair} = -(12\pm3)\cdot A^{-(0.32\pm0.05)}, \bar{\chi}^2 = 3.0;$$

$$C_1 = E_1 - E_{pair} = -(6.6\pm2.7)\cdot A^{-(0.32\pm0.08)}, \bar{\chi}^2 = 6.2.$$

The back shift C_o of the constant temperature formula is a smoother function of A than C_1.
The experimental level spacing distributions $P(S_i/D(E_i'))$ are given in Figs. 5 and 6. These distributions are compared with exponential (Poisson),Wigner and Brody distributions. Brody's parameter ω was fitted. The spacing including all levels is close to the exponential curve as expected. The distribution of levels with the same spin and parity lies definitely between an exponential and Wigner distribution (Fig. 5).
Fig. 6 demonstrates that the spacing distribution is close to an exponential function for even-even nuclides and different for odd and odd-odd nuclides. This is a clear indication that either very different structures exist or that at least one further good quantum number plays a role at low excitation energies in addition to spin and parity, in particular in even-even nuclides. This quantum number is less good in odd and odd-odd nuclides. The K-quantum number or the seniority may be relevant in this context. These results are at variance with previous findings [14] and spacings distributions of neutron[15] and proton[16] resonances which are reproduced by Wigner distributions.

Acknowledgement: We wish to thank A.H. Weidenmüller for stimulating discussions and P. Stoeckel for graphical help.

References

1 T. Ericson, Adv.Phys. 9 (1960) 425
2 T.A. Brody et al., Rev.Mod.Phys. 53 (1981) 385
3 A. Gilbert and A.G.W. Cameron,Can.J.Phys. 43 (1965) 1446
4 W. Dilg et al., Nucl.Phys. A217 (1973) 269
5 G. Rohr, Z. Physik A 318 (1984) 299
6 E.P. Wigner, Oak Ridge Nat.Lab.Rep. ORNL-2309 (1957) p.59
7 G. Rohr et al., Geel, Belgium, private communication 1979
8 T.v.Egidy, A.N.Behkami and H.H.Schmidt, Nucl.Phys.A,accepted
9 T. Ericson, Nucl. Phys. 11 (1959) 481
10 H.A. Bethe, Rev.Mod.Phys.9 (1937) 69
11 T.A. Brody, Lett.Nuovo Cimento 7 (1973) 482

12 R. Fischer et al., Phys.Rev. C30 (1984) 72, and references therein
13 A.H. Wapstra, G. Audi, Nucl. Phys. A 432 (1985) 55
14 T.A. Brody et al., Nucl.Phys. A259 (1976) 87
15 H.I. Liou et al., Phys. Rev. C11 (1975) 462
16 W.A. Watson III et al., Z. Physik A300 (1981) 89

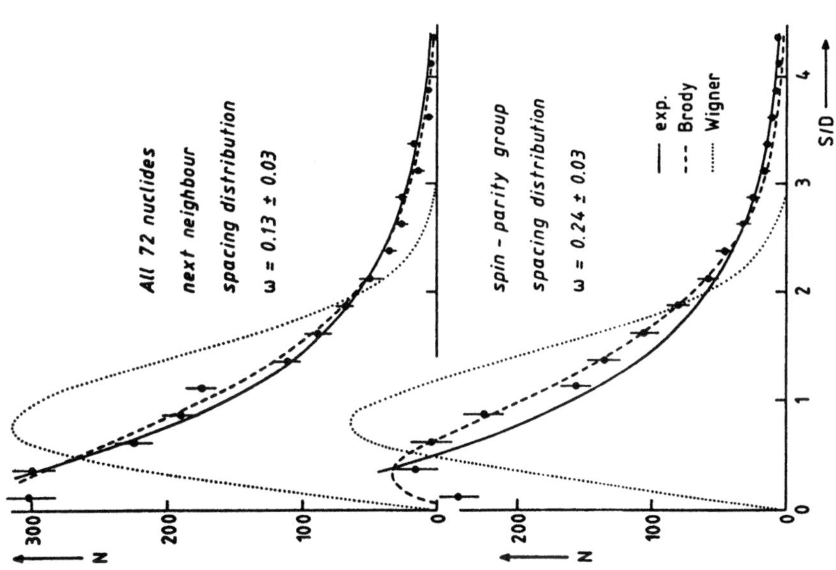

Fig. 5: Level spacing distributions $P(S_i/D(E'_i))$ for all nuclei without and with equal spin and parity.

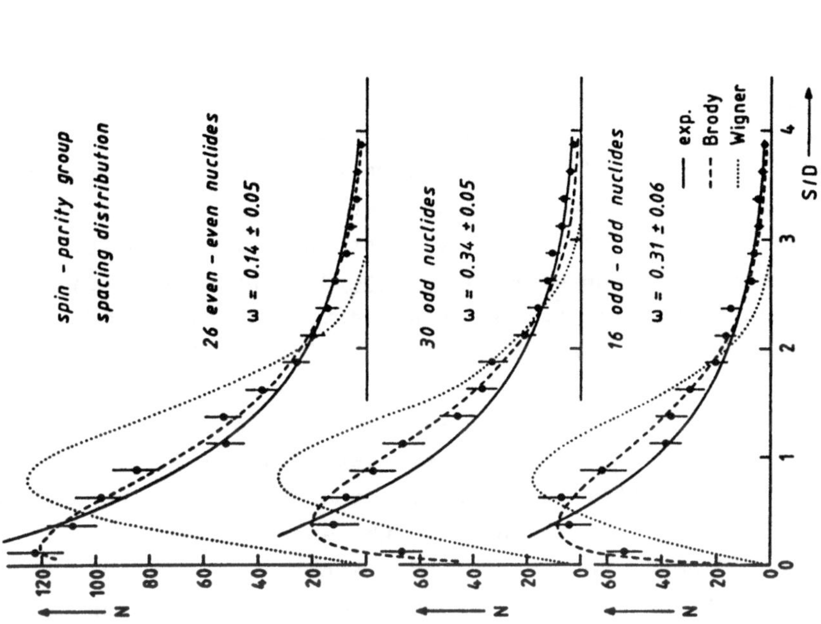

Fig. 6: Spacing distribution $P(S_i/D(E'_i))$ of levels with equal spin and parity for even-even, odd and odd-odd nuclides.

RECENT PROGRESS IN THE THEORY OF RANDOM-MATRIX MODELS

Akhilesh Pandey
Department of Physics & Astronomy
University of Rochester, Rochester, NY 14627 USA

1. Introduction

The notion that the energy levels of complex systems should be modeled, for their (local) fluctuation properties, by the eigenvalues of asymptotically-large random matrices is due to Wigner [1,2]. The three "classical" matrix ensembles, invariant under the transformations indicated, are: (i) the Gaussian orthogonal ensemble (GOE) of real symmetric matrices, (ii) the Gaussian unitary ensemble (GUE) of complex Hermitian matrices, and (iii) the Gaussian symplectic ensemble (GSE) of quaternion-real Hermitian matrices. They apply respectively to: (i) time-reversal invariant (TRI) systems with angular momentum either good or integer-valued, (ii) time-reversal noninvariant (TRNI) systems, and (iii) TRI systems with angular momentum broken but having half-integral values. In all three cases the energy levels under consideration should have the same set of exact quantum numbers.

The main predictions of these ensembles are the von-Neumann-Wigner [3] level repulsion and the Dyson-Mehta [4] long-range spectral rigidity, both increasing with β where β (= 1,2,4 respectively for real, complex and quaternion-real) counts the number of independent "sites" in the off-diagonal matrix elements. For example, the repulsion between nearest-neighbor levels is linear, quadratic and quartic respectively for the three ensembles.

Of the three, the GOE has received the most attention, applying, as it does, to spectra of complex nuclei, atoms and molecules. By the 1970's a broad agreement with GOE was established for the nuclear [5,2] and atomic [6] spectra. The subsequent theoretical studies focused on the "ergodic" properties [7] of the ensembles, as well as on new ensembles which explicitly deal with the few-body nature of the interactions [8,9], ignored in the GOE, or which contain large non-random components [10]. These studies, reviewed in [2], indicated that the fluctuation properties might be "universal" with the GOE defining the most significant universality class.

Three important developments of the past few years are: (i) demonstration of a remarkably close agreement between nuclear spectra and GOE by combining the nuclear data into a nuclear data ensemble

(NDE) of approximately 1800 levels and then using new spectrally-averaged measures for comparison [11], (ii) derivation of a bound on the TRNI part of the nuclear interaction by random-matrix methods [12], this resulting from combining the data analysis of (i) with a good understanding of the random-matrix theory for partial TRI-breaking [12,13], and (iii) studies of spectral fluctuations in quantum systems, with two degrees of freedom, whose classical analogs are chaotic, establishing the above-mentioned universality as a manifestation of quantum chaos [14,15]. These topics are discussed in detail in other talks of this conference. The primary purpose of this talk is to review the underlying random-matrix theory, stressing in particular the theory for the "transition to quantum chaos" as a symmetry of the system is gradually broken.

2. Matrix Ensembles with Partial Symmetry Breaking

Dyson [16] has emphasized the "all-or-nothing" character of the classical ensembles. They do not apply, for example, to systems for which a symmetry is only partially conserved, leading either to approximate quantum numbers or partial TRI (or both). We consider therefore the more general ensembles* [2,10]: $\{H(\alpha)=H(0)+\alpha V\}$, where $\{V\}$ is a classical ensemble $G_\beta(v^2)$ and $H(0)$, to be chosen according to the problem, may be a fixed (Hermitian) matrix, or may form another independent ensemble, with the constraint that, when transformed to the $H(0)$-diagonal representation, $\{V\}$ remains invariant. Here the decomposition is such that a symmetry of interest may be preserved for $\alpha=0$ but irrelevant for large α. For approximate quantum numbers, the $H(0)$-spectrum may be a superposition of several subspectra, whereas, for partial TRI, β is necessarily 2 but $H(0)$ may be a characteristic member of a GOE or GSE; $\alpha(\geq 0)$ thus determines the relative norm of the (global) symmetry-breaking.

2.1 Response Theory: The H_α-eigenvalues move as α increases and the motion is decomposable into infinitesimally small random components.

---

* Notation: The symbol $\{M\}$ stands for an ensemble of matrices M, and $\overline{W}$ for an ensemble average of the quantity W. $G_\beta(v^2)$ denotes a classical ensemble with variance v^2 for each of the β off-diagonal sites. The matrices are d-dimensional for $\beta=1,2$, with the primary interest being in large d; for $\beta=4$ the dimensionality is 2d, but, since the $H(\alpha)$-eigenvalues are doubly-degenerate, only the d distinct eigenvalues need usually be considered.

Let $\{V'\}$ be another independent $G_\beta(v^2)$. Since the sum of two independent Gaussian variables is a Gaussian with the summed variance, the perturbed ensemble $\{H(\alpha)+\delta\alpha V'\}$ is statistically equivalent to $\{H(\sqrt{\alpha^2+\delta\alpha^2})\}$ for any $\delta\alpha$. Moreover, in the first ensemble, transforming the matrices to $H(\alpha)$-diagonal representation leaves $\{V'\}$ invariant. It follows then that, for every $H(\alpha)$, the level motion, $\delta E_j(\alpha) = E_j(\sqrt{\alpha^2+\delta\alpha^2}) - E_j(\alpha)$, arising from the infinitesimal perturbation, can be described, to order $\delta\alpha^2$, by

$$\overline{\delta E_j(\alpha)} = \beta\ \delta\alpha^2\ v^2 \sum_{i(\neq j)} \frac{1}{E_j(\alpha)-E_i(\alpha)} \quad , \tag{1}$$

$$\overline{\delta E_i(\alpha)\delta E_j(\alpha)} = 2\ \delta\alpha^2\ v^2\ \delta_{ij} \quad , \tag{2}$$

all higher-order correlations being zero to the same order. This is equivalent to the Brownian motion of a two-dimensional Coulomb gas constrained to move on a line, and gives a differential equation for the joint-probability distribution of the eigenvalues [16]. The main problem is to derive the density-correlation functions, and the observable fluctuation measures therefrom, for this Brownian motion.

2.2 Level Density: As an example of the response theory we consider the ensemble average of the level density $\rho(E;\alpha) = d^{-1}\ \mathrm{tr}\ \delta(E-H(\alpha))$, or equivalently the ensemble average of the Stieltjes transform (or the Green function) $G(z;\alpha) = d^{-1}\ \mathrm{tr}\ (z-H(\alpha))^{-1}$. The effect of the infinitesimal perturbation can be derived from (1) or more directly by an expansion in powers of $\delta\alpha V'$. For large d we find [17], with $\beta v^2 d=1$,

$$\left(\frac{\partial}{\partial\alpha^2} + \overline{G}(z;\alpha)\ \frac{\partial}{\partial z}\right)\ \overline{G}(z;\alpha) = 0 \tag{3}$$

whose solution is the Pastur equation [18,10]

$$\overline{G}(z;\alpha) = \overline{G}(z - \alpha^2\overline{G}(z;\alpha);0). \tag{4}$$

For an illustration, take $H(0)=0$ so that $G(z;0)=z^{-1}$. Eq. (4) then yields the Stieltjes transform of the semicircular density.

2.3 Transition Parameter for the Fluctuations: We now show [10,12] that the transition in the fluctuation properties occurs when $\alpha^2 \sim d^{-1}$, or more precisely for finite values of the (energy-dependent) local parameter, $\Lambda=\alpha^2 v^2/D^2$, which measures, to within a multiplicative constant, the mean-square-symmetry-breaking matrix element in units of the local average spacing $D(E) = (\overline{\rho}(E)d)^{-1}$. (This incidently implies that, for $d\to\infty$, the transition as a function of α is discontinuous at

α=0, there being thus an extreme sensitivity to global symmetry-breaking and relatively smooth behavior with respect to local symmetry-breaking.) For the fluctuation properties of the levels in an interval [E-rD/2, E+rD/2], the effect of the αV-perturbation can be divided into two parts: (i) the effect of the interaction with the distant levels, say the levels outside the interval [E-pD/2, E+pD/2] where d>>p>>r, and (ii) the effect of the interaction with the levels inside the p-interval (including the levels of the r-interval). For $\alpha^2 \lesssim d^{-1}$ the first can be dealt with in second-order perturbation theory, and produces a shift of the centroid and rescaling of the average spacing by a factor $(1+o(d^{-1}))$. For the second, one must diagonalize the local matrix $\hat{H}(0)+\hat{V}$ where $\hat{H}(0)$ gives the H(0)-levels contained in the p-interval, (locally) rescaled to unit average spacing and shifted to zero centroid, and $\{\hat{V}\}$ is a $G_\beta(\Lambda)$. The fluctuations in the r-interval become those of G_β when $\hat{H}(0)$ and $\hat{V}$ have comparable norms, i.e., when $p \simeq 12\beta\Lambda >> r$. The effect of further increments in α, $(\alpha^2 >> d^{-1})$, is to rescale the average spacing without any change in the fluctuation properties. For the locally renormalized spectra $e_j(\Lambda) = (E_j(\alpha)-E)/D$, with $\delta e_j(\Lambda) = e_j(\Lambda+\delta\Lambda)-e_j(\Lambda)$, $\delta\Lambda=\delta\alpha^2 v^2/D^2$, the Brownian-motion equations (1,2) become [17]

$$\overline{\delta e_j}(\Lambda) = \beta\, \delta\Lambda \sum_{i(\neq j)} \frac{1}{e_j(\Lambda) - e_i(\Lambda)} , \tag{5}$$

$$\overline{\delta e_i(\Lambda)\delta e_j(\Lambda)} = 2\, \delta\Lambda\, \delta_{ij} , \tag{6}$$

where the sum extends over the levels of the p-interval with $p\to\infty$.

2.4 <u>Hierarchical Relations for the Correlation Functions</u>: All the fluctuation measures (see Section 3) derive from the set of n-level correlation functions $R_n(r_1,\ldots,r_n)$ which give the probability density of observing a level each at $r_1,\ldots,r_n$, or equivalently from the n-level cluster functions $Y_n(r_1,\ldots,r_n)$ which follow from a cumulant expansion of the R_n [19]; $R_1=Y_1=1$, $R_2=1-Y_2$ etc; for example for the Poisson ensemble $R_n=1$ and $Y_n=0$ for all n>1. For locally stationary ensembles the functions depend only on the relative coordinates (r_i-r_j). The H(α)-ensemble is locally stationary for all α if it is for α=0. It can be shown [17] that (5,6) yield, for n>1, BBGKY-like hierarchical relations among the correlation functions:

$$\frac{\partial R_n}{\partial\Lambda} = \sum_{j=1}^{n} \frac{\partial}{\partial r_j}\left(\Delta_n^\beta \frac{\partial}{\partial r_j}\left(\frac{R_n}{\Delta_n^\beta}\right) - \beta\, P\!\int_{-\infty}^{\infty} \frac{R_{n+1}}{r_j - r_{n+1}}\, dr_{n+1}\right) \tag{7}$$

where P denotes the principal value of the integral and $\Delta_n=\Pi|r_k-r_\ell|$

with $n \geq k > \ell \geq 1$. These relations should give uniquely the correlation functions for any Λ in terms of those for $\Lambda=0$ (i.e., for the H(0)-eigenvalues).

2.5 Classical Ensembles: The "stationary" solution of (7), $\partial R_n/\partial\Lambda=0$, obtains for $\Lambda=\infty$ (i.e., for the classical ensembles) and yields, after one integration,

$$\Delta_n^\beta \frac{\partial}{\partial r_j}\left(\frac{R_n}{\Delta_n^\beta}\right) - \beta\, P\int_{-\infty}^{\infty} \frac{R_{n+1}}{r_j - r_{n+1}}\, dr_{n+1} = 0 \tag{8}$$

valid for all $j=1,\ldots,n$. Two significant features follow immediately from (8). For small Δ_n, the integral term can be ignored and, then, $R_n \propto \Delta_n^\beta$, defining the degree of level repulsion. Moreover, multiplying (8) by r_j and then letting $r_j\to\infty$ yields, for $n\geq 1$,

$$\int_{-\infty}^{\infty} \big(R_{n+1}(r_1,\ldots,r_{n+1}) - R_n(r_1,\ldots,r_n)\big)dr_{n+1} = -n\, R_n(r_1,\ldots,r_n) \tag{9}$$

which, for n=1, is an essential ingredient for spectral rigidity; see Section 3.1. Both features are also implied if we extend β, in (8), to all values ≥ 0; then $\beta=0$ yields the Poisson ensemble, whereas $\beta=\infty$ describes the uniform spectrum, in line with Dyson's interpretation [16] of β as the inverse of a "temperature".

The solution of (8) for classical β is given by Dyson and Mehta [19]:

$$R_n = \mathrm{Qdet}\,[\Phi(r_i-r_j)]_{i,j=1,\ldots,n} \quad . \tag{10}$$

Here Φ, a quaternion function, is represented by a two-dimensional matrix,

$$\begin{aligned}
\Phi(r) &= \begin{pmatrix} S(r) & D(r) \\ J(r) & S(r) \end{pmatrix} \quad , \beta=1 \ , \\
&= \begin{pmatrix} S(r) & 0 \\ 0 & S(r) \end{pmatrix} \quad , \beta=2 \ , \\
&= \begin{pmatrix} S(2r) & D(2r) \\ I(2r) & S(2r) \end{pmatrix} \quad , \beta=4 \ ,
\end{aligned} \tag{11}$$

$$\begin{aligned}
S(r) &= (\sin \pi r)/\pi r \quad , \quad D(r) = dS/dr \quad , \\
I(r) &= \int_0^r S(r')dr' \quad , \quad J(r) = I(r) - \epsilon(r) \quad , \\
\epsilon(r) &= 1/2,\ 0,\ -1/2 \ , \quad \text{for } r > \ , = \ , < 0 \ .
\end{aligned} \tag{12}$$

Qdet, the quaternion-determinant, is the square root of the determinant of the self-dual 2n-dimensional matrix $[\Phi]$, and is a scalar polynomial in Φ; for example $Y_2 = \Phi(r_1-r_2)\Phi(r_2-r_1)$.

2.6 TRNI Interpolations: The solutions of (7), for β=2, with GOE and GSE as boundary (Λ=0) conditions, are also now known [13]. We have again (10) with Λ-dependent Φ given by

$$\Phi(r;\Lambda) = \begin{pmatrix} S(r) & D(r;\Lambda) \\ J(r;\Lambda) & S(r) \end{pmatrix} ,$$

$$D(r;\Lambda) = -\pi^{-1} \int_0^{\pi} k \sin(kr) \exp(2\Lambda k^2) dk ,$$

$$J(r;\Lambda) = -\pi^{-1} \int_{\pi}^{\infty} k^{-1} \sin(kr) \exp(-2\Lambda k^2) dk \qquad (13)$$

for GOE→GUE transition, and

$$\Phi(r;\Lambda) = \begin{pmatrix} S(r) & K(r;\Lambda) \\ I(r;\Lambda) & S(r) \end{pmatrix}$$

$$K(r;\Lambda) = -\pi^{-1} \int_{\pi}^{\infty} k \sin(kr) \exp(-2\Lambda k^2) dk ,$$

$$I(r;\Lambda) = -\pi^{-1} \int_0^{\pi} k^{-1} \sin(kr) \exp(-2\Lambda k^2) dk \qquad (14)$$

for GSE→GUE transitions. Eq. (9) is valid for these ensembles also. Note that, for $\Lambda \to \infty$, the products JD and IK are both zero, yielding thereby (11) for β=2. Moreover, Λ=0 in (13) corresponds to β=1 in (11), whereas the same in (14) yields the β=4 form of (11) after the double-degeneracy of the eigenvalues is incorporated into the latter.

2.7 General Solution: The general solution of (7) with arbitrary boundary condition is not yet known. However approximate expressions for Y_2 have been derived by the binary-correlation method [10,12]; for further results the replica trick and supersymmetry methods [20] may be worth exploring. Besides that some new results have been obtained directly from (7); for example small-Λ results may be derived by replacing the integral term in (7) by its value at Λ=0; see (17) ahead. We also remark that Eqs. (5,6) provide a simple procedure for making Monte-Carlo simulations of the ensembles for any boundary condition. Finally, the response theory applied to more complicated ensembles implies a universality of fluctuations which we consider in Section 4.

3. Fluctuation Measures

3.1 Two-Level Measures: The important measures are the variance $\Sigma^2(r)$ of the number of levels in intervals of length rD, and the average least square deviation $\bar{\Delta}_3(r)$ of the spectral staircase function from the best-fit straight line. Both derive from Y_2[4,7]:

$$\Sigma^2(r) = r - 2 \int_0^r (r-s)Y_2(s)ds \quad , \tag{15}$$

$$\bar{\Delta}_3(r) = 2r^{-4} \int_0^r (r^3-2r^2s+s^3)\Sigma^2(s)ds \quad . \tag{16}$$

For Poisson, $\Sigma^2(r)=r$ and $\bar{\Delta}_3(r)=r/15$, whereas for the classical ensembles $\Sigma^2(r)=2(\beta\pi^2)^{-1}\ln r$ + const. and $\bar{\Delta}_3(r)=(\beta\pi^2)^{-1} \ln r$ + const. for large r; see [2] for detailed expressions. Similarly, for the TRNI interpolations of Section 2.6, $\Sigma^2(r;\Lambda) = \Sigma^2(r;\infty)+(2\pi^2)^{-1} \ln(r^2+16\pi^2\Lambda^2) + \xi(r;\Lambda)$ where ξ depends on the boundary condition (GOE or GSE) but becomes independent of r for large r [12,13]. The important result here is that, if $\int Y_2(r)dr=1$ as in (9), Σ^2 and $\bar{\Delta}_3$ do not contain terms linear in r for large r, and then the logarithmic expressions derive from the asymptotic form of Y_2, resulting in a high degree of long-range rigidity in the spectrum. For other boundary conditions, the perturbation method, mentioned in Section 2.7, yields $o(\Lambda)$-terms in Σ^2. For example, for Poisson→classical transitions we have [12]

$$\Sigma^2(r;\Lambda) = r - 2\,\beta\,\Lambda\,\ln(r^2/2\Lambda) + 2\,\beta\,\Lambda\,(\psi(\beta/2+1)-1) \tag{17}$$

valid to order Λ for $r \gg \Lambda^{1/2}$ with ψ the digamma function.

3.2 <u>Higher-Order Measures</u>: The (discrete) probability function $E_n(r)$, giving the probability of observing n levels in intervals of length rD, derives from the correlation functions of all orders [21]:

$$E_n(r) = 1+ (n!)^{-1} \sum_{k=1}^{\infty} (-1)^k (k!)^{-1} \hat{R}_{n+k}(r) \quad , \tag{18}$$

$$\hat{R}_k(r) = \int_0^r \cdots \int_0^r R_k(r_1,\ldots,r_k)dr_1\ldots dr_k \quad . \tag{19}$$

The average and variance of the $E_n(r)$-distribution are respectively r and $\Sigma^2(r)$. Its skewness $\gamma_1(r)$ and excess $\gamma_2(r)$, deriving from two-, three-, and four-level functions, are used as higher-order measures in the data analysis [11]; see Fig. 1. For Poisson $E_n(r)=(r^n/n!)\exp(-r)$, whereas, for the ensembles satisfying (10), we have [21]

$$E_n(r) = \frac{(-1)^n}{n!} \left(\frac{d}{dz}\right)^n \prod_{\ell=0}^{\infty} (1-z\mu_\ell)\Big|_{z=1} \tag{20}$$

where the μ are the distinct eigenvalues of the integral equation

$$\int_{-r/2}^{r/2} \Phi(x-y)\, f(y)dy = \mu\, f(x) \quad . \tag{21}$$

3.3 <u>Spacing Distributions</u>: The probability density $p_k(S)$, $k\geq 0$, of the (k+1)'th-nearest-neighbor spacing (having the mean k+1), as well as its variance $\sigma^2(k)$, are also used often. The former follows from the E_n[21]:

$$p_k(S) = \sum_{n=0}^{k+1} (k+1-n) \frac{d^2 E_n(S)}{dS^2} \quad , \tag{22}$$

but becomes a Gaussian density for large k. For Poisson, $p_k(S) = (S^k/k!)\exp(-S)$. For the classical ensembles the Wigner surmise $AS^\beta \exp(-BS^2)$, with A, B adjusted to yield unit normalization and mean, provides excellent approximation for $p_0(S)$. For the variance, the difference $\Sigma^2(k+1)-\sigma^2(k)$ is usually small, zero for Poisson but 1/6 for large k for the ensembles satisfying (9) for n=1,2.

3.4 Sum Rules and "Uncorrelated" Ensembles: More generally joint-distributions of nearest-neighbor spacings could be considered. They are related to the R_n by sum rules. Eqs. (18,22) together provide one such example. Two other important sum rules are:

$$\sum_{k=0}^{\infty} p_k(S) = R_2(S) \quad , \tag{23}$$

$$\sigma^2(0)\left(1+2\sum_{k=0}^{\infty} C_k\right) = 1-\int_{-\infty}^{\infty} Y_2(r)dr \quad . \tag{24}$$

Here, in (24), C_k is the correlation coefficient between nearest-neighbor spacings separated by k levels, and derives from $2\sigma^2(0)C_k = \sigma^2(k+1) - 2\sigma^2(k)+\sigma^2(k-1)$. For Poisson the C_k are all zero, whereas for the classical ensembles they are all negative but rapidly approach zero for large k. One can construct other ensembles with given $p_0(S)$ but $C_k=0$ for all k. For example, the uncorrelated Wigner (UW) ensemble, with $p_0(S)$ the Wigner surmise for $\beta=1$, has been used in data analysis to illustrate spectral rigidity and higher-order effects [11]; see Fig. 1. For such ensembles, $\sigma^2(k+1) = (k+1)\sigma^2$ and, from (15,23), $\Sigma^2(r) = r\sigma^2 + (3(1+\sigma^2)^2-2M_3)/6$ for large r where σ^2 and M_3 are respectively the variance and third moment of $p_0(S)$.

3.5 Superposition of Independent Spectra: When several independent subspectra (say, ℓ of them with f_i the fraction of levels belonging to the i'th subspectrum, $\Sigma f_i=1$, $i=1,\ldots,\ell$) are randomly superposed, an interval of length rD contains Σn_i levels where n_i is the number belonging to the i'th subspectrum. Since the n_i are independent, and $\overline{n}_i=f_i r$, we have for the superposed spectrum

$$\Sigma^2_{\text{sup.}}(r) = \sum_{i=1}^{\ell} \Sigma^2(f_i r) \quad . \tag{25}$$

Similarly for $\overline{\Delta}_3$ and higher-order measures; see for example [7]. A significant result here is that, for $f_i \sim \ell^{-1}$ and $\ell >> r$ (but not for $\ell \lesssim r$), Poisson results are obtained. These results are useful when spectra belonging to different sets of the exact quantum numbers (for example, J^π for the nuclear and atomic spectra) are overlapping.

3.6 Effect of Spurious and Missing Levels: Spectra with high degree of rigidity are extremely sensitive to spurious and missing levels. Both these effects arise in data due to experimental uncertainties and are usually random in nature. We have

$$\Sigma^2_{spur.}(r) = fr + \Sigma^2 \left((1-f)r\right) \quad , \tag{26}$$

$$\Sigma^2_{miss.}(r) = fr + (1-f)^2\, \Sigma^2 \left(r/(1-f)\right) \quad , \tag{27}$$

f being the fraction of levels spurious or missed.

3.7 Bias, Ergodicity and Sample Errors: The fluctuation measures are given as ensemble averages whereas, in actual applications with samples containing say p levels, they are estimated as spectral averages. The question is whether the difference goes to zero for large p for most members of the ensemble. The answer is in the affirmative, if the ensemble average and variance of the difference go to zero. A non-zero value for the first indicates a biased estimator, whereas the same for the second shows lack of ergodic behavior in the ensemble. The main source of bias, especially in small samples, arises from the uncertainty associated with the estimation of the average spacing. As for the second, the above ensembles have been shown to be ergodic [7]; the ensemble variance, which is the square of the sample error, is usually proportional to p^{-1} for large p and therefore easily estimated from Monte-Carlo simulations [11]; the proportionality constant, small for rigid spectra, increases with the correlation length of the measure (for example with r for $\Sigma^2(r)$) and with its level-rank (for example it is larger for γ_1 and γ_2 than Σ^2 for the same value of r).

3.8 Comparison with Experiment: Extensive tests of the above-mentioned NDE, for comparison with GOE, have been recently carried out [11]. Shown in Fig. 1 are the results for Σ^2, γ_1 and γ_2 as functions of r. The GOE agrees with NDE, with a very high precision for the two-level measures, but also with significant precision for the higher-order ones. For the former, by fitting $\Sigma^2(r)$ with $ar+2b\ell nr+c$ for $1 \le r \le 25$, it is found that a = -0.007, b = 0.11 and c = 0.45, agreeing with the GOE predictions (0±0.005, 0.10±0.02, 0.44±0.02) but disagreeing with Poisson (1,0,0), UW (0.273,0,0.17), and GUE (0,0.05,0.34) values; note that (1-a) is an estimator of the total integral of Y_2.

An extension of the above analysis, using (26,27), shows that only a few percent of the levels may be missed or spurious. Moreover, comparison with the TRNI ensemble of GOE-GUE interpolation yields a bound of 0.1 on $\Lambda^{1/2}$ at 99.7% confidence level [12]; see next section for the significance of this bound.

The atomic spectral data may be classified into three parts according to whether the L,S quantum numbers are good, approximate, or completely broken. In the first case Poisson is obtained (because of mixed spectra; Section 3.5 above), in the second Poisson-GOE interpolation, and in the third the GOE [6]. New analysis of the same data, with the new improved measures, has not yet been carried out. In particular a comparison of the second with (17) (or its extension to larger values of Λ) should yield a value of Λ for the LS-breaking.

The quality of the molecular spectral data is improving, but the fraction of missed and spurious levels is still believed to be large and consequently no clean example of GOE has yet been found [22].

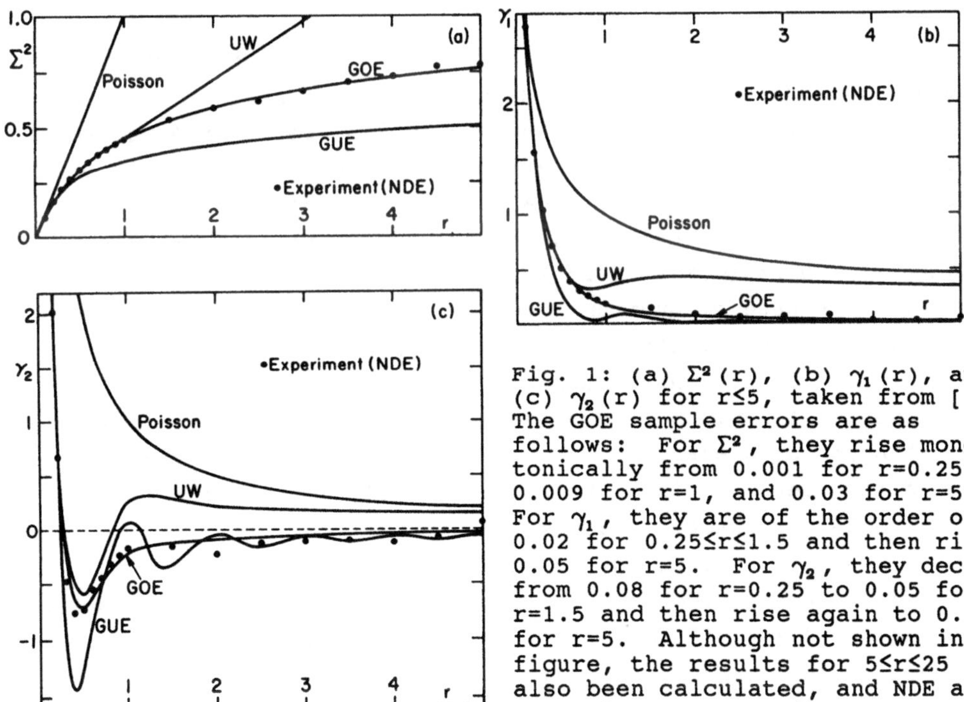

Fig. 1: (a) $\Sigma^2(r)$, (b) $\gamma_1(r)$, and (c) $\gamma_2(r)$ for $r \leq 5$, taken from [11]. The GOE sample errors are as follows: For Σ^2, they rise monotonically from 0.001 for r=0.25, to 0.009 for r=1, and 0.03 for r=5. For γ_1, they are of the order of 0.02 for $0.25 \leq r \leq 1.5$ and then rise to 0.05 for r=5. For γ_2, they decrease from 0.08 for r=0.25 to 0.05 for r=1.5 and then rise again to 0.08 for r=5. Although not shown in the figure, the results for $5 \leq r \leq 25$ have also been calculated, and NDE and GOE values are consistent.

4. Embedded Ensembles and Other Systems

The response theory in Section 2.1 relies on the Gaussian distribution of the V′-matrix elements as well as on the invariance of {V′}

in the H_α-diagonal representation for any α. However, as is clear from the discussion of Section 2.3, this is far too restrictive and only a locally-generated classical ensemble $\{\hat{V}'\}$ is needed for (5,6) to hold, and hence for the resultant correlation functions to be valid. On the other hand, if the diagonal $\hat{V}'$-matrix elements are (locally) independent Gaussian variables and if the off-diagonal $\hat{V}'$ is a sparse matrix with mostly zero elements, then (5,6) would hold with $\beta=0$. We have thus obtained criteria to decide, for more general ensembles, whether, as Λ increases, the fluctuation properties undergo a transition toward classical or Poisson.

As an example of this, consider the embedded Gaussian ensemble (EGE) [8,2] of H_α-operators of $k(\geq 1)$-particle rank acting in an $m(\geq k)$-particle space. For m=k this is the H_α-ensemble of Section 2, and the question is whether embedding, which introduces correlations among matrix elements for m>k, affects the fluctuation properties. For k=1, the V' matrices in H_α-diagonal representation are sparse, having a direct-product structure and hence generating transition to Poisson as Λ increases. On the other hand, for k>1, a local classical ensemble is generated; this is because V' introduces simultaneous interaction among all the states of H_α and because, in the H_α-diagonal representation, the matrix-element correlations, which are of order $d^{-1}(m)$, are locally of no consequence; here d(t) is the dimensionality of the t-particle space and, for the local p-interval, $d(m) \gg p$. These predictions have been confirmed, for k=2, by Monte-Carlo simulations of EGOE [9] and of EGOE-EGUE interpolation [12].

In the language of quantum chaos a local classical ensemble should be associated with complex systems whose classical analogs are chaotic, and a local sparse matrix with integrable systems. Moreover, for a given α, Λ would be energy-dependent, increasing with energy. Then, at asymptotic energies, Poisson-like fluctuation would be obtained for "generic" integrable systems [23] and classical-like (such as GOE) for chaotic systems [14,15,24].

The theory for the energy dependence of Λ has recently been developed for the nucleon-nucleon interaction, and this, combined with the bound on TRNI Λ of Section 3.7, yields a bound of 0.35% on the TRNI part of the interaction [12].

Finally we mention that the embedded ensembles differ from the classical ones for very long-range fluctuation measures. For all $m>k\geq 1$, a Fourier decomposition of $(\delta(r)-Y_2(r))$ shows that excitations of wavelengths $\gtrsim d^{-1}(k)d(m)$ are strongly suppressed [8]. This implies a saturation for $\Sigma^2(r)$ and $\bar{\Delta}_3(r)$ for $r \gtrsim d^{-1}(k)d(m)$, the saturation values depending, among other things, on energy as well as on k. This has not been observed in data because of experimental limitations, but a similar saturation has been found in spectral fluctuations of integrable and chaotic systems with two degrees of freedom [15,25].

Acknowledgements: The author is grateful to J. B. French and S. Tomsovic for a critical reading of the manuscript, and to Ms. C. Murdoch for preparing the manuscript. This work has been supported in part by the U.S. Department of Energy.

References

1. C. E. Porter, Ed., Statistical Theories of Spectra: Fluctuations, Academic Press, New York (1965); M. L. Mehta, Random Matrices, Academic Press, New York (1967); E. P. Wigner, Symmetries and Reflections, Ox Bow Press, Woodbridge (1979), p. 203.
2. T. A. Brody, J. Flores, J. B. French, P. A. Mello, A. Pandey and S. S. M. Wong, Rev. Mod. Phys. 53, 385 (1981).
3. J. Von Neumann and E. P. Wigner, Phys. Z. 30, 462 (1929).
4. F. J. Dyson and M. L. Mehta, J. Math. Phys. 4, 701 (1963).
5. H. I. Liou, G. Hacken, J. Rainwater, and U. N. Singh, Phys. Rev. C11, 462 (1975); W. M. Wilson, E. G. Bilpuch, and G. E. Mitchell, Nucl. Phys. A245, 285 (1975); C. Coceva and M. Stefanon, Nucl. Phys. A315, 1 (1979); and other references therein.
6. N. Rosenzweig and C. E. Porter, Phys. Rev. 120, 1698 (1960); H. S. Camarda and P. D. Georgopulos, Phys. Rev. Lett 50, 492 (1983).
7. A. Pandey, Ann. Phys. (N.Y.) 119, 170 (1979).
8. K. K. Mon and J. B. French, Ann. Phys. (N.Y.) 95, 90 (1975).
9. S. S. M. Wong and J. B. French, Nucl. Phys. A198, 188 (1972).
10. A. Pandey, Ann. Phys. (N.Y.) 134, 110 (1981).
11. R. U. Haq, A. Pandey, and O. Bohigas, Phys. Rev. Lett. 48, 1086 (1982); O. Bohigas, R. U. Haq, and A. Pandey, Phys. Rev. Lett. 54, 1645 (1985).
12. J. B. French, V. K. B. Kota, A. Pandey, and S. Tomsovic, Phys. Rev. Lett. 54, 2313 (1985), and to be published.
13. A. Pandey and M. L. Mehta, Commun. Math. Phys. 87, 449 (1983); M. L. Mehta and A. Pandey, J. Phys. A16, 2655 (1983).
14. O. Bohigas, M. J. Giannoni, and C. Schmit, Phys. Rev. Lett. 52, 1 (1984).
15. M. V. Berry, Proc. Roy. Soc. London A400, 229 (1985).
16. F. J. Dyson, J. Math. Phys. 3, 1191 (1962).
17. A. Pandey, to be published.
18. L. A. Pastur, Th. Math. Phys. 10, 67 (1972).
19. F. J. Dyson, Commun. Math. Phys. 19, 235 (1970); M. L. Mehta, Commun. Math. Phys. 20, 245 (1971).
20. J. J. M. Verbaarschot and M. R. Zirnbauer, Ann. Phys. (N.Y.) 158, 78 (1984); K. B. Efetov, Adv. in Phys. 32, 53 (1983).
21. M. L. Mehta and J. des Cloizeaux, Ind. J. Pure Appl. Math. 3, 329 (1972); M. L. Mehta and A. Pandey, J. Phys. A16, L601 (1983).
22. E. Haller, H. Koppel, and L. S. Cederbaum, Chem. Phys. Lett. 101, 215 (1983); E. Abramson, R. W. Field, D. Imre, K. K. Innes, and J. L. Kinsey, J. Chem. Phys. 80, 2298 (1984); S. Mukamel, J. Sue, and A. Pandey, Chem. Phys. Lett. 105, 134 (1984).
23. M. V. Berry and M. Tabor, Proc. Roy. Soc. London A356, 375 (1977).
24. P. Pechukas, Phys. Rev. Lett. 51, 943 (1983); T. Yukawa, Phys. Rev. Lett. 54, 1883 (1985).
25. G. Casati, B. V. Chirikov, and I. Guarneri, Phys. Rev. Lett. 54, 1350 (1985); T. H. Seligman, J. J. M. Verbaarschot, and M. R. Zirnbauer, Phys. Rev. Lett. 53, 215 (1984).

SPECTRAL FLUCTUATIONS AND TIME-REVERSAL INVARIANCE+

J. B. French, V. K. B. Kota++, A. Pandey and S. Tomsovic
Department of Physics and Astronomy
University of Rochester
Rochester, NY 14627 USA

Abstract

The effects of time-reversal non-invariance on spectral fluctuations of chaotic systems are considered, and a bound on non-invariance in the nucleon-nucleon interaction is deduced.

1. Introduction

This talk will make connections with both topics of this conference. For a system with a time-reversal non-invariant (TRNI) Hamiltonian we shall give first an analysis of energy-level fluctuations in a chaotic region of the spectrum. We shall then use fluctuation-free theoretical spectroscopy to reduce the information which the fluctuations in nuclei carry (about the complex system) to a bound on TRNI in the nucleon-nucleon interaction. The spectral-fluctuation problem, suggested by Wigner [1], relies on the fact that spectra become more rigid when TRI is broken; Wigner's parallel topic, that of studying TRI via the strength distributions, will be discussed separately. The solution of the quite distinct reduction problem will enable us to assess the fundamental significance of other weakly broken symmetries (parity, isospin,...) for which the data do not give rise to statistical measures (as they do with TRI). We shall not deal with that subject, nor with other problems in the theory of complex nuclei which can be treated similarly.

The TRI symmetry is fundamental, both in classical and quantum systems, and well-defined bounds on its breaking are almost unknown. The present analysis will give information about TRI, first in the form of a bound on a certain complex-system "transition parameter" Λ whose value, once the fluctuation theory is available, can be determined directly from experimental data. See [2] for a first account. It is essential to understand that, while a non-zero value

---

+ Presented by the first-named author. This work has been supported in part by the U.S. Department of Energy.

++ On leave from Physical Research Laboratory, Ahmedabad, India.

or Λ would be of interest in itself (indicating the existence of RNI), a bound on Λ tells us nothing of consequence until the reduc- ion has been made, either by theoretical analysis, or, conceivably, y comparison with the data and results for other broken symmetries. his point, clearly stated by Simonius [3] ("manifestations of ymmetries and symmetry breaking in complex systems...in most ases...are not of great help...due to uncertainty in the complex uclear physics part of their analysis"), has been well appreciated y some but not considered at all by many others. We shall of ourse have to resolve the uncertainty.

Wigner's general prescription [4] for dealing, via the GOE, ith complicated many-body systems which have TRI Hamiltonians mplies that spectral fluctuations are in general of GOE type. As hould be known the GOE involves only physically unreasonable many- ody interactions, all particles interacting simultaneously. hether this has any effect on the fluctuations, a subject of onfusion and controversy at the 1971 Albany Conference, has long ince been answered in the negative (though, despite some contrary laims, a rigorous formal proof is still lacking). This, coupled ith the remarkable agreement [5] of the experimental data with GOE redictions, implies that Wigner's universality is very general ndeed. By the same token, fluctuations carry little information; t seems in fact that, if we rule out trivial cases (where, for xample, the spectrum decomposes into a set of non-interacting pectra - mixed J is the best case) the principal information that pectral fluctuations in a complex system can carry is about the eak breaking of a symmetry.

It is not safe to assume that general properties of the lassical ensembles, GOE, GUE and GSE, apply to ensembles in which n underlying fundamental symmetry is broken. We shall indeed see hat the (large) matrix dimensionality d, which is a "harmless" arameter in the classical ensembles, becomes an important physical arameter when, for studying TRI, we use the GE(α) ensemble which nterpolates between GOE and GUE; ignoring this fact has rendered rrelevant the early Monte-Carlo studies of TRNI. In the same way hen we should consider the much more complicated EGE(α) (embedded E(α)) which, defined, as it is, in a two-particle space but acting n the many-particle space, avoids the multibody-interaction roblem. It turns out however that the same insight (namely that he embedded GOE, like the interaction in the spectrally chaotic egions of real systems, acts locally like a GOE), which indicates the significance of d for TRNI systems, also demonstrates that embedding is an unnecessary complication in determining the TRNI bound. Thus, on the one hand, we can deal with the simpler GE(α) and, on the other, can determine Λ directly from data. It is worthwhile noting that a strong central limit theorem (CLT) acts in the EGOE, and in real many-body systems, resulting in a remarkably sharp separation in the power spectrum; the result is that, except for a small number of long wavelength modes (λ~spectrum span), the only modes excited are of short wavelength (λ~level spacing), intermediate excitations being completely suppressed. This has implications for fluctuation measures of long range.

2. The TRI→TRNI Ensemble and Transition Parameter

The ensemble is GE(α), i.e.

$$\{H_\alpha(v^2)\} = \{H(S:v^2) + i\alpha H(A:v^2)\} \quad (1)$$

where $\{H(S):v^2\} = GOE(v^2)$, i.e. the GOE with off-diagonal matrix-element variances v^2, and $\{H(A:v^2)\}$ is the corresponding ensemble of real antisymmetric matrices. The parameter α measures the relative TRNI to TRI nucleon-nucleon interaction. Fig. 1 shows, for various α's but the same H(S) and H(A), 100 levels taken from the central regions of large matrices.

Fig. 1 100-level segments taken from the central region of large GE(α) spectra. Reading from the left they correspond to α=0 (GOE), 0.03, 0.09 and 1 (GUE). The arrowheads indicate pairs of levels with spacing less than D/4.

If the implicit assumption made in earlier calculations were correct, namely that the dimensionality d is irrelevant as long as it is large enough, we would conclude that the smallest detectible α would be a disappointing few percent (say 3%). However the assumption is wrong. The subspectra of Fig. 1 could well be found with matrices of dimensionality d=21,000 rather than d=210, as is the case; then the corresponding α would be smaller by a factor 10 so that the limit above would become 0.3%. This comes about because, for fixed v, the GOE spectral width $[d^{-1}\mathrm{Tr}(H^2)]^{1/2}$, essentially the semicircular radius, is proportional to $d^{1/2}$ so that the locally-averaged level spacing $D \propto d^{-1/2}$; the effect of the distant states is thereby to enhance by a factor $d^{1/2}$ the imaginary amplitudes $i\alpha H_{ij}(A)/(E_j-E_i)$ which for small α are admixed into the real H(S) eigenstate ψ_j. In other words the transition parameter is not α^2 but $\sim d\alpha^2$; more generally we have, where i, j are close-lying states,

$$\Lambda = \overline{|H_{ij}(A)|^2/D^2} \xrightarrow{\text{GOE}} d\alpha^2/\pi^2 \qquad (2)$$

where, by virtue of the local GOE result referred to above, the first form may be applied directly to the data (once the Λ value is derived from the fluctuation analysis). The second (ensemble) form derives from the GOE density at the center and displays the enhancement referred to above. While the transition parameter is determined via the wave-function admixing, we can take for granted that $\Lambda \sim 1$ marks the completion of the TRI→TRNI transition* for other things as well, this because the uniformity of the ensemble (which accurately describes the experimental data) scarcely permits another parameter. We shall verify that ahead. On the other hand note carefully that there is no such transition parameter for the weak admixing of a TRI amplitude into a TRNI state. Finally observe that for each nucleus there must be an effective dimensionality which measures, in some appropriate way, the number of interacting states, and eventually determines the sensitivity of the analysis. A formal definition of this is possible, but we shall approach the reduction problem rather differently.

3. Fluctuation Theory

The problem is to calculate TRI→TRNI "transition curves", i.e.

---

* In defining Λ we have taken account of the admixing of a single nearest-neighbor state. To take account of all admixing states we should multiply by $\sim 2\Sigma n^{-2} = \pi^2/3$ so that the "effective transition parameter" is $\Lambda^* = \pi^2\Lambda/3$ and $\Lambda \sim 3/\pi^2$ should more accurately mark the completion.

statistical measures - quantities directly derivable from data - as functions of Λ. Important measures are the number variances, $\Sigma^2(r;\Lambda)$, variances of the number of levels in an energy interval of length rD. The problem for two-point measures has been solved at four different levels which we illustrate by considering $\Sigma^2(r;\Lambda)$.

3.1 Guess plus Transition Parameter: Taking for granted that $\Sigma^2(r:\Lambda)$ can be expanded in powers of Λ we have a natural guess that $\Sigma^2(r;\Lambda) \sim \Sigma^2(0:\Lambda)-\pi^2\Lambda/3$ where the negative sign comes by considering spectral rigidity; note that $\Sigma^2(1)=0.446$ and 0.344 for GOE and GUE respectively. Similarly for other measures.

3.2 Exact First-order Theory: This is based on two facts: (a) that the sum of two independent GOE's is a GOE, the variances v^2 adding; (b) exact to first order in Λ for the eigenvalues, $iH(A)\equiv\hat{H}(A)$ where the real symmetric $\hat{H}(A)$ is defined by $\hat{H}_{ij}(A)=\hat{H}_{ji}(A)=H_{ij}(A)$ while $\hat{H}_{ii}(A)=0$. Then $\{iH(A)\}$ is equivalent to an independent $H(S:v^2)$ except for the missing diagonal elements; these form an ensemble of real diagonal matrices $\{W\}$ which are independent of H_α and whose elements w_{ii} are distributed as independent zero-centered Gaussians of variances $2v^2$. Manipulating these obvious results we see that $\{\hat{H}_\alpha = H(S)+\alpha[\hat{H}(A)+W]\}$ is a $GOE(1+\alpha^2)v^2$, and this leads, for the variance of the motion of the j'th level in $\{H_\alpha(v^2)\}$, to

$$V[E_{\alpha:j}(v^2)] = V[E_{0:j}((1+\alpha^2)v^2)] - 2\alpha^2v^2 + \ldots \qquad (3)$$

the $2\alpha^2v^2$ terms being all independent. With this result we can write the slopes, at $\Lambda=0$, of all spectral measures and correlation functions of all orders. Inspection shows that the k'th-nearest-neighbor spacing variances are $\sigma^2(k;\Lambda) = \sigma^2(k;0)-4\Lambda+\ldots$. For the number variances it turns out that $\Sigma^2(r;\Lambda) = \Sigma^2(r;0)-4[1-Y_2(r)]\Lambda$ where $Y_2(r)$ $\left(\simeq 1/\pi^2r^2 \text{ for } r\gtrsim 1\right)$ is Dyson's [6] two-level GOE cluster function. These results, which are adequate for deriving the TRNI bound, confirm our identification of the transition parameter and justfy our casual treatment of it in section 3.1.

3.3 Transition Curves via Binary-Correlation Approximation: The above methods give only the slopes of fluctuation measures at $\Lambda=0$. The complete transition curves are of considerable interest representing, as they do, the solution of a complicated problem over the whole range $(0 \le \Lambda \le \infty)$ of the transition parameter, and suggesting that the techniques used in generating them may have a wider application.

The binary-correlation method [7] which is simple, widely applicable and often (as in the present case) quite accurate, effectively represents the microscopic eigenvalue density $I(x) = d \times \rho(x)$ in terms of a sequence (labeled by $\varsigma=1,2,\ldots d$) of Chebyshev-polynomial excitations of the averaged (semicircular) density $\overline{\rho}(x)$. The polynomials, $v_{\varsigma}(x)$, are orthonormal with $\overline{\rho}(x)$ as weight function. The outcome of this is the representation of the distribution function as a random function with specified first and second-order statistical properties, and eventually to

$$\Sigma^2(r;\Lambda) = \Sigma^2_{GUE}(r) + \frac{1}{2\pi^2}\ln\left\{1 + \frac{\pi^2 r^2}{4(\tau + 2\pi^2\Lambda)^2}\right\} \qquad (4)$$

in which τ=0.615 is fixed from the GOE(Λ=0) - GUE(Λ=∞) difference. A plot is given in Fig. 2.

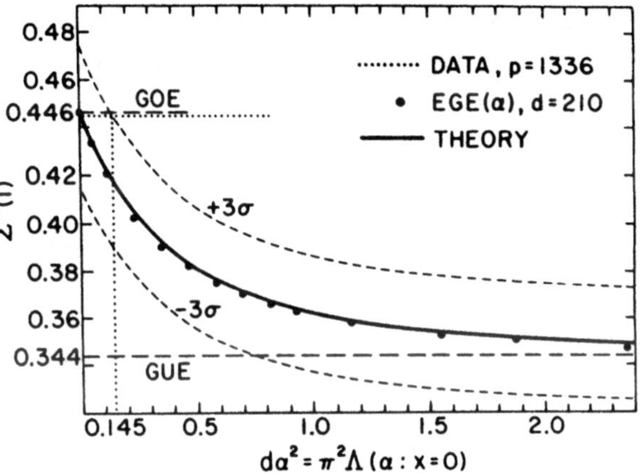

Fig. 2 The GOE→GUE transition curve for $\Sigma^2(1)$, as given by Eq. (4) and by exact theory; the two curves coincide. Shown also, as described in the text, are the $\pm 3\sigma$ lines drawn for a set of 1336 neutron-resonance levels, and the lines showing the datum point (0.445) and the corresponding $\pi^2\Lambda$ bound (0.145). Points are also indicated for a fifty-member EGE(α) of two-body interactions acting in a four-particle 210-dimensional space.

3.4 Exact Theory: The exact joint-probability distribution for the $\{H_\alpha\}$ eigenvalues has been given by Pandey and Mehta [8]. The exact forms for the two-point measures of interest to us have, among other things, enabled us to verify and better understand the approximate methods described above.

4. Data Analysis

The analysis will yield an upper bound on Λ. We begin with the recent data analysis [5] based on a TRI assumption, and ask for the largest Λ value compatible with the agreement found therein for the $\Sigma^2(1)$ measure which turns out to be optimal for the present purpose. The condition for agreement is defined by a 3σ criterion which then defines a confidence interval about the theoretical $\Sigma^2(1)$ transition curve of Fig. 2. The sample error, derived [5] for $\alpha=0$ (TRI) by a combination of theory and Monte-Carlo calculations, and since verified for TRNI, is $\sigma(p;\Lambda)\sim 0.35\ p^{-1/2}$ where p is the number of levels in the sample. For a run of 100 levels $3\sigma \simeq 0.11$ so that the confidence band may span the entire GOE-GUE difference. One therefore groups the data, 1762 levels from 32 nuclei, into larger sets [2]. The outcome of the analysis is that $\Lambda^{1/2} \lesssim 0.1$, indicating that, in the slow neutron (or proton) resonance regions of the nuclei considered, the RMS admixing matrix element of H(TRNI)$\lesssim$D/10 where D is the resonance-region level spacing.

5. The Reduction Problem: General Procedure

The question is: what does the Λ bound tell us about TRNI in the nucleon-nucleon interaction. Take

$$H = h(1) + V(2) + i\alpha U(2) \quad ; \quad Tr^{(2)}(U^2) = Tr^{(2)}(V^2) \tag{5}$$

where $i\alpha U$ = H(TRNI) and U is normalized in the two-particle space as shown, so that α plays the same role as with the ensembles and gives the TRNI/TRI interaction ratio. We have

$$\alpha^2 = \overline{|H_{ij}(TRNI)|^2} \ / \ \overline{(U_{ij})^2} \tag{6}$$

where we have an experimental bound on the numerator and must calculate the denominator.

In principle now we take a reasonable form for V, and an adequate model space and one-body Hamiltonian h (e.g. from a zero-deformation Nilsson model). V_0, the magnitude parameter in V, is chosen so that a calculation fits the level spacing in the resonance region. The stability is verified by recalculating with a larger model space and for neighboring nuclei. Finally we calculate the numerator of (8), which is a strength, generated by a complicated (two-body) transition operator and connecting neighboring H-eigenstates in the chaotic spectrum. We see then that we must be able, for interacting particle systems, to calculate level densities and strengths at high excitations. We briefly describe methods for dealing with these non-trivial problems.

6. Level Densities

The Bethe formula for the level density, $I(E) \sim \exp\{(bE)^{1/2}\}$, derived for non-interacting particles (NIP), being simply related to the Hardy-Ramanujan (HR) partition formula, is essentially combinatorial. But with interactions the energies are not additive and this feature is lost. During the past dozen years methods [9] which rely on the existence of CLT's in IP spaces have successfully dealt with the extended problem of calculating IP level densities as explicit functions of the Hamiltonian matrix elements. But these are inelegant in our huge spaces (characteristically $d \gtrsim 10^{15}$) and make no direct connection with Bethe-HR.

The key to a much superior theory is to decompose V according to its local actions (i.e. actions in the neighborhood of a given multiparticle energy E) on the spherical orbits which are active (neither essentially empty nor filled) at E. To do that we introduce a set of N=N(E) single-particle (s.p.) states, adequate to define the Fermi surface at E, and the corresponding group U(N) of unitary transformations in the s.p. space. Then V is decomposed according to: (a) The U(N) irreps; there are three of them, $\nu=0,1,2$ where $[N-\nu,\nu]$ defines the Young columnar shape; (b) The direct-sum U(N) subgroup defined by individual spherical orbits. The label is $[\sigma]$ where $\sigma=[0,0,0,\ldots]\equiv[0]$ defines the scalar and $[\sigma]=[\Gamma_1,\Gamma_2,\ldots]\equiv[\Gamma]$ with every $\Gamma_i \neq 0$ refers to the sum of all the other contributing irreps; (c) The harmonic-oscillator excitations, $t\hbar\omega$, which V generates. t=0 is dominant; t=2,4,...may be ignored or treated perturbatively. Then we have

$$V = \sum V_t^{\nu,[\sigma]} \longrightarrow V^{0,[0]} + V^{1,[0]} + V^{2,[0]} + V_0^{2,[\Gamma]} + \sum_{t>0} V_t^{2,[\Gamma]} \qquad (7)$$

in which the first term is a harmless (number-dependent) constant, the second renormalizes the single-particle energies, thereby shifting the spherical-orbit configuration centroids, while the third shifts the centroids by a pattern which does not correspond to renormalization of the s.p. energies. By evaluating norms of the separate parts of V, and their correlation coefficients with h, we find the general result that $V^{2,[0]}$ is negligible for realistic interactions so that the entire effect of $V^{[0]}$ (which does not spread the states of a given configuration) may be incorporated into h, generating simply a new NIP density. Since we have argued that the last term is ineffective the spreading and admixing are produced by $V_0^{2,[\Gamma]}$. It turns out here that, if we label the configurations by their harmonic oscillator excitations, $s\hbar\omega$, the spreading variances are very closely constant for given s, increasing however with s (or effectively with excitation energy), as we would expect. Since the spreading will be Gaussian we have immediately that

$$I(E) = \sum_s I_{NIP}^{(s)} \otimes \rho_G(E_s=0,\ \sigma_s^2) \tag{8}$$

where $\otimes$ denotes convolution, I_{NIP} is the Bethe-HR level density, and the spreading variances are directly calculable from V. For $E \leq 30$ MeV we would need $s \lesssim 5$. Note that we have automatically a parity decomposition.

The Bethe-HR density has survived as a useful object. However we still need its s-decomposition, given moreover as a function of the renormalized s.p. energies. This problem has been solved completely by J. Smith (to be published) who finds that $I_{NIP}^{(s)}$ is itself quite close to Gaussian, so that the convolution above is immediate and we are left with an elegant solution of the entire problem. That solution obviously has many applications besides the present one.

7. Strength Distributions

The strength for a transition operator O is defined as $R(E',E) = |\langle E'|O|E\rangle|^2$, its distribution being then given by $I(E')R(E',E)I(E) = \langle\langle O^+\delta(H-E')O\delta(H-E)\rangle\rangle^m$ where $\langle\langle\ \rangle\rangle^m$ is the m-particle trace. For an NIP system (V=0) the strength is easily evaluated exactly in terms of the two level densities indicated. It turns out that, just as with the level density, we can incorporate the interactions by means of convolutions. If we could assume the interaction spreadings at E′ and E to be uncorrelated we would separately convolute the V=0 density with the two spreading Gaussians appropriate to E′ and E. More generally we need the correlation coefficient, defined in terms of $\langle O^+VOV\rangle^m$, which can be evaluated by ensemble averaging or for fixed operators. Then the convolution leads to a general bivariate Gaussian form, to which we can add Edgeworth corrections. The details of the procedure sketched here have not yet been published.

8. TRNI Bound for the Nucleon-Nucleon Interaction

For ^{169}Er, for which a reasonable stable shell model has d~6×10^{14}, and whose level spacing in the slow-neutron region is ~100 eV, we find, for the relative norm of the TRNI interaction,

$$\alpha \lesssim 3.5 \times 10^{-3}$$

We shall shortly report on TRNI bounds as derived from more complex nuclei (D~10 eV) and less complex ones (D~10 keV).

The spectral TRNI bound, and the bound determined by the neutron strengths, discussed separately in this conference, are coupled because the levels are detected via their strengths. The strength analysis shows that, if we ignore the smallest strengths (5% of the total according to GOE), the data are consistent with GOE and give rise to the same TRNI bound. There is however a deficiency in the small-strength distribution. The level analysis, fully consistent with GOE, does not permit us to assign the complete deficiency to missing levels. There is a non-zero residual effect, defining an "effective" $\alpha \sim 1.5\times10^{-3}$, which may be due to non-statistical nuclear effects, or to TRNI, or to a combination of the two.

Acknowledgement: We thank J. Smith for giving us access to his unpublished results.

References:

1. E. P. Wigner, SIAM Rev. 9, 1 (1967).
2. J. B. French, V. K. B. Kota, A. Pandey and S. Tomsovic, Phys. Rev. Lett. 54, 2313 (1985).
3. M. Simonius, in Proc. of Amsterdam International Conf. on Nuclear Structure, A. Van der Woude and B. J. Verhaar, editors, Nucl. Phys. A396, 203c (1983).
4. E. P. Wigner (1961), reprinted in Symmetries and Reflections, Ox Bow Press (1979) p.203.
5. R. U. Haq, A. Pandey, and O. Bohigas, Phys. Rev. Lett. 48, 1086 (1882).
6. F. J. Dyson, J. Math. Phys. 3, 166 (1962).
7. K. K. Mon and J. B. French, Ann. Phys. (N.Y.) 95, 90 (1985).
8. A. Pandey and M. L. Mehta, Commun. Math. Phys. 87, 449 (1983).
9. J. B. French and V. K. B. Kota, Ann. Rev. Nucl. Part. Sci. 32, 35 (1982).

ANTIUNITARY SYMMETRIES AND ENERGY LEVEL STATISTICS

M.Robnik
H.H.Wills Physics Laboratory
University of Bristol
Tyndall Avenue
Bristol BS8 1TL, U.K.

INTRODUCTION

One surprising development in the search of quantum chaos was the finding that the statistics of random matrix theories apply not only to the spectra of sufficiently complex systems with many coupled degrees of freedom, but also to relatively simple systems having a few degrees of freedom, provided that their classical dynamics is ergodic. A firm numerical evidence for this has been given for the first time by Bohigas, Giannoni and Schmit /7/, who studied the spectral fluctuations of the Sinai billiard which is a classically ergodic system. They have formulated the following conjecture supported by the numerical level spacings distribution and the spectral rigidity: spectral fluctuations of classically ergodic systems with time reversal symmetry obey the statistics of the Gaussian Orthogonal Ensemble (GOE) of random matrix theories, while in absence of time reversal invariance the statistics is that of the Gaussian Unitary Ensemble (GUE). (This was the stronger version of the conjecture, the weaker one requiring in addition the Kolmogorov entropy to be positive.)

The spectral fluctuations of classically ergodic systems fall into two universality classes namely GOE and GUE. Given a Hamiltonian H, how can we decide which universality class applies? In random matrix theories /16,6/ the classification is due to the time reversal T: if H has the symmetry T, TH=HT, then GOE applies and GUE otherwise (we ignore spin). Robnik and Berry /23/ have pointed out that this classification scheme is incomplete and that the non-trivial generalization is in terms of antiunitary symmetries A which admit real representations of H. To explain this and to further extend our results is the main purpose of this lecture.

The confidence in the conjecture put forward by Bohigas, Giannoni and Schmit has been strengthened by the numerical results of Seligman and Verbaarschot /26/ and by Berry and Robnik /5/, who have demonstrated that in absence of an antiunitary symmetry the spectral fluctuations are indeed consistent with the GUE statistics. There are also theoretical arguments in support of the conjecture, originally due to Pechukas /15/, and recently by Yukawa /31/. (The results of Pechukas and of Yukawa can be easily generalized to include GUE case by allowing complex valued matrix elements V_{mn} (or f_{mn} in Yukawa's notation). For example, the exponent in the formula (9) of Yukawa's paper becomes 1 rather than 1/2, yielding the GUE level spacings distribution in the limit $N\to\infty$. The condition for the realness of $f_{mn}(t)$ is that there is an antiunitary A such that $AH(t)=H(t)A$ for all t, admitting a real representation of H(t), in which case the exponent is 1/2 and when $N\to\infty$ one gets the GOE level spacings distribution.) Semiclassical theory can be a helpful tool in understanding the link between the ergodicity of classical dynamics and the statistics of spectral fluctuations of a given Hamiltonian H. The periodic orbit sum formula gives the semiclassical spectrum in terms of all classical periodic orbits/9-13,-1,2/. It is a rather useless formula in calculating individual levels, but can be the decisive tool to express the statistics of spectral fluctuations in terms of the statistical laws governing the distribution of the classical closed orbits/21/. Berry /3/ has calculated the semiclassical spectral rigidity which turns out to be the same as the leading term in the asymptotic expansion of the GOE or GUE spectral rigidity, depending on whether H has a discrete symmetry or not, no matter whether unitary or antiunitary!

We emphasize that we will deal only with systems whose classical dynamics is ergodic. For integrable systems one observes locally the Poisson statistics. For the generic systems in transition region between integrability and chaos one finds a smooth continuous transition of the spectral statistics/19,14,28,29,4,25/.

THE ROLE OF ANTIUNITARY SYMMETRIES

The significance of antiunitary symmetries is that if a Hamiltonian H has such a symmetry A, i.e. HA=AH, then there exists a large family of (complex) basis such that H has a real representation. (It is important that these so called A-adapted basis $\{\psi_\alpha\}$ defined by $A\psi_\alpha = \psi_\alpha$, are not related to the eigenbasis of the Hermitian operator H by an orthogonal transformation. So the realness is <u>not</u> a trivial one.) In fact an A-adapted basis can be constructed from arbitrary(complex) basis in Hilbert space if A satisfies certain conditions.

The time reversal T is a particular case of an antiunitary transformation. The general theory of such operators was given by Wigner /30/,Dyson/8/, and pp2-87 in Porter/16/. We review the essentials. In absence of spin, and using the coordinate representation, T can be represented simply by the complex conjugation operator K. This follows from the definition

$$TqT^{-1} = q,\ TpT^{-1} = -p, \qquad (1)$$

so that in the coordinate representation, $p=-\frac{i}{\hbar}\frac{\partial}{\partial q}$, we find T=K.

Now we discuss the antiunitary operators more generally. An operator A is antilinear if for any states ψ and ϕ and any complex numbers a and b

$$A(a\psi+b\phi) = a^*A\psi + b^*A\phi \ , \tag{2}$$

where * denotes complex conjugation. Since AK is a linear operator, it follows that every antilinear operator A can be written as a product of a linear operator L and the complex conjugation operator K, i.e.

$$A = LK. \tag{3}$$

Clearly $K^2=1$, and K is Hermitian, i.e. $K=K^+$. Thus $A^+=KL^+$. For antilinear operators we cannot define the eigenvalues. An operator S is called a symmetry of H if it commutes with H, HS=SH, and if it preserves all transition probabilities, i.e. $|\langle\phi|\psi\rangle| = |\langle S\phi|S\psi\rangle|$ for any ϕ and ψ (S is an isometry). Thus, S preserves the physics of H.

Now there is a theorem by Wigner, discussed in Porter /16/, that every isometry S must satisfy $SS^+=1$ and either S is linear or antilinear, with no other possibility left. In the first case the symmetry S is termed unitary and in the latter case antiunitary. A unitary operator U preserves scalar products, i.e. $\langle\phi|\psi\rangle=\langle U\phi|U\psi\rangle$ for any ϕ and ψ. Any antiunitary A can be written, because of (3), as a product of some unitary operator U and K,

$$A = UK. \tag{4}$$

Clearly, $AA^+=UKKU^+=UU^+=1$. Therefore for any antiunitary A we have

$$\langle A\phi|A\psi\rangle = \langle UK\phi|UK\psi\rangle = \langle K\phi|K\psi\rangle = \langle\phi|\psi\rangle^* \tag{5}$$

for arbitrary ϕ and ψ.

Let H have an antiunitary symmetry A. In any basis $\{\psi_\alpha\}$ defined by

$$A\psi_\alpha = \psi_\alpha \ , \tag{6}$$

and called A-adapted basis, the matrix elements of H are real. The very simple argument is given in Porter/16/,pp2-87:

$$\langle\psi_\alpha|H\psi_\beta\rangle = \langle A\psi_\alpha|AH\psi_\beta\rangle^* = \langle A\psi_\alpha|HA\psi_\beta\rangle^* = \langle\psi_\alpha|H\psi_\beta\rangle^* \tag{8}$$

The question is whether an A-adapted basis exists for the given A. It seems that it does but I am able to prove it only for two classes of A.

The first class is defined by

$$A^n=1, \text{ n integer.} \tag{9}$$

We generalize the arguments given by Porter/16/. Let $\{\phi_\alpha\}$ be any basis. Define the operator

$$R_+(A) = \sum_{\ell=1}^{n} A^\ell \ , \tag{10}$$

which has the property

$$AR_+=R_+ \quad , \tag{11}$$

as a consequence of (9). Further define the first basis vector ψ_1 as follows

$$\psi_1 = R_+(A)a_1\phi_1 \ , \ A\psi_1 = \psi_1 \quad , \tag{12}$$

where a_1 is a complex number (chosen to achieve normalization). Next arrange that ϕ_2 and ψ_1 are orthogonal, i.e.

$$\langle\psi_1|\phi_2\rangle = 0. \tag{13}$$

Now define

$$\psi_2 = R_+(A)a_2\phi_2 \tag{14}$$

where a_2 is a complex constant. Obviously $A\psi_2 = \psi_2$, and even orthogonality is satisfied as can be easily verified:

$$\langle\psi_1|\psi_2\rangle = \langle\psi_1|\sum_{\ell=1}^{n} A^\ell a_2\phi_2\rangle = \langle\psi_1|a_2\phi_2\rangle = 0. \tag{15}$$

By appropriate choice of a_2 we can achieve $||\psi_2||=1$. By induction we can thus construct an orthonormal A-adapted basis $\{\psi_\alpha\}$ which is complete if $\{\phi_\alpha\}$ was complete.

The second class of antiunitary symmetries admitting an A-adapted basis consists of all A such that

$$A=UK, \ U^n=1, \ n \text{ even}. \tag{16}$$

To construct the A-adapted basis we start from any real orthonormal basis $\{\phi_\alpha\}$. (If ϕ_α are complex, then we decompose them in real and imaginary parts and consider their union which is made orthonormal.) We define the operators

$$R_+(U) = \sum_{\ell=1}^{n} U^\ell \ , \tag{17}$$

$$R_-(U) = \sum_{\ell=1}^{n} (-U)^\ell \ . \tag{18}$$

Clearly,

$$UR_+ = R_+ \ , \ UR_- = -R_- , \tag{19}$$

where the evenness of n is essential for the second equality. Now let us define

$$\phi_\alpha^+ = R_+(U)\phi_\alpha , \phi_\alpha^- = R_-(U)\phi_\alpha \ , \tag{20}$$

so that ϕ_α^+ and ϕ_α^- are eigenvectors of U,

$$U\phi_\alpha^\pm = \pm\phi_\alpha^\pm \ . \tag{21}$$

They are orthogonal, or more generally, for any α and β

$$\langle\phi_\alpha^+|\phi_\beta^-\rangle = \langle U\phi_\alpha^+|U\phi_\beta^-\rangle = -\langle\phi_\alpha^+|\phi_\beta^-\rangle = 0. \tag{22}$$

The complete A-adapted basis is constructed as follows

$$\psi_\alpha = \phi_\alpha^+ \pm i\phi_\alpha^- \ . \tag{23}$$

Indeed we verify

$$A\psi_\alpha = UK\psi_\alpha = U\phi_\alpha^+ \mp iU\phi_\alpha^- = \phi_\alpha^+ \pm i\phi_\alpha^- = \psi_\alpha \ . \tag{24}$$

The orthogonality

$$< \psi_\alpha | \psi_\beta > = \begin{cases} 0 \text{ if } \alpha \neq \beta \\ \text{nonzero if } \alpha = \beta \end{cases} \tag{25}$$

is satisfied only for n=2. This is easy to show. From (22) follows

$$< \psi_\alpha | \psi_\beta > = <\phi_\alpha^+ \pm i\bar{\phi}_\alpha | \phi_\beta^+ \pm i\bar{\phi}_\beta > = <\phi_\alpha^+ | \phi_\beta^+ > + <\phi_\alpha^- | \phi_\beta^- >. \tag{26}$$

Now using (22) and $<U^\ell \phi_\alpha | U^\ell \phi_\beta> = 0$ for any ℓ we get

$$<\phi_\alpha^+ | \phi_\beta^+> = \sum_{\substack{\ell,k=1 \\ \ell \neq k}}^{n} <U^\ell \phi_\alpha | U^k \phi_\beta > =$$

$$= \sum_{\substack{\ell,k=1 \\ \ell<k}}^{n} \{<U^{k-\ell}\phi_\alpha | \phi_\beta > + <\phi_\alpha | U^{k-\ell}\phi_\beta>\}. \tag{27}$$

On the other hand

$$<\phi_\alpha^- | \phi_\beta^-> = \sum_{\substack{\ell,k=1 \\ \ell \neq k}}^{n} <(-U)^\ell \phi_\alpha | (-U)^k \phi_\beta> =$$

$$= \sum_{\substack{\ell,k=1 \\ \ell<k}}^{n} \{ <(-U)^{k-\ell}\phi_\alpha | \phi_\beta> + <\phi_\alpha | (-U)^{k-\ell}\phi_\beta>\}. \tag{28}$$

Those terms in (27,28) with $(k-\ell)$ odd will cancel in (26) and the remainder is

$$<\psi_\alpha | \psi_\beta> = 2 \sum_{\substack{\ell,k=1 \\ \ell<k,(k-\ell)\text{even}}}^{n} \{<U^{k-\ell}\phi_\alpha | \phi_\beta> + <\phi_\alpha | U^{k-\ell}\phi_\beta>\}, \tag{29}$$

which can be nonzero if $\alpha \neq \beta$. If n=2, then $k-\ell \leq 1$, and the orthogonality (25) is satisfied. By choosing appropriate real constants a_α we can also normalize ψ_α so that $||\psi_\alpha||=1$. For n>2 the orthogonality (25) is not satisfied. However, because $<\psi_\alpha | \psi_\beta >$ is real for any α and β according to (26) we can construct an orthonormal basis from the A-adapted basis by linear combinations of ψ_α with real coefficients. Real linear combinations do not spoil the A-adaptedness, and consequently we obtain a complete,orthonormal A-adapted basis denoted by $\{\psi_\alpha\}$ again.

Whenever an ensemble of Hamiltonians possesses an antiunitary symmetry A belonging to one of the two classes, we predict GOE spectral fluctuations. If A belongs to neither of the two classes, I am not able to construct an A-adapted basis, and the representation argument cannot be used.

But there are invariance arguments. The distribution of the Hamiltonians of the given ensemble must be invariant with respect to the canonical group of transformations that preserve the symmetry structure of the ensemble. A similarity transformation SHS^{-1} must preserve the Hermitian property of H,

$$(SHS^{-1})^{+} = SHS^{-1}, \tag{30}$$

from which follows

$$SS^{+}H - HSS^{+} = 0. \tag{31}$$

Since H is arbitrary SS^{+}=constant and can be chosen unity. Therefore the most general canonical group is the group of unitary transformations.

If every member H of the ensemble has the antiunitary symmetry A=UK, U unitary, then we require that SAS^{+} is also of the form U'K,U'unitary i.e.

$$SAS^{+} = SUKS^{+} = SUS^{+}K=U'K, \tag{32}$$

which can be satisfied only if

$$KS^{+}= S^{+}K, \tag{33}$$

whence $S^{+} = KS^{+}K = \tilde{S}$ = transpose of S, and thus S is real, i.e.

$$S=S^{*} \tag{34}$$

The conclusion is that, if the ensemble has an antiunitary symmetry, S must be real and the canonical group of transformations preserving the symmetry structure is the group of real orthogonal transformations. On these grounds we arrive at the most general classification scheme: the spectral fluctuations of an ensemble with arbitrary antiunitary symmetry and ergodic classical dynamics are described by the GOE statistics, and by the GUE statistics if no antiunitary symmetry exists.

The role of antiunitary symmetries is quite different from that of unitary symmetries. The antiunitary symmetries for which eigenvalues cannot be defined dictate the type of spectral fluctuations. The unitary symmetries have eigenvalues and imply decomposition of energy level sequences into subsequences corresponding to different eigenvalues. They do not affect the type of spectral fluctuations for each subsequence.

CHARGED PARTICLE IN A MAGNETIC FIELD: ANTICANONICAL SYMMETRY

Consider a charged particle with mass m and charge e moving in the $\underline{r}=(x,y)$ plane under the influence of a scalar potential $V(\underline{r})$ and a magnetic field $\underline{n}B(\underline{r})$ directed perpendicularly to the plane. The Hamiltonian is

$$H(\underline{r},\underline{p}) = \frac{1}{2m}(\underline{p} - \frac{e}{c}\underline{A})^2 + V(\underline{r}), \tag{35}$$

where the vector potential $\underline{A}$ satisfies

$$\underline{\nabla}\times \underline{A}(\underline{r}) = \underline{n}B(\underline{r}). \tag{36}$$

It is clear that in discussing the symmetries of (35) we must remove the arbitrariness of gauge. The symmetry of the kinetic energy of (35) can be changed rather arbitrarily by the (physically insignificant) gradient transformations

$$\underline{A} \to \underline{A} + \underline{\nabla}f, \tag{37}$$

where $f = f(\underline{r})$ is any function. Therefore we decompose $\underline{A}$ into its curl and gradient parts and require vanishing of the gradient part. So

$$\underline{A} = \underline{\nabla}\times \underline{F}, \tag{38}$$

implying the Coulomb gauge

$$\underline{\nabla}\cdot\underline{A} = 0. \tag{39}$$

In (38) we can choose

$$\underline{F} = \underline{n}F(\underline{r}), \tag{40}$$

and obtain the Poisson equation for F,

$$B = -\left(\frac{\partial^2 F}{\partial x^2} + \frac{\partial^2 F}{\partial y^2}\right) = -\Delta F. \tag{41}$$

For the uniform magnetic field we can choose $F=-\frac{1}{2}Bx^2$, so that $A_x=0$, $A_y=Bx$. But there is further a continuum of other choices, namely e.g. $F=-\frac{1}{2}B(x^2\cos^2\alpha+y^2\sin^2\alpha)$, where α is arbitrary real angle. Thus to make (38) unique we must impose conditions on the solution F of the Poisson equation (41). We require that F is the 'purely inhomogeneous' solution (thereby removing the freedom of adding to F any harmonic function F', $\Delta F'=0$.) Specifically our requirement is that

$$F(\underline{r}) = \frac{1}{2\pi}\int d^2\underline{r}' B(\underline{r}')\,\ell n|\underline{r}-\underline{r}'|. \tag{42}$$

This is the only natural gauge that we can have. According to (42) F and with it the vector potential (38) vanish if the field B is identically zero. Also $F(\underline{r})$ has the same symmetry as the field $B(\underline{r})$. In fact the gauge (42) maximises the symmetry of F relative to B. In the Hamiltonian (35) we shall henceforth assume that $\underline{A}$ is given by (38) and (42).

Normally time reversal T is defined as the reversal of velocities and not of the canonical momenta p. Nevertheless we define T as the reversal of canonical momenta. No difficulty arises because with the gauge (42) the system does or does not have the invariance with respect to the reversal of velocities if H is or is not invariant under T defined as p-reversal. Namely, if B=0 then $\underline{A}=0$ and H has T and also has invariance with respect to the reversal of velocities. If $B\neq 0$ then $\underline{A}\neq 0$ and H neither has T nor is it invariant with respect to the reversal of velocities.

Now we consider the following symmetry operations in the classical phase space/22,24/.

Inversion P: $(x,y,p_x,p_y) \rightarrow (-x,-y,-p_x,-p_y)$

Time reversal T: $(x,y,p_x,p_y) \rightarrow (x,y,-p_x,-p_y)$

Reflection S_x: $(x,y,p_x,p_y) \rightarrow (-x,y,-p_x,p_y)$. (43)

These maps satisfy $P^2=T^2=S_x^2=1$. For any symmetry U we will say that H has symmetry U if HU=H. It follows from (35), (38) and (42) that

(i) H has P if both V and B have P;

(ii) H has T if B=0;

(iii) H has S_x if B=0 and V has S_x;

Remark: If both V and B have S_x this does not imply that H has S_x, but we have instead

(iv) H has TS_x if B and V have S_x.

There is an important difference between (i,iii) and (ii,iv). The operations (i) and (iii) are canonical, i.e. they leave invariant both the Hamiltonian H and the Hamilton equations, for the Poisson brackets remain unchanged. On the other hand the transformations (ii) and (iv) leave H invariant, but they change the sign of the Hamilton equations, this change of sign being equivalent to the time inversion. We call them anticanonical as they preserve the Poisson brackets apart from the change of sign. The canonical symmetries establish the equivalence between forward orbits, whereas the anticanonical symmetries relate a forward orbit to a backward orbit. Anticanonical symmetries correspond to the antiunitary symmetries in quantum mechanics.

When the periodic orbit sum formula is used to calculate the spectral statistics the degeneracy of classical orbits is very important. Degeneracy is either due to canonical or anticanonical symmetry and implies equivalence of pairs (or of n-tuples) of closed orbits, such that their periods and amplitudes are the same, and in such a case they must be combined coherently, but incoherently in absence of symmetries. At this level of semiclassical approximation the anticanonical symmetries affect the spectral statistics in much the same way as the canonical symmetries. The effect in the spectral rigidity $\Delta(L)$ is much less pronounced and the leading asymptotic term of $\Delta(L)$ is the same as for the GOE ensembles if the Hamiltonian has a discrete symmetry (no matter whether canonical or anticanonical), or the same as for the GUE ensemble if it does not have a symmetry. On the other hand, the distribution of level spacings P(S) is known to depend very strongly on the unitary or antiunitary nature of the symmetry: in the unitary case we get superposition of independent sequences/4/, while the antiunitary symmetry merely changes the type of fluctuations from GUE to GOE. From this we conclude that the semiclassical theory based on the usage of the periodic orbit sum formula cannot explain the level repulsion laws ($P(S) \sim$ const, S, S^2 as $S \rightarrow 0$ for integrable, GOE and GUE, respectively) and even less so the global aspects of level spacings distributions.

There is a large class of systems corresponding to the case (iv), where H has neither T nor S_x, but $TS_x=S_xT$. For all such Hamiltonians we predict GOE spectral fluctuations if the classical dynamics of H is ergodic. Important examples are atoms in uniform magnetic fields. The hydrogen atom was studied by Robnik /17,18/ and was shown to become chaotic above a certain critical energy. Other examples have been given by Berry and Robnik/5/, for the case of Aharonov-Bohm billiards, analyzed by a conformal method as developed in Robnik/19/. Due to the limited space of these proceedings I am not able to present these results but the interested reader is referred to /5/ and /23/.

DISCUSSION AND CONCLUSIONS

The conclusion of this lecture is that the discrete energy spectra of Hamiltonians that have ergodic classical dynamics fall into two classes as to their fluctuation properties. The spectral fluctuations of the first class are described by the statistics of the Gaussian Orthogonal Ensemble (GOE) of random matrix theories, and those of the second class are described by the statistics of the Gaussian Unitary Ensemble (GUE). The criterion necessary to classify a given Hamiltonian H is in terms of the antiunitary symmetries A. The representation arguments relying on the existence of an A-adapted basis guarantee that H belongs to the GOE class if there is an antiunitary A such that AH=HA and either $A^n=1$ for an integer n, or A=UK, and the unitary operator U satisfies $U^n=1$ for some even n. The invariance arguments, however, suggest that the existence of an antiunitary symmetry A is sufficient for GOE statistics to apply.

Both assertions - the existence of exactly two universality classes (we ignore spin) - and the classification criterion - are firmly supported by the numerical evidence of Bohigas, Giannoni and Schmit/7/, Seligman and Verbaarschot/26,27/, Berry and Robnik/5/, and Robnik and Berry/23/. In the present lecture I have given theoretical arguments for the classification criterion. In particular, the invariance arguments suggest that an A-adapted basis exists for arbitrary antiunitary symmetry A, but this is yet to be proven and explicitly constructed. I have further pointed out that arguments in the papers by Pechukas/15/ and Yukawa/31/ lead to the conclusion that there are two universality classes. Let us rediscuss these arguments.

Following Yukawa/31/ we study a one parameter family of Hamiltonians

$$H(t) = H_o + tH_1 \tag{44}$$

which may be general complex Hermitian. We use the notation: $x_n(t)$ for eigenvalues, $\phi_n(t)$ for eigenfunctions (generally complex), $p_n = \langle \phi_n(t)|H_1|\phi_n\rangle$, and $f_{mn} = |x_m - x_n| \langle \phi_m(t)|H_1|\phi_n(t)\rangle$, $m \neq n$, the matrix elements. The equations of motion follow from the (nondegenerate) perturbation theory.

$$\frac{dx_n}{dt} = p_n \tag{45}$$

$$\frac{dp_n}{dt} = 2 \sum_{m\neq n} \frac{|f_{mn}|^2}{(x_n-x_m)^3} \tag{46}$$

$$\frac{df_{mn}}{dt} = (x_n-x_m) \sum_{\ell\neq m,n} \frac{f_{m\ell}f_{\ell n}}{|x_n-x_\ell||x_\ell-x_m||} \left\{ \frac{1}{x_n-x_\ell} + \frac{1}{x_m-x_\ell} \right\} \tag{47}$$

Clearly, x_n and p_n are always real, while f_{mn} are complex in general. It can be easily seen that the flow defined by (45-47) preserves volume in the phase space $\{x_n, p_n, f_{mn}\}$. Further, the 'energy'

$$E = \frac{1}{2} \sum_n p_n^2 + \frac{1}{2} \sum_{n\neq m} \frac{|f_{mn}|^2}{(x_n-x_m)} \tag{48}$$

is a constant of the motion. For a large enough t one can assume that the distribution in phase space is canonical with probability one, so

$$P\{x_n, p_n, f_{mn}\} = \text{const} \exp(-\text{const } E), \tag{49}$$

and after integrating over p_n and $|f_{mn}|$ we get the probability density for the distribution of the eigenvalues $\{x_n\}$,

$$P\{x_n\} = \text{const} \prod_{m>n} |x_m-x_n|^\beta, \tag{50}$$

where $\beta=2$ in the general case of complex f_{mn}, and $\beta=1$ if f_{mn} are real.

Now, when are f_{mn} real? Suppose there is an antiunitary symmetry A such that $H_oA = AH_o$ and

$$H_1A = AH_1 \tag{51}$$

and suppose that $\phi_n(0)$ is an A-adapted basis, i.e.

$$A\phi_n(0) = \phi_n(0), \tag{52}$$

for each n. Then f_{mn} are clearly real and their realness is preserved by the flow, for the perturbation theory implies:

$$\frac{d\phi_n}{dt} = \sum_{\ell\neq n} \frac{\langle\phi_\ell|H_1|\phi_n\rangle}{x_n - x_\ell} \phi_\ell , \tag{53}$$

which was used to obtain (47). As has been argued by Yukawa, $\beta=1$ leads to the GOE level spacings distribution, while $\beta=2$ yields GUE level spacings distribution. The agreement of these results with our previous conclusions will be perfect once we can show that an A-adapted basis exists for any A.

REFERENCES

1 R.Balian and C.Bloch, Ann.Phys. (N.Y.) 69, 76-160 (1972)
2 R.Balian and C.Bloch, Ann.Phys.(N.Y.) 85, 514-545 (1974)
3 M.V.Berry, Proc.Roy.Soc.Lond. A400, 229-251 (1985)

4 M.V.Berry and M.Robnik, J.Phys.A17, 2413-2421 (1984)
5 M.V.Berry and M.Robnik, J.Phys.A19 (1986) in press
6 O.Bohigas and M.J.Giannoni, Lecture Notes in Physics 209,1-99 (1984)
7 O.Bohigas, M.J.Giannoni, and C.Schmit, Phys.Rev.Lett. 52, 1-4 (1984)
8 F.J.Dyson, J.Math.Phys.3, 1191-1198 (1962)
9 M.C.Gutzwiller, J.Math.Phys.8, 1979-2000 (1967)
10 M.C.Gutzwiller, J.Math.Phys. 10, 1004-1020 (1969)
11 M.C.Gutzwiller, J.Math.Phys. 11, 1791-1806 (1970)
12 M.C.Gutzwiller, J.Math.Phys. 12, 343-358 (1971)
13 M.C.Gutzwiller, in Path Integrals and their applications in quantum statistical and solid state physics (ed. G.J.Papadopoulos and J.T.Devreese) pp 163-200 (New York: Plenum) (1978)
14 H.D.Meyer, E.Haller, H.Köppel and L.S.Cederbaum, J.Phys.A. 17 L831-836 (1984)
15 P.Pechukas, Phys.Rev.Lett. 51, 943-946 (1983)
16 C.E.Porter, Statistical Theories of Spectra: Fluctuations (Academic Press: New York) (1965)
17 M.Robnik, J.Phys.A.14, 3195-3216 (1981)
18 M.Robnik, J.Physique C2, 43, 45-61 (1982)
19 M.Robnik, J.Phys.A.17, 1049-1074 (1984)
20 M.Robnik, J.Phys.A.17, 109-130 (1984)
21 M.Robnik, in Proceedings of the NATO Advanced Study Jnstitute "Photophysics and Photochemistry in the Vacuum UV" , eds; S.P.McGlynn, G.L.Findley and R.H.Huebner, (dordrecht: Reidel) pp579-629) (1985)
22 M.Robnik, Regular and Chaotic Billiard Dynamics in Magnetic Fields, in Nonlinear Phenomena and Chaos, ed.S.Sarkar (Bristol: Adam Hilger) (1986)
23 M.Robnik and M.V.Berry, J.Phys.A.19 (1986) in press
24 M.Robnik and M.V.Berry, J.Phys.A.18, 1361-1378 (1985)
25 M.Robnik (1986) to be published
26 T.H.Seligman and J.J.M.Verbaarschot, Phys.Lett. 108A, 183-187 (1985)
27 T.H.Seligman and J.J.G. Verbaarschot J.Phys.A 18, 2227-2234 (1985)
28 T.H.Seligman, J.J.M.Verbaarschot and M.R.Zirnbauer, Phys.Rev. Lett. 53, 215-217 (1984)
29 T.H.Seligman, J.J.M.Verbaarschot and M.R.Zirnbauer, J.Phys. A.18,2751-2770 (1985)
30 E.Wigner, "Group Theory and its Applications to the Theory of Atomic Spectra" (New York: Academic Press) (1959)
31 T.Yukawa, Phys.Rev.Lett. 54, 1883-1886 (1985)

SPECTRAL STATISTICS OF SCALE INVARIANT SYSTEMS

T.H. Seligman

Instituto de Fisica, University of Mexico in Cuernavaca (UNAM)
Apdo. Post. 20364, 01000 Mexico DF, Mexico

and

J.J.M. Verbaarschot

Department of Physics, University of Illinois at Urbana-Champaign
Loomis Laboratory, 1110 W Green Street, Urbana, Il 61801

Abstract

We give both numerical and analytical results for the spectral statistics of two dimensional systems with a homegeneous polynomial as potential. When the hamiltonian is time reversal invariant we find a gradual transition from the Poisson ensemble to the GOE going from the classically integrable to the classically chaotic case. When the time reversal symmetry is broken we find the statistics of the GUE. Most features observed in the numerical calculations, with inclusion of the 'kink' in the Δ_3 statistic for integrable integrable systems and the asymptotic logarithmic behaviour for chaotic systems, are described by the semiclassical limit of the fluctuating part of the level density. It is shown that the transition to the chaotic regime can be described in the semiclassical limit. Finally, we construct a random matrix model that describes well the short range behaviour of all statistics studied.

1. Introduction.

The study of classically chaotic systems has made a great deal of progress during the last few years. Since the investigations of the quantum Sinai billiard by Berry [1] and Bohigas et al. [2] numerous other investigations have confirmed the conjecture that the level statistics of time reversal invariant chaotic systes are identical with those of the Gaussian Orthogonal Ensemble (GOE). The study of level statistics of integrable systems started more than a decade ago with the work of Berry and Tabor [3]. They found that the eigenvalues were distributed as independent random variables (also called the Poisson ensemble because the nearest neighbour spacing distribution has the form $\exp(-S)$).

We have been interested in the level statistics of chaotic systems in the transition between order and chaos [4]. For these studies the class of systems with a homegeneous polynomial as potential appeared to be particularly useful [5]. The reason is that due to a scale invariance [6] the topological structure of the classical phase space does not depend on the energy (see section 2 of this lecture). For example, the fraction of phase space that is chaotic (from now on called the chaotic volume) does not depend on energy. This fact greatly simplified the search for classical order parameters that characterize the fluctuations of the energy levels in the quantum system.

In section 4 we discuss numerical results and find that the spectral statistics are a monotonous function of the chaotic volume.

Later on we have extended our studies to scale invariant systems with broken time reversal invariance [7]. This was an important test for the concept that ergodic systems are described by the invariant random matrix ensembles. One expects that the level statistics of this class of hamiltonians coincides with the statistics of the ensemble of hermitean matrices (the Gaussian Unitary Ensemble (GUE)). This expectation has been confirmed by our numerical studies for cases with a true time reversal violation (see section 4).

The success of random matrix theory for ergodic systems urged us to construct a random matrix ensemble that interpolates between the Poisson ensemble and the GOE. We can no longer fall back on the invariant random matrix ensembles. Of course, we have to construct an ensemble whose statistics are universal, i.e. the functional dependence of the statistics does not depend on the details of the ensemble. This topic will be discussed in section 5 of this lecture.

The study of level statistics suggests that we are considering the limit in which a very large number of quantum levels are available (e.g. the semiclassical limit). In the latter limit Berry [8, 9] was able to prove that the asymptotic behaviour of the Δ_3 statistic coincides with the GOE (or GUE) result. The semiclassical limit of the Δ_3 statistic for integrable systems has been understood competely. Not only the linear behaviour predicted by Berry and Tabor [3] but also the appearance of the 'kink' discovered by Seligman et al. [4]. These issues are discussed in section 6. In the same section we discuss the semiclassical limit of the transition between integrability and ergodicity.

2. Description of the Model.

We have studied the quantum spectra in the transition from order to chaos for the class of models characterized by the hamiltonian [5,7,10,11,12]

$$H = \frac{1}{2}(p_1 + \alpha x_2^q)^2 + \frac{1}{2}(p_2 - \alpha x_1^q)^2 + \left(\frac{x_1}{\sigma_1}\right)^{2q} + \left(\frac{x_2}{\sigma_2}\right)^{2q} + \lambda(x_1 - x_2)^{2q}. \tag{2-1}$$

For $\lambda = 0$ and $\alpha = 0$ this hamiltonian is serapable and thus integrable. For $\alpha = 0$ it is time reversal invariant. When λ varies from 0 until the value λ_{max} for which the potential gets unbound and $\alpha = 0$ numerical calculations show the chaotic volume to vary from zero until one. In the case of broken time reversal invariance ($\alpha \neq 0$) and $\lambda = 0$ the chaotic volume ranges from zero to one as a function of α. In all our studies we have chosen $\sigma_1 \neq \sigma_2$ (for historical reasons). Consequently, the invariance under the permutation of 1 and 2 is broken. However, the invariance under the parity transformation is not affected.

The hamiltonian (2-1) has the property that the momentum independent terms and the terms linear in the momenta are homogeneous polynomials in the x_i. This has the important consequence that the hamilton equations of motion are invariant under the following scale transformation

$$\begin{aligned} p_i &\rightarrow \lambda^q p_i, \\ x_i &\rightarrow \lambda x_i, \\ H &\rightarrow \lambda^{2q} H, \\ t &\rightarrow \lambda^{1-q} t. \end{aligned} \tag{2-2}$$

This implies that for each trajectory $(x_i(t), p_i(t))$ at energy E we can construct a trajectory at energy $\lambda^{2q}E$: $(\lambda x_i(t\lambda^{1-q}), \lambda^q p_i(t\lambda^{1-q}))$. The Lyapunov exponent for the two trajectories is the same. This follows by using the following scale invariant definition for the Lyapunov exponent [11]

$$\Lambda = \lim_{t \rightarrow \infty, d(0) \rightarrow 0} \frac{\tau_c}{t} \log\left(\frac{d(t)}{d(0)}\right) \tag{2-3}$$

where

$$d(t) = \sqrt{(x_1(t) - x_1'(t))^2 + (x_2(t) - x_2'(t))^2 + \tau_c^2((p_1 - p_1')^2 + (p_2 - p_2')^2)}. \tag{2-4}$$

The cycle time τ_c has been introduced in order to be able to add terms of the same physical dimension. It is defined by

$$\tau_c = \frac{x_{av}}{p_{av}} \tag{2-5}$$

where the averages are defined by

$$x_{av} = \left(\frac{1}{2} < (x_1^2 + x_2^2) > -(\frac{1}{2} < (x_1 + x_2) >)^2\right)^{\frac{1}{2}}, \tag{2-6a}$$

$$p_{av} =< \frac{1}{2}(p_1^2 + p_2^2) >^{\frac{1}{2}} \tag{2-6b}$$

and the phase space average $< . >$ is defined as

$$< A >= \frac{\int dx_1 dx_2 dp_1 dp_2 A \delta(E - H)}{\int dx_1 dx_2 dp_1 dp_2 \delta(E - H)}. \tag{2-7}$$

The Lyapunov exponent defined in eq. (2-3) is invariant under the scale transformation given in (2-2). Consequently the distribution of the Lyapunov exponents is the same at all energies. In particular the fraction of phase space that is chaotic does not depend on energy. These results greatly facilitate the study of the corresponding quantum systems. For given value of the coupling constant the chaotic properties do not change over the quantum spectrum and can be related unambiguously to the level statistics. When the surface of section is defined as the (x_1, p_1)-plane, for trajectories related by a scale transformation give rise to the section points related by a scale transformation. Therefore the Poincaré-sections at different energies are simply related by a scale transformation.

3. Analysis of classical phase space.

In all our model studies we have determined the distrubution of the Lyapunov exponents by first sampling a given number (say N) of initial conditions according to a unifrom distribution on the energy-shell $(\delta(E-H)dx_1dx_2dp_1dp_2)$. For each initial condition we chose a nearby point and integrate the equations of motion in order to obtain the Lyapunov exponent defined in eq. (2-3). We use a rescaling technique due to Benettin et al. [13]. This technique provides us with both the chaotic volume μ and the Kolmogorov entropy K:

$$\mu = \frac{1}{N}\{number\ of\ \Lambda_i \neq 0\}, \tag{3–1}$$

$$K = \frac{1}{N}\sum_{i=1}^{N}\Lambda_i. \tag{3–2}$$

The errors in μ and K, essentially due to the Monte-Carlo algorithm, are of the order $\frac{1}{\sqrt{N}}$.

4. Analysis of the quantum spectra.

The Schrödinger equation corresponding to the hamiltonian eq. (2-1) has been solved by diagonalizing the quantum hamiltonian in a finite harmonic oscillator basis. By adapting the frequency of the basis such that the eigenvalues are minimized we were able to obtain ≈500 reliable eigenvalues from a ≈2000×2000 matrix [5]. The reliability has been tested by several independent methods. A particular useful tool appeared to be the comparison of the integrated level density obtained from matrix diagonalization and the Weyl formula given by

$$N(E) = \frac{1}{h^2}\int_{H<E} dx_1dx_2dp_1dp_2. \tag{4–1}$$

As a general rule we found that when to two level densities coincided the errors in the eigenvalues did not exceed our required accuracy of 1% of the average level spacing.

We have performed a statistical analysis of the fluctuations of the eigenvalues around their average position as given by the Weyl formula. The following statistics have been used: the Δ_3-statistic, the skewness and the excess of the number of levels in a given interval and the nearest neighbour spacing distribution. For a definition of these quantities we refer to the talk of O. Bohigas [14].

As a first example we show in Fig. 1 the nearest neighbour spacing distribution $P(S)$ and the Δ_3 statistic $\Delta_3(L)$. The spacing S is expressed in units of the average level spacing and the argument of $\Delta_3(L)$ is equal to the average number of levels in the interval under consideration. In each figure we have superimposed the even- and the odd-parity states. In this way the statistical errors have been reduced by a factor $\sqrt{2}$. The exact form of the potential is given by

$$H = \frac{1}{2}p_1^2 + \frac{1}{2}p_2^2 + 50\Big(\big(\frac{x_1}{0.8}\big)^4 + \big(\frac{x_2}{1.2}\big)^4 - \lambda(x_1-x_2)^4\Big). \tag{4–2}$$

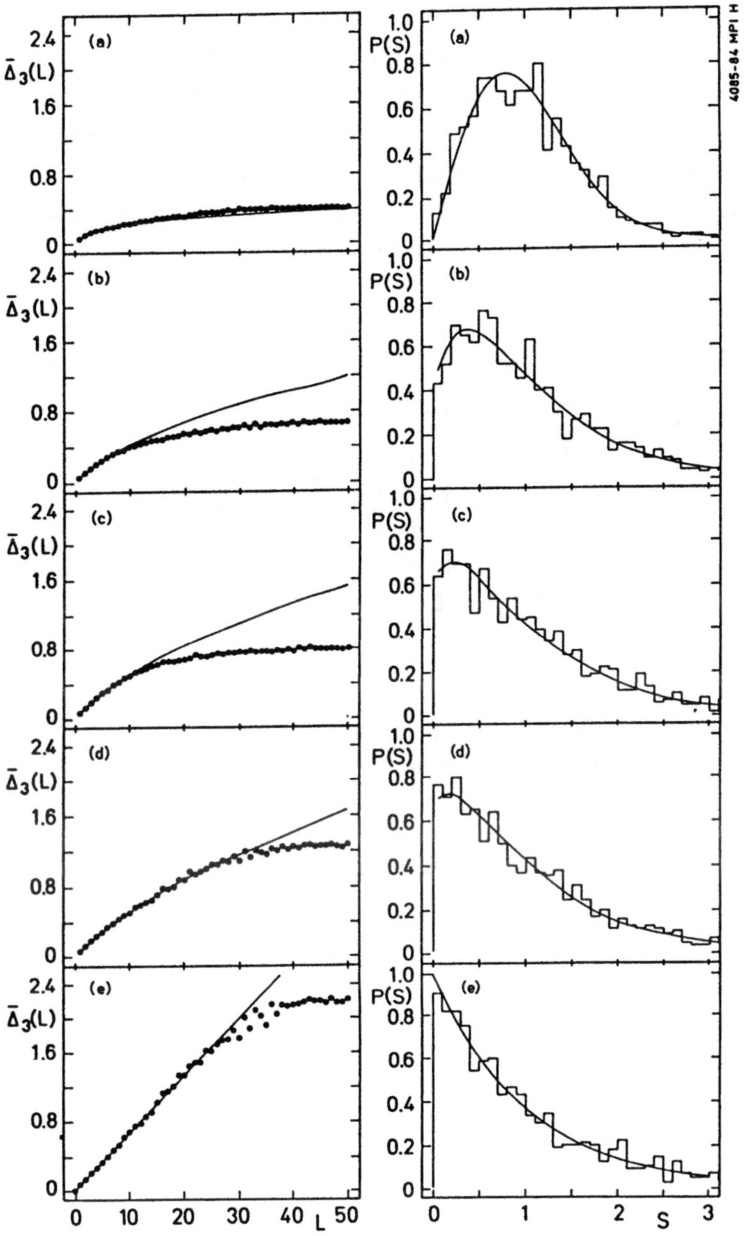

Fig[1]: Numerical results for the Δ_3 statistic and the distribution of nearest neighbour spacings, $P(S)$. Dots and the histograms represent the results obtained for the hamiltonian (4-2). The lines were obtained from a random matrix model defined in section 5. Figures $(a)-(e)$ correspond to interaction strengths $\lambda = 0.10$, 0.04, 0.02, 0.01 and 0.0 and chaotic volumes 1.00, 0.95, 0.50, 0.25 and 0.0, respectively.

The values of λ and the chaotic volume μ are given in the figure caption. We observe a gradual transition from the integrable case (lower figures) to the ergodic case (upper figures). The full lines in these figure correspond to the Poisson ensemble and the GOE, respectively. The full lines in the other figures have been obtained from a random matrix ensemble to be discussed in the next section. Another striking result from these figures is the appearance of a 'kink' in the Δ_3 statistic: after a certain value of L, $\Delta_3(L)$ remains constant. This effect which is particularly obvious in the integrable case will be discussed in the next section.

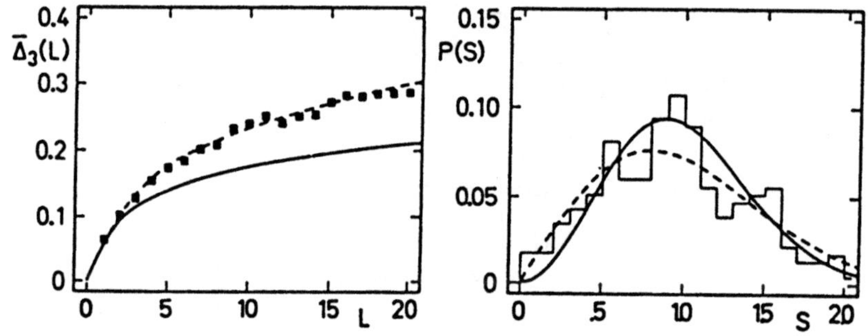

Fig [2]: The dots on the left and the histogram on the right give the numerical result for the Δ_3 statistic and the nearest neighbour spacing distribution with $\alpha = 3.0$ and $\lambda = 0$. The full and dashed lines give the corresponding results for the GUE and the GOE, respectively.

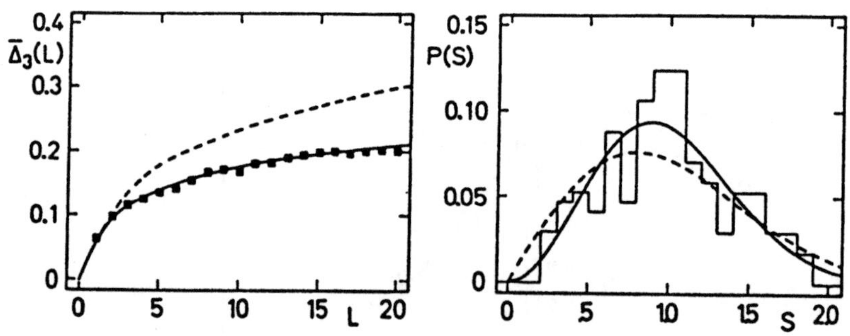

Fig [3]: The dots on the left and the histogram on the right give the numerical result for the Δ_3 statistic and the nearest neighbour spacing distribution with $\alpha = 3.0$ and $\lambda = 0.167$. The full and dashed lines give the corresponding results for the GUE and the GOE, respectively.

As a second example we show a case where the time reversal symmetry is broken. The hamiltonian given by

$$H = \frac{1}{2}(p_1 - \alpha x_2^3)^2 + \frac{1}{2}(p_2 - \alpha x_2^3)^2 + \frac{25}{3}\left(\left(\frac{x_1}{1.2}\right)^6 + \left(\frac{x_2}{0.8}\right)^6\right) - \lambda(x_1 - x_2)^6 \qquad (4\text{–}3)$$

was studied as a function of α and λ. When we switch on the parameter α until its chaotic value and leave $\lambda = 0$ the level statistics approach the value of the GOE (see Fig. 2). This has been identified by Berry and Robnik as a case of false time reversal violation [15, 16]. As a consequence of the reflection symmetry it is possible to choose a so called adapted basis in which the hamiltian is represented by a real symmetric matrix. Only, when we also switch on λ until its chaotic value (see Fig. 3) we obtain the statistics of the GUE, the ensemble appropriate for hermitean matrices.

5. A random matrix model.

In order to parametrize the spectral statistics inbetween the integrable and ergodic case we have proposed a random matrix model. After some trial and error we arrived at the following random matrix ensemble

$$H_{ij} = (H_{GOE})_{ij} \frac{1}{\sqrt{2\pi\sigma}} \exp\left(-\frac{(i-j)^2}{2\sigma^2}\right). \qquad (5\text{–}1)$$

Here, H_{GOE} is a gaussian orthogonal ensemble. By choosing different values of σ this ensemble interpolates between the Poisson ensemble ($\sigma = 0$) and the GOE ($\sigma = \infty$). Of course, the results for the statistics depend on the size of the matrix. However, this is not true for the family of *all* curves obtained for the Δ_3-statistic and the nearest neighbour spacing distribution. We also have studied ensembles with a nongaussian cutoff. By reparametrizing the cutoffs we were able to map the resulting spectral statistics to those of eq. (5-1). An ensemble that does *not* show the behaviour observed in Fig. 1, is surprisingly given by the superposition of a Poisson ensemble H_P and a Gaussian Orthogonal Ensemble H_{GOE}

$$H = \alpha H_P + (1 - \alpha) H_{GOE}. \qquad (5\text{–}2)$$

The reason is the nonuniformity of this ensemble: for small distances in the spectrum this ensemble always follows the GOE and it starts deviating at larger distances [17]. The detailed form of the statistics depends on the value of α.

For a more detailled discussion of the ensemble defined in eq. (5-1) we refer to Roman et al. [22] elsewhere in this volume.

6. Semiclassical interpretation.

Recently, Berry succeeded [8] to derive the functional form of the Δ_3 statistic in the semiclassical limit. For integrable systems he was able to give a semiclassically exact expression, whereas for ergodic systems only the asymptotic logarithmic behaviour could be extracted. In the first part of this section we review the general arguments given by Berry and in the second part we apply them to homogeneous potentials.

In the semiclassical limit the fluctuating part of the level density is given by

$$\rho(\epsilon) = \frac{1}{\hbar^{\frac{d+1}{2}}} Re \sum_j A_j \exp(\frac{i}{\hbar}(S_j(E) + \epsilon \frac{dS_j(E)}{dE})) \tag{6–1a}$$

for *integrable* systems and

$$\rho(\epsilon) = \frac{1}{\hbar} Re \sum_j A_j \exp(\frac{i}{\hbar}(S_j(E) + \epsilon \frac{dS_j(E)}{dE})) \tag{6–1b}$$

for *ergodic* systems. The sum is over all periodic orbits and d is the dimension of the system. Note that the coefficients A_j in eq. (6-1a) and eq. (6-1b) have a different physical dimension. We have expanded S_j around E. The local energy variable is denoted by ϵ. This expression for $\rho(\epsilon)$ can be used to calculate the two-point function

$$< \rho(\epsilon)\rho(\epsilon') >= \frac{2}{\hbar^{2\mu+2}} Re \sum_j A_j^2 \exp(\frac{i}{\hbar}\frac{dS_j}{dE}(\epsilon - \epsilon')). \tag{6–2}$$

We have introduced the symbol μ which is equal to $\frac{d-1}{2}$ for *integrable* systems and *zero* for *ergodic* systems. The brackets $< . >$ denote local averaging on a scale of many level spacings. Berry has shown that in the limit $h \to 0$ only the diagonal terms contribure to the l.h.s. of eq. (6-2). We can rewrite eq. (6-2) as

$$< \rho(\epsilon)\rho(\epsilon') >= \frac{2}{\hbar^{2\mu+2}} Re \int_0^\infty dT \phi(T) \exp(\frac{i}{\hbar}T(\epsilon - \epsilon')) \tag{6–3}$$

where

$$\phi(T) = \sum_j \delta(T - \frac{dS_j}{dE}) A_j^2. \tag{6–4}$$

The large T limit of the function $\phi(T)$ has been determined analytically by Hannay and Ozorio de Almeida [18]:

$$\phi(T) = \frac{1}{(2\pi)^{d+1}} \frac{d\Omega(E)}{dE} \tag{6–5a}$$

for *integrable* systems and

$$\phi(T) = \frac{T}{2\pi^2} \tag{6–5b}$$

for *ergodic* systems. $\Omega(E)$ is the volume of the energy shell at energy E. By inserting this result in eq. (6-4) we immediatly find that th density correlation function is given by $\rho(E)\delta(\epsilon - \epsilon')$

for *integrable* systems and by $-\frac{1}{\pi^2}\frac{1}{(\epsilon-\epsilon')^2}$ for *ergodic* systems. In terms of the Δ_3 statistic this corresponds to a linear and a logarithmic behaviour, respectively. For the discussion of the subtleties in the aforementioned arguments we refer to Berry [8].

For scaling systems the sum rules for the coefficients A_j can be understood in a particularly simple way. Under the scale transformation (2-2) the coefficients A_j transform as

$$\frac{A_j^2}{T_j} \rightarrow \lambda^{d(1+q)-2}\frac{A_j^2}{T_j} \tag{6–6a}$$

for *integrable* systems and

$$\frac{A_j^2}{T_j} \rightarrow \lambda^{1-q}\frac{A_j^2}{T_j} \tag{6–6b}$$

for *ergodic* systems. When we apply this transformation to the sumrule eq. (6-4) we obtain the functional dependence given in eqs. (6-5). The proportionality constants *cannot* be obtained by scaling arguments.

For integrable scaling systems the amplitudes A_j and the actions S_j can be obtained in a particularly simple way by starting from the WKB-approximation for the eigenvalues. The WKB-level density is given by

$$\rho(\epsilon) = \sum_{n_i}\delta(E - \sum_{i=1}^{2}\gamma_i(n_i+\alpha_i)^p). \tag{6–7}$$

The power p depends on the order of the polynomial in the scaling potential and the constants α_i will be taken equal to $\frac{1}{2}$. By performing a Poisson resummation [19] we can rewrite the sum over the n_i as

$$\rho(\epsilon) = \int_0^\infty\int_0^\infty dn_1 dn_2 \sum_{M_1,M_2}\exp(2\pi i(\sum_{j=1}^{2}M_j n_j - \alpha_j))\delta(E - \sum_{i=1}^{2}\gamma_i n_i^p). \tag{6–8}$$

We transform the variables n_i to polar coordinates. The integration over the delta function can be carried out and we are only left with the angular integrations. The result is

$$\begin{aligned}\rho(\epsilon) =& const. \sum_{M_1,M_2}\int_0^{\frac{\pi}{2}} d\phi(\cos\phi\sin\phi)^{\frac{2}{p}-1}\times \\ &\exp(-2\pi i(\sum_{i=1}^{2}M_i\alpha_i - E^{\frac{1}{p}}(M_1c_1\cos^{\frac{2}{p}}\phi + M_2c_2\sin^{\frac{2}{p}}\phi))).\end{aligned} \tag{6–9}$$

Here c_1 and c_2 are constants that can be evaluated easily. The explicit expressions are given in ref. [12]. The term with M_1 and M_2 equal to zero yields the Weyl formula for the average level density:

$$\rho_{sm}(\epsilon) = \int dn_1 dn_2 \delta(E - \gamma_1 n_1^p - \gamma_2 n_2^p). \tag{6–10}$$

When both M_1 and M_2 are different from zero we can make a stationary phase approximation. This provides us with the amplitudes A_j and the actions S_j in closed orbit sum for the fluctuating part of the level density. By using this result as an input for the general expressions given by Berry [8] we obtain a semiclassically exact expression for the $\Delta_3(L)$ statistic. As an example we show in Fig. 4 a comparison of the exact values of the Δ_3 statistic for the WKB-approximation of the spectrum of a homogeneous fourth order potential and its semiclassical approximation.

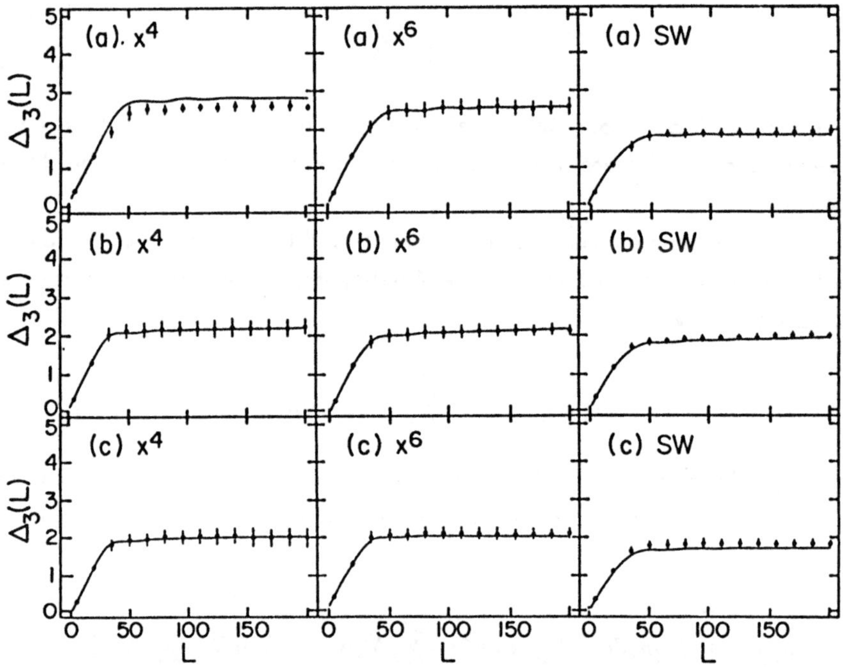

Fig [4]: Comparison of the Δ_3 statistic calculated from the WKB-eigenvalues (dots with error bars) for the 500th eigenvalue up to the 1500th eigenvalue and the analytical expression that can be obtained from eq. (6-9) for an homogeneous fourth order potential. The ratio of γ_1/γ_2 is equal to π.

One concludes that the kink in the Δ_3 can be explained entirely by the semiclassical approximation to the fluctuating part of the level density.

The semiclassical limit of the nearest neighbour spacing distribution $P(S)$ in the transition region between the Poisson statistics and the GOE statistics has been given by Berry and Robnik [20]. They argued that each disconnected chaotic region in phase space give rise to a sequence of levels with GOE-statistics and a level density proportional to the volume of the region, whereas the integrable part of phase space corresponds to a sequence of levels with Poisson statistics and level density proportinal to its volume. By superimposing all sequences they were able to derive an analytical expression for $P(S)$ in terms of the volumina μ_i of the different regions. The same analysis can be carried through for the Δ_3-statistic [10]. Using the result that the Δ_3-statistic is the integral transform of the number variance [21], which is an additive quantity, we arrive at the result

$$\Delta_3(L) = \Delta_{3P}(\mu_1 L) + \sum_{i=1}^{k} \Delta_{3GOE}(\mu_i L), \tag{6–11}$$

where the sum of over the k disconnected chaotic region with volumes μ_i. The volume of the integrable region is equal to μ_1. The subscripts P and GOE refer to the Poisson ensemble and the GOE, respectively. Now it becomes clear that in order to see a sizeble deviation from the Poisson ensemble in Fig. 1 we need a large chaotic fraction of phase space: the total chaotic volume is divided up into many different disconnected pieces which only merge when the total chaotic volume is near to 1. The net effect of the superposition of independent GOE sequences is a sequence of levels with Poisson statistics.

In Fig. 5 we give an example of the application of the aforementioned semiclassical result. The parameters of the homogeneous x^4-potential and the partioning of phase space can be found in the figure caption.

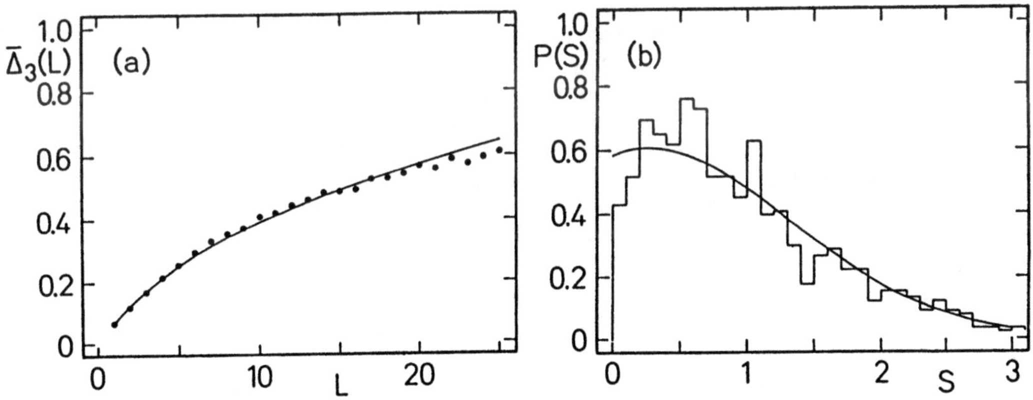

Fig [5]: The Δ_3 statistic (a) and the nearest neigbour spacing distribution (b) for the hamiltonian given in eq. (4-2) with $\lambda = 0.04$. The dots and the histogram correspond to the exact quantum mechanical calculation. The semiclassical results corresponding to the full line have been calculated with the partioning (0.09, 0.45, 0.56) that has been obtained from an analysis of the distribution of the Lyapunov exponents. The first number in the sequence is the ordered fraction of phase space.

We observe an excellent agreement for the Δ_3-statistic and the tail of the nearest neighbour spacing distribtion. As a consequence of tunneling between the different chaotic regions the semiclassical limit of the nearest neighbour spacing distribution deviates from the exact quantum results for small spacings.

7. Conclusion.

The basic conclusion of this paper is that the fluctuations of the energy levels of quantum systems which are classically ergodic are described by the invariant random matrix ensemble (either GOE or GUE) whereas the fluctuations of the spectra of systems in the transition between integrable and ergodic can be parametrized by a one parameter random matrix model. An important point is that all statistics are described by one and the same value of the parameter. However, the random matrix model does not provide the long range stiffness of the integrable part of the spectrum.

On the other hand the quantum results can be related directly to the structure of classical phase space via the semiclassical limit of the correlation function. This completely explains the 'kink' in the Δ_3 statistic for integrable systems and the asymptotic logarithmic behaviour for ergodic systems. Even the transition region between order and chaos can be described satisfactorily in the semiclassical limit. The most important open problem is the semiclassical limit of the nearest neighbour spacing distribution for small spacings which a lack of level repulsion.

Acknowledgements.

This work was supported by grant NSF-PHY-84-15064 and by Consejo Nacional de Ciencia y Tecnologia Mexico. Among the many people we would like to thank we mention in particular: M.V. Berry, O. Bohigas, M.J. Gianonni, H.A. Weidenmüller and M.R. Zirnbauer.

References.

1. M.V. Berry, Ann. of Phys. NY **131** 163 (1981)
2. O. Bohigas, M.J. Giannoni, C. Schmit, Phys. Rev. Lett. **52** 1 (1984)
3. M.V. Berry, M. Tabor, Proc. Roy. Soc. A **356** 375 (1977)
4. T.H. Seligman, J.J.M. Verbaarschot. M.R. Zirnbauer, Phys. Rev. Lett. **53** 215
5. T.H. Seligman, J.J.M. Verbaarschot. M.R. Zirnbauer, J. of Phys. **A 18** 2751 (1985)
6. L. Landau and E.M. Lifschitz, Mechanics (Pergamon, Oxford, 1969)
7. T.H. Seligman and J.J.M. Verbaarschot, Phys. Lett. **108A** 183 (1985)
8. M.V. Berry, Proc. Roy. Soc. **A400** 229 (1985)
9. M.V. Berry, Proceedings of this Conference
10. T.H. Seligman, J.J.M. Verbaarschot, J. Phys. **A18** 2227 (1985)
11. T.H. Seligman, J.J.M. Verbaarschot, M.R. Zirnbauer, Phys. Lett. **110a** 231 (1985)
12. T.H. Seligman, J.J.M. Verbaarschot, J. Phys. **A** (1986) (in press)
13. G.L. Benettin, L. Galgani, A. Giorgilli and J.M. Strelcyn, Meccanica **21** (1980)
14. O. Bohigas, Proceedings of this conference
15. M.V. Berry and M. Robnik, to be published
16. M. Robnik, Proceedings of this Conference
17. T.A. Brody, J. Flores, J.B. French, P.A. Mello, A.Pandey, S.S.M. Wong, Rev. Mod. Phys. **53** 385 (1982)
18. J.H. Hannay and A.M. Ozorio de Almeida J. Phys **A17** 3429 (1984)
19. M.V. Berry and M. Tabor, Proc. Roy. Soc. **A349** 101 (1976)
20. M.V. Berry and M. Robnik, J. Phys. **A17** 2413 (1984)
21. A. Pandey, Ann. Phys. NY **119** 170 (1979)
22. C. Roman, T.H. Seligman and J.J.M. Verbaarschot, Proceedings of this Conference

Spectral Fluctuations in the Three-Level LMG Model

D.C. Meredith and S.E. Koonin
W.K. Kellogg Radiation Laboratory
California Institute of Technology
Pasadena, California 91125 U.S.A.

"Spectra of time reversal invariant systems whose classical analogs are K systems show the same fluctuation properties as predicted by GOE." [1] This hypothesis of Bohigas, Giannoni, and Schmidt is supported by numerical studies of spectral fluctuations for several Hamiltonians that describe particles in a potential well: Sinai's billiard, the stadium billiard,[1] and the two-dimensional harmonic oscillator perturbed by fourth and sixth order terms.[2] Our work extends these studies to the three-level Lipkin-Meshkov-Glick model (LMG),[3] an essentially quantum Hamiltonian mimicking a nuclear shell model.

The Model.

The two-level LMG model has only one degree of freedom (e.g., the number of particles in the upper level), therefore the classical analog cannot exhibit chaotic behavior at any value of the coupling constant. To study chaotic behavior we must turn to a generalization of the model that describes a system of N particles in three N-fold degenerate levels, which has with two degrees of freedom (e.g., the numbers of particles in the two highest levels). With this additional degree of freedom comes the possibility of seeing chaotic behaviour in the classical limit. This three-level Hamiltonian is written using single-particle creation and annihilation operators:

$$H = \sum_{k=0}^{2} \epsilon_k (\sum_{n=1}^{N} a_{kn}^{\dagger} a_{kn}) - \frac{1}{2} V \sum_{k,l=0}^{2} (1 - \delta_{kl}) (\sum_{n=1}^{N} a_{kn}^{\dagger} a_{ln})^2$$

where the single-particle states are labeled by k (= 0, 1, 2 denoting the three levels) and n (=1, 2, 3, ..., N denoting the different degenerate states within each level).

It is sufficient to look only at the totally symmetric states of the three- level model.[1] These states are denoted by $|l,m\rangle$, where m is the number of particles in the third level and l the number in the second level. We note that since the basis is finite we do not have to introduce an ad-hoc truncation of the basis to find the eigenvalues. We choose the single- particle energies to be $-\epsilon, 0$, and ϵ. In this symmetric basis the Hamiltonian is written[4]

$$\langle l'm'|H|lm\rangle = \epsilon(-N+l+2m)\delta_{ll'}\delta_{mm'} - \frac{V}{2}A_{l'm',lm}$$

where

$$\begin{aligned} A_{l'm',lm} = &\sqrt{l(l-1)(N-l-m+1)(N-l-m+2)}\delta_{l-2,l'}\delta_{mm'} \\ &+\sqrt{(l+1)(l+2)(N-l-m)(N-l-m-1)}\delta_{l+2,l'}\delta_{mm'} \\ &+\sqrt{m(m-1)(N-l-m+1)(N-l-m+2)}\delta_{ll'}\delta_{m-2,m'} \\ &+\sqrt{(m+1)(m+2)(N-l-m)(N-l-m-1)}\delta_{ll'}\delta_{m+2,m'} \\ &+\sqrt{(l+1)(l+2)m(m-1)}\delta_{l+2,l'}\delta_{m-2,m'} \\ &+\sqrt{(m+1)(m+2)l(l-1)}\delta_{l-2,l'}\delta_{m+2,m'}. \end{aligned}$$

Because the interaction moves *pairs* of particles, H conserves the parities of l and m separately. This conservation principle creates four symmetry classes and thereby reduces the size of the Hamiltonian matrix and the labor needed to find the spectrum.

The Classical Limit.

The classical Hamiltonian is obtained in the limit $N \to \infty$ with $\chi = (N-1)V/\epsilon$ fixed.[5] It has been shown that there exists an overcomplete set of coherent states that become orthogonal in this limit. Also in this limit, matrix elements of "reasonable" operators between two different states vanish, products of operators factorize, expectation values of commutators become Poisson brackets, and quantum equations of motion become Hamilton's equations of motion.

In our model the coherent states are[6]

$$|z\rangle = \exp(\sum_{k=1}^{2}\sum_{n=1}^{N} z_k a_{kn}^{\dagger} a_{0n})|0\rangle$$

where $|0\rangle$ is the "vacuum" state with all N particles in the hole (k=0) level, and z_k is a complex number. The subscripts k=1,2 designate particle levels. We introduce the variables

$$q_k = \mathrm{Re}\frac{\sqrt{2}z_k}{\sqrt{1+|z_1|^2+|z_2|^2}}, \qquad p_k = \mathrm{Im}\frac{\sqrt{2}z_k}{\sqrt{1+|z_1|^2+|z_2|^2}}.$$

The equations of motion for these variables, which are obtained using a variational principle,[7] are Hamilton's equations; therefore p and q can be interpreted as coordinates and momenta. The classical Hamiltonian is then the expectation value of H between normalized coherent states:

$$\frac{H}{N\epsilon} = -1 + \frac{1}{2}q_1^2(1-\chi) + \frac{1}{2}q_2^2(2-\chi) + \frac{1}{2}p_1^2(1+\chi) + \frac{1}{2}p_2^2(2+\chi)$$

$$+\frac{\chi}{4}[(q_1^2+q_2^2)^2 - (p_1^2+p_2^2)^2 - (q_1^2-p_1^2)(q_2^2-p_2^2) - 4q_1q_2p_1p_2]$$

If χ=0 (i.e., there is no two-body interaction) we obtain a two-dimensional harmonic oscillator, an integrable system. Therefore by varying the coupling constant we may be able to produce a transition from integrability to chaos.

The total phase space is limited to a four-ball of radius $\sqrt{2}$ (z_k is free to take on all values, but the p_k and q_k are restricted). Because the phase space is compact, solutions to the equations of motion exist only in a finite energy interval.

Numerical Calculations.

We have done numerical calculations for four values of the coupling constant: χ=0.75, 2, 10, 100, which correspond to classical potentials with one, two, four and again four minima. Balancing computer time constraints and the desire for large N, we chose N=85, and diagonalized the Hamiltonian matrices corresponding to each of the four symmetries (one of dimension 903 and three of dimension 946).

The classical level density needed for unfolding the spectra was obtained by Monte Carlo integration of

$$\rho(E) = \int \delta(E - H(q_1, q_2, p_1, p_2))dq_1dq_2dp_1dp_2.$$

It agreed qualitatively with the level density obtained directly from the eigenvalues.

We have done a preliminary classification of phase space by integrating the classical equations of motion for all four values of χ, each at several energies, and for several initial conditions. The classification was based on the generated surfaces of section. Toroidal surfaces of section indicated an integrable or near-integrable system, while those sections that were covered densely by one orbit indicated a chaotic system. We plan a more careful survey of the phase space, calculating Lyapunov exponents and the chaotic volume.

We found that χ=0.75 showed toroidal surfaces of section over the entire energy range, indicating near-integrability. For χ=2 and χ=10 there was a wide range of behavior from near-integrable to apparently totally chaotic, but there was no long stretch of energy values for which the classical behavior was constant. Because the classification was not clear, we did not do further calculations for these values of χ. For χ=100 we found a large energy range for which the motion seems totally chaotic. This range encompassed about 400 eigenvalues from each symmetry class. Further computations were restricted to these "chaotic" levels for χ=100, while we used all levels for χ=0.75.

After unfolding the spectra, we calculated six different spectral measures. This was done separately for the four different symmetries, then the results were combined to obtain better statistics. These measures were the nearest neighbor spacing distribution ($P(s)$), the average number of levels ($\bar{n}(L)$) in an interval of length L, the variance, excess, and skewness of the number of levels in an interval of length L, and $\Delta_3(L)$. This latter is a measure of the rigidity of the spectrum, and is defined by

$$\Delta_3(L,\alpha) = \frac{1}{L}\text{Min}_{A,B}\int_{\alpha}^{\alpha+L}[N(x) - Ax - B]^2dx,$$

where $N(x)$ is the number of eigenvalues less than x (a staircase function).

Because the classical motion was near-integrable for χ=0.75 we expected Poisson statistics for the corresponding eigenvalues. The nearest neighbor spacing distribution was almost Poisson, although the maximum did not occur at the origin, but near 0.4. The agreement with $\bar{n}(L) = L$ was good, indicating that the unfolding was done correctly. The other four statistics for this χ did not agree at all with values predicted for a Poisson spectrum. We concluded that this is another instance of the "harmonic oscillator anomaly" noticed by others.[2,8] For this value of χ the system is close to a harmonic oscillator with frequencies in the ratio of 1:5. Indeed, the spectrum of such an oscillator was also calculated to have spectral fluctuations qualitatively similiar to our three-level LMG spectrum, and similarly far from Poisson predictions.

Because the classical motion was chaotic for χ=100, we expected GOE statistics. The nearest neighbor spacing distribution and $\bar{n}(L)$ were in good agreement with GOE predictions. The results for the Δ_3 calculation gave a relatively smooth curve, but were consistently higher than the GOE values. The remaining measures did not show smooth behavior, and they differed significantly from the GOE predictions, but the general trend of the data did agree with GOE. The scatter in the numerical data may have arisen from poor statistics due to the small sample size. The smallest matrix gave a spectrum that was signficantly less stiff than the other three, while the spectral measures of the three equal size matrices varied only slightly.

We can interpret our results for χ=100 based on the work by Seligman *et al.*,[2] which shows that there is a smooth transition between Poisson and GOE statistics as the classical motion changes from integrable to chaotic. Based on their conclusions, we see that our calculations for the χ=100 system are qualitatively consistent with classical motion that is nearly, but not totally, chaotic. This gives a less stiff spectrum than GOE.

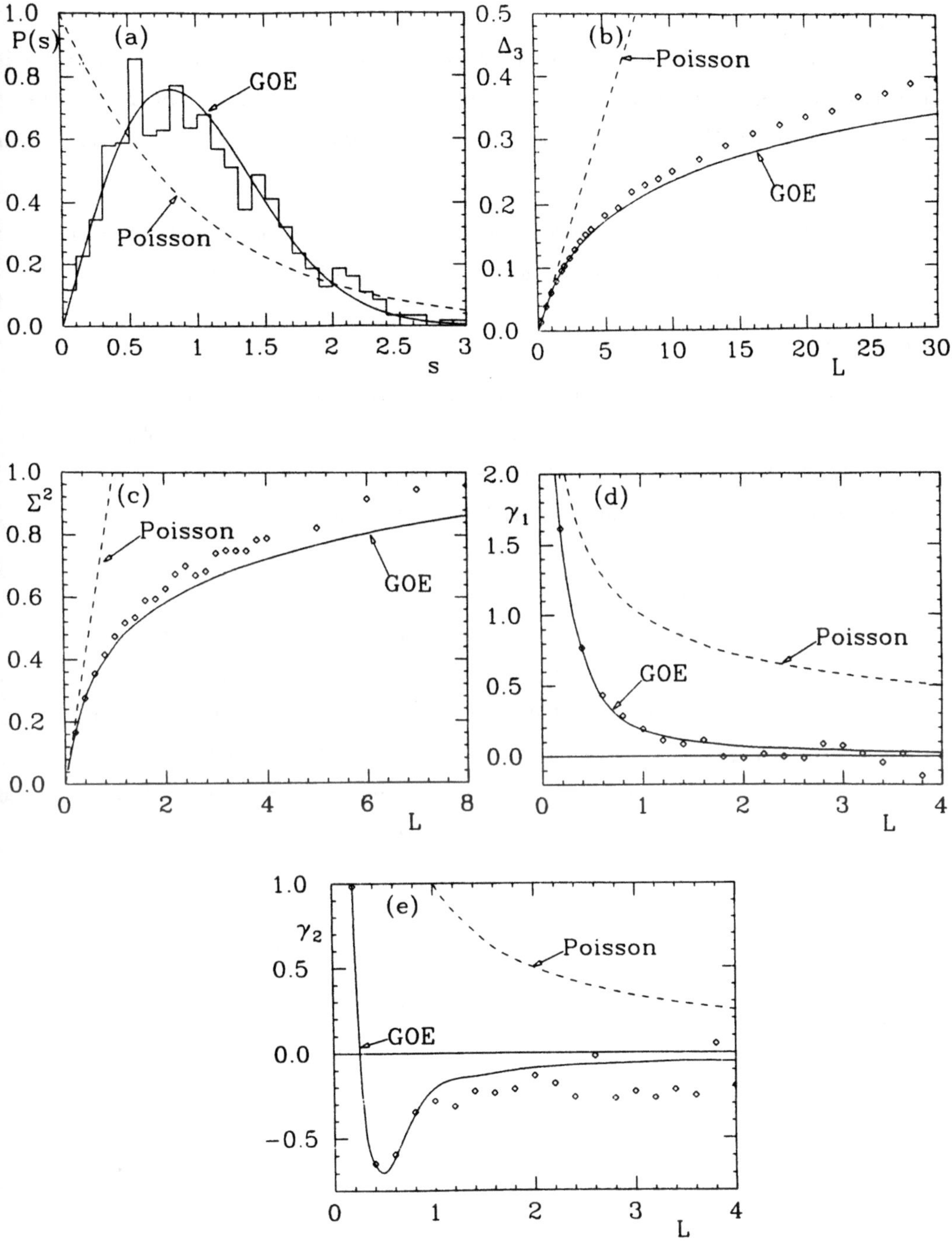

Spectral fluctuations of the three-level LMG model with N=85 and χ=100. A total of 1190 levels from three different symmetry classes (viz., those classes giving equal sized matrices) were included in the analysis. We show a) the nearest neighbor spacing distribution, b) spectral rigidity, c) variance of the number of levels ($n(L)$) in an interval of length L, d) excess of $n(L)$, and e) skewness of $n(L)$. The curves for GOE and Poisson were obtained from references 9-11.

Conclusion.

Our preliminary results for the spectral fluctuations of the three-level LMG model are in qualitative agreement with results obtained for other Hamiltonian systems. Final results await a more careful classification of classical phase space, and better statistics achieved by diagonalizing larger matrices.

We would like to thank M.R. Zirnbauer for many useful discussions. This work was supported in part by the National Science Foundation, grants PHY85-05682 and PHY82-07332.

References

1) O. Bohigas, M.J. Giannoni, and C. Schmit, Phys. Rev. Lett. **52**, 1 (1984).
2) T.H. Seligman, J.J.M. Verbaarschot, and M.R. Zirnbauer, Phys. Rev. Lett. **53**, 215 (1984).
3) H.J. Lipkin, N. Meshkov, and A.J. Glick, Nucl. Phys. **62**, 188 (1965).
4) B. Flanders, Ph.D. Thesis, California Institute of Technology, 1981.
5) L.G. Yaffe, Rev. Mod. Phys. **54**, 407 (1982).
6) R.D. Williams and S.E. Koonin, Nucl. Phys. **A391**, 72 (1982).
7) K.-K. Kan, J.J. Griffin, P.C. Lichtner, and M. Dworzecka, Nucl. Phys. **A332**, 109 (1979).
8) M.V. Berry and M. Tabor, Proc. Roy. Soc. (Lond.) **A356**, 375 (1977).
9) M.L. Mehta and J. des Cloizeaux, Indian Journ. Pure and Appl. Math. **3**, 329 (1972).
10) R.V. Haq, A. Pandey, and O. Bohigas, Phys. Rev. Lett. **16**, 1086 (1982).
11) O. Bohigas and M.-J. Giannoni, in *Mathematical and Computational Methods in Nuclear Physics, 1983*, edited by J.S. Dehesa (Springer-Verlag, 1984).

FLUCTUATION PROPERTIES OF REGULAR AND IRREGULAR SPECTRA

V.B. Sheorey
Physical Research Laboratory
Navrangpura, Ahmedabad 380 009
India

Although many aspects of regular and chaotic motion in classical systems are fairly well understood the way in which they are expressed in the corresponding quantal systems is not always equally clear [Zaslavsky, 1981 and Bohigas and Giannoni, 1984]. In order to understand chaos in quantal systems there have been several studies of quantal energy eigenvalues for potentials which are known to exhibit chaotic classical motion. This is with a view of obtaining a "signature" in the eigenvalue spectrum of the underlying classical chaotic motion. The main suggestions in this respect have included (i) an examination of the properties of individual energy eigenvalues and (ii) an examination of the fluctuation properties of a sequence of levels.

As regards the latter suggestion, spectral measures for a sequence of eigenvalues have long ago been introduced to describe the statistical properties of spectra of many-particle systems such as nuclei [Brody et al., 1981]. These spectral measures have been employed for bound systems with a few degrees of freedom to understand ways in which these spectral measures for a sequence of levels reflect integrability or chaos in the underlying classical trajectories and thus, perhaps, provide a signature of "quantum chaos". It must be mentioned that in a recent paper Berry [Berry, 1985] has obtained, in the semi-classical limit ($\hbar \to 0$) formulae for the spectral rigidity $\Delta^3(n)$, where n is the energy interval in units of the mean spacing. In particular, it has been shown that $\Delta_3(n)$ displays certain universal behaviour as a result of properties of very long classical orbits.

As regards suggestion (i), Percival [Percival, 1977] has predicted that the appearance of "avoided crossings" in the quantal eigenvalue spectrum corresponds to irregular classical trajectories which are associated with unstable orbits, i.e., orbits which do not lie on invariant tori. Such irregular quantal eigenvalues are sensitive to a parameter in the potential.

We have considered both these suggestions taken together. We briefly present our calculational procedure and some of the results for the quartic potential,

$$V(x,y) = (x^2 + y^2)/2 + \alpha x^2 y^2$$

This potential has a totally bound eigenvalue spectrum and the Hamiltonian has a C_{4v} point group symmetry.

For this potential Pullen and Edmonds [1981] have examined both the quantal eigenvalues and the classical trajectories. The well known Poincare surface of section technique has been utilized to study classical motion and it is clear from their results that for low energies the motion is almost regular while for high energies it is almost totally irregular. The quantal eigenvalues, correspondingly, exhibit avoided crossings, which are characterized by large second differences with respect to small variations in the coupling constant α.

$$\Delta_i^2 = | \{E_i(\alpha + \delta\alpha) - E_i(\alpha)\} - \{E_i(\alpha) - E_i(\alpha - \delta\alpha)\}|$$

The procedure used for the calculation of eigenvalues for this potential is similar to the one followed by Pullen and Edmonds. The C_{4v} symmetry of the Hamiltonian was taken into account and eigenvalues belonging to the irreducible representations A_1, A_2, B_1 and B_2 for several values of α in the range 0.04500 to 0.05250 with $\delta\alpha = 0.00125$, were obtained by diagonalizing appropriate truncated Hamiltonian matrices. All eigenvalues in the range 0.0 to 50.0 were thus obtained. The total number of levels for the four symmetries considered and for five values of α were about 2000. In order to calculate the Δ_3-statistic it is important to decompose the distribution function into its secular variation part and its fluctuating part. This was done using an unfolding procedure discussed by Brody et.al. [1981]. The mapping function for the potential considered was obtained to first order in the coupling constant α following a procedure discussed by Jennings et.al. [1975]. For each value of α all eigenvalues were tagged according to symmetry type and a regular or irregular type according to the second difference criteria. For the quartic potential all eigenvalues, for the four symmetries considered, for which the second difference was $\geq 8.10^{-3}$ were tagged as irregular. Thus about 20% - 25% of all levels were labelled as irregular. The Δ_3-statistic for each of the regular and irregular sequences was then calculated for each α value and each symmetry type and the ensemble average taken.

In Table I we present our results for Δ_3, as a function of n. Physically Δ_3 provides a measure of the departure of the exact eigenvalue distribution function (staircase curve) from a straight line. It is seen that for the regular sequence and irregular sequence, as defined above, the calculated values of $\Delta_3(n)$ are in agreement with the Gaussian orthogonal ensemble (GOE) form and the Poisson form respectively. Our results for the regular sequence differs from those of others [see, for example, Seligman et.al. 1984]. One possibility is that, compared to our case, they may have a much larger number of uncorrelated sequences in their calculation. We had earlier obtained a similar result for the Henon-Heiles potential [Parikh and Sheorey, 1985]. The sequence of irregular levels has gaps since the intervening regular levels are not present by construction and this may give to the Poisson-like form. We have also found that the values of Δ_3 calculated ignoring the second difference criteria were of the GOE form. This finding is again similar to that for the Henon-Heiles eigenvalues. On the basis of these calculations we feel that the second difference criteria for defining the irregular sequence of levels may not be useful.

Table I: Values of $\Delta_3(n)$ for the quartic potential, GOE (from Monte-Carlo calculations of Bohigas et.al.) and Poisson spectra ($\Delta_3(n)$= n/15)

n	Regular Levels	Irregular Levels	GOE	Poisson
0.25	0.01600	0.01714	0.0165	0.0167
0.50	0.03119	0.04200	0.0325	0.0333
0.75	0.04512	0.06624	0.0471	0.0500
1.00	0.05806	0.09419	0.0605	0.0667
2.00	0.10157	0.22181	0.1023	0.1333
3.00	0.13369	0.30476	0.1320	0.2000
4.00	0.16088	0.44148	0.1549	0.2667
5.00	0.18571	0.53476	0.1735	0.3333
6.00	0.20942	0.61698	0.1839	0.4000
7.00	0.23121	0.69107	0.2028	0.4667
8.00	0.25248	0.77384	0.2148	0.5333
9.00	0.27272	0.84754	0.2255	0.6000
10.00	0.29155	0.92889	0.2356	0.6667

We would like to thank Dr.J.C.Parikh for valuable discussions and Drs.O.Bohigas, R.Haq and A.Pandey for providing us with some of their computer programmes and for results of Monte-Carlo calculations prior to publication.

References

Berry M.V., 1985, Proc. Roy. Soc. **A400**, 229.

Bohigas, O. and Giannoni, M.J., 1984, Chaotic Motion and Random Matrix Theories, in Mathematical and Computational Methods in Nuclear Physics (ed. J.S.Dehesa, J.M.G.Gomez and A.Polls), Lecture Notes in Physics, **Vol.209**, 1 (New York: Springer-Varlag).

Brody, T.A., Flores, J., French, J.B., Mello, P.A., Pandey, A., and Wong, S.S.M., 1981, Rev. Mod. Phys. **53**, 385.

Jennings, B.K., Bhaduri, R.K. and Brack, M., 1975, Nucl. Phys. A **253**, 29.

Parikh, J.C., and Sheorey, V.B., 1985, Pramana, **24**, 39.

Percival, I.C., 1977, Adv. Chem. Phys., **36**, 1.

Pullen, R.A. and Edmonds, A.R., 1981, J. Phys. A, **14**, L477.

Seligman, T.H., Verbaarschot, J.J.M. and Zirnbauer, M.R., 1984, Phy. Rev. Lett., **53**, 215.

Zaslavsky, G.M., 1981, Phys. Rep. **80**, 157.

NUMERICAL CALCULATION OF CORRELATION PROPERTIES OF SPARSE REAL SYMMETRIC RANDOM MATRIX *

R. JOST, SNCI/CNRS, B.P. 166 X
38042 - GRENOBLE CEDEX, FRANCE

I - INTRODUCTION.

Random matrices are strongly related with "quantum chaos" in the sense that they model very complex hamiltonians. Only the statistical correlations properties of the eigenvalues for these random matrices are significant. As an example, consider the hamiltonian of a multi-oscillators system as model for the vibrations of a polyatomic molecule. We can write $H = H_0 + H_1$, where H_0 represents the hamiltonian of n uncoupled oscillators and H_1 represents the coupling between them. In the basis of H_0, the diagonal matrix elements are mainly related with the properties of H_0, and the off diagonal elements V_{ij} are related with H_1. The large size of the H matrix usually prohibits direct diagonalization. We consider only a submatrix of H, corresponding to a set of basis states in the range of ΔE, around an energy E_0. The density of states of H_0 within this range is assumed to be roughly constant and assume a homogeneous coupling. We are concerned only with the statistical properties (i.e. Nearest Neighbor Distribution - N.N.D. -and Δ_3 function (1)) of a set of eigenvalues and not with a detailed understanding of the individual eigenvalues of H.

II - THE MODEL.

These real symmetric random matrices are constructed with a uniform distribution of diagonal elements, corresponding to a constant density of states. The mean spacing of diagonal values defines a unit of energy. The off diagonal elements are uniformly distributed in the matrix (i.e. the distribution of V_{ij} is independent of i and j). Their magnitudes are determined by various distributions (see fig. 2). When there is no coupling ($V_{ij} = 0$), the N.N.D. of the eigenvalues follow POISSON statistics. We want to define how the correlations properties, especially those of the N.N.D., evolve with increasing coupling, i.e. for different parameters of the distribution of V_{ij} (see below).

III - METHOD OF ANALYSIS.

For each type of distribution of V_{ij}, and for each size of matrix, an ensemble averaging is performed over 18,000 eigenvalues in order to obtain good statistics. The chosen distributions of V_{ij} (see fig. 2) give a N.N.D. intermediate between POISSON and WIGNER. The BRODY distribution does not provide the best fit of the observed N.N.D. ; however, the BRODY parameter, q, remains a good description of the observed N.N.D. and we use it. This point will be discussed elsewhere (2). The mean spacing of eigenvalues in the central part of the spectrum (i.e. in region B of figure 1), remains very close to the mean spacing of diagonal elements (from H_0). Consequently, only the B part of the spectrum (11,600 eigenvalues) is used in the calculation of the statistical properties. As a result, the coupling, H_1, can induce strong correlation properties (N.N.D. close to WIGNER, Δ_3 close to G.O.E.) without altering the initial density of states. This situation can be called "low coupling", in contrast with the G.O.E. model, which corresponds to "strong coupling". Submatrices of dimension d = 25, 50, 100, 200, 400, 800 of a very large uniform matrix are diagonalized. We find that the statistical properties like N.N.D. and Δ_3 are independent of the size of submatrix for $d \gtrsim 100$ (except when the initial matrix is too sparse, see below). The truncation of a large uniform matrix alters the eigenvalues and eigenstates, but does not change the statistical correlation properties.

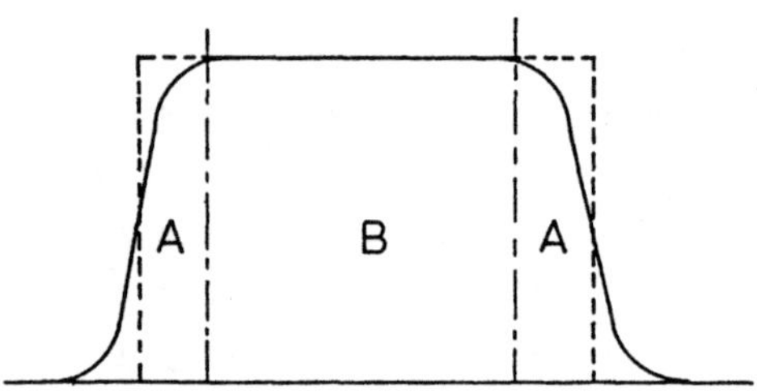

Fig. 1 : Shape of histogram of eigenvalues.

- dashed line for diagonal elements (H_0 only).
- solid line for $H_0 + H_1$.

IV - RESULTS.

A - N.N.D. versus "coupling parameter".

Consider, for example, 200 x 200 random matrices with different types of distributions of off diagonal elements, as shown in figure 2. These distributions are chosen in order to change the ratio of $\overline{|V_{ij}|}$ (the mean of the absolute value of the off diagonal elements) and the standard deviation of V_{ij} : $\sigma(V_{ij})$.

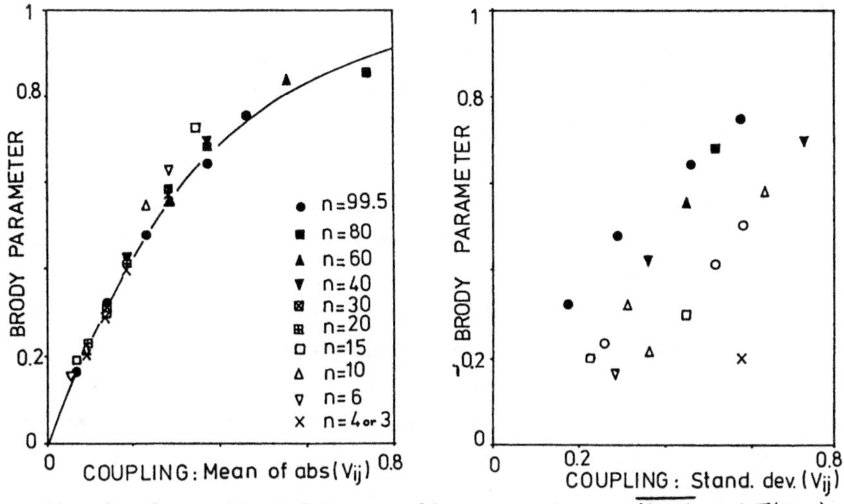

Fig. 3 : Comparison of two coupling parameters : $\overline{|V_{ij}|}$ and $\sigma(V_{ij})$. n defines the sparseness of the matrix (n = 99.5.α(see fig.2)).

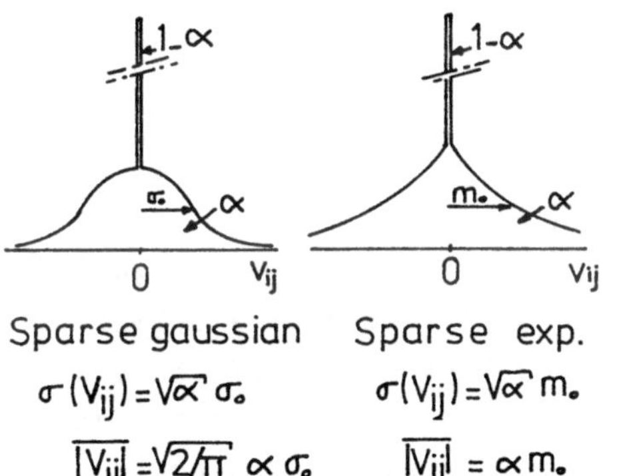

Fig. 2 : Distribution of off diagonal matrix elements (V_{ij}) : the sparseness parameter α range from $\simeq$ 0 to 1.

Each N.N.D. is obtained numerically and is fit to a BRODY distribution. Figure 3 shows a plot of the BRODY parameter q versus $\overline{|V_{ij}|}$ and versus $\sigma(V_{ij})$ for many distributions. This figure shows that q (and the corresponding N.N.D.) is a function of $\overline{|V_{ij}|}$ and not of σ or σ^2. Thus, it is appropriate to call $\overline{|V_{ij}|}$ the "coupling parameter" for this "low coupling" situation. Qualitatively, the same relationship (q = f($\overline{|V_{ij}|}$)) holds when the N.N.D. is fit with the BERRY and ROBNIK distribution (3) or other distributions (2). This result is independent of the size of the matrix for $d \gtrsim 100$. A coupling parameter as low as unity (in units of the mean spacing of diagonal terms) is sufficient to obtain strong level repulsion. This result has been found implicitly by J. VERBAARSCHOT et al. (4). Long range correlation properties are examined with the Δ_3 function and also with a new method based on the fourier transform of the spectrum (5). We only mention that, for a specified "coupling parameter", the larger the variance of V_{ij}, the longer the length of the correlation. Like the N.N.D., the long range correlation properties (Δ_3) are independent of the size d of the matrix (however, $\Delta_3(L)$ can't be determined for L larger than d !).

B - Truncation effect (extremely sparse matrices).

A matrix which is too sparse may be decomposed into submatrices. Each submatrix may produce sets of strongly correlated eigenvalues. But, as the sets overlap, correlations disappears. This effect is matrix size dependent, and therefore acts as a truncation effect. To avoid this effect, the number, N, of non zero randomly distributed V_{ij} (for $i > j$) of a d x d matrix, should be larger than approximately 3d. In other words, as the total number of matrix elements is $d(d-1)/2$ (for $i > j$), the degree of sparseness, f, ($f = 2N/(d-1)d$) must be larger than $6/d$. Conversely, consider a very large uniform random matrix, with a given degree of sparseness. Only a submatrix with d larger than $6/f$ will yield correct correlation properties of the very large initial matrix.

V - SUMMARY AND CONCLUSIONS.

The N.N.D. or a real symmetric sparse random matrix is governed by the "coupling parameter" $\overline{|V_{ij}|}$.

Meaningful statistical properties of a such uniform random large matrix can be obtained from the diagonalization of a submatrix ; however, this submatrix must not be too small (function of the sparseness of the large matrix). This approch will be used in the study of vibrations levels of the ground state of polyatomic molecules like formaldehyde and glyoxal. To apply this model to molecular systems, one must check that the assumption of a uniform distribution of off diagonal elements holds.

REFERENCES

1 - T.A. BRODY, J. FLORES, J.B. FRENCH, P.A. MELLO, A. PANDEY and SS. WONG, Rev. Mod. Phys. 53, 385 (1981) and references cited within.
2 - R. JOST, to be published.
3 - M.V. BERRY and M. ROBNIK, J. Phys. A : Math. Gen. 17, 2413.
4 - J. VERBAARSHOT, H.A. WEINDENMULLER and M. ZIRNBAUER, Ann. of Phys. 158 , 78 (1984).
5 - L. LEVIANDIER, M. LOMBARDI, R. JOST, J.P. PIQUE, accepted in Phys. Rev. Lett.

* Computational facilities are provided by the scientific counsel of the "Centre de Calcul Vectoriel pour la Recherche" (France).

ENERGY LEVEL FLUCTUATIONS FOR 'DEFORMED' ENSEMBLES

P. Bauer and R. Haq
Department of Physics
Laurentian University
Sudbury, Ontario
Canada, P3E 2C6

Energy level fluctuations using random matrix methods are calculated for the Quadrupole-Quadrupole interaction augmented by a random two-body Hamiltonian. The rapid transition from almost Poisson to GOE fluctuations found provides some understanding as to why realistic nuclear interactions show GOE fluctuations.

The study of fluctuation properties of energy spectra for a wide class of quantum systems has shown agreement with the predictions of the Gaussian Orthogonal Ensemble (GOE)[1]. A detailed analysis of nuclear data consisting of slow-neutron resonances of medium and heavy nuclei and of proton resonances of some light nuclei has shown excellent agreement with theory[2]. The same conclusion is reached in analysing atomic[3] and molecular systems [4]. Further, Monte Carlo calculations of ensembles of two-body operators embedded in many particle shell model spaces (two-body random ensemble TBRE[5]) and its generalisations (embedded ensembles EGOE[6]) as well as eigenvalues of large shell model matrices using realistic nuclear interactions give GOE fluctuations [1].

While this 'universality' of GOE fluctuations is pleasing, the reasons behind are not obvious. The Gaussian Orthogonal Ensemble takes no account of the specific properties of the Hamiltonian, its (1+2)-body nature, its dominant components and range etc. The only requirement is rotation and time reversal invariance resulting in real symmetric matrices. The question then is--Why are the fluctuations not affected by the special features of the Hamiltonian that exhibit themselves in observed spectra?

In order to understand the generality of GOE fluctuations, one can proceed as follows: Consider a Hamiltonian matrix of the form

$$H(\alpha)=H_o + \alpha H_R \quad (1)$$

where H_o is a non-random matrix arising from a specific Hamiltonian which takes into account the dominant features of the system, H_R is a random matrix corresponding either to GOE or TBRE and α is a mixing parameter. It has been shown[7] that as α increases, there is a rapid transition from the fluctuation pattern characteristic of H_o to GOE fluctuations. Recently, by making appropriate choices of the matrices involved, this method has been used[8] to determine a bound on time-reversal noninvariance in the nuclear Hamiltonian.

In this work, we apply the method to gain some insight as to why realistic nuclear interactions give GOE fluctuations. The choice of H_o is governed by the following criterion-a) it must constitute a significant part of any realistic interaction and b) it must have its own characteristic fluctuations different from GOE. The Quadrupole-Quadrupole (Q.Q)- interaction[9] satisfies the above requirements since most nuclear interactions commonly used have a large Q.Q component and because of the inherent symmetries (SU(3)and L),the fluctuations, as shown later, are close to those obtained for a Poisson (randomly spaced levels) spectrum. The two-body matrix elements of H_R are randomly chosen from a normal distribution with zero mean and variance equal to that of H_o.

The basis space chosen is $(sd)^8$, J=0, T=2. This leads to a matrix dimensionality d=287. The many-particle matrices of H_o and H_R were obtained using the shell model program. An ensemble is constructed by making several choices for the two-body matrix elements corresponding to H_R. Our ensemble consisted of 45 different trials.

The transition parameter Λ, in this model is defined[8] as $\Lambda=\alpha^2 v^2/D^2$ where v^2 is the variance of the off-diagonal matrix elements of H_R in the basis in which H_o is diagonal and D is the mean spacing of the H_o spectrum. The matrices of H_R were therefore transformed using the eigenvectors of H_o and v^2 determined. By choosing α appropriately for different members of the ensemble, a unique value of Λ can be maintained. This ensures that for a given Λ, the strengths of various H_R matrices is the same relative to H_o. Further, the parameter Λ is independent of the dimensionality of the matrix.

The eigenvalues of H were obtained for several choices of Λ. These were then unfolded to a constant density and various fluctuation measures calculated. Details of the method of calculation for the measures used are given in ref.2. The nearest-neighbour spacing distribution for H_o (Λ=0) is shown in Fig. 1. It is clear that the distribution is close to Poisson. When the random part of the interaction H_R is switched on, the level repulsion comes into play and the distribution tends to GOE as Λ increases becoming almost GOE for Λ=0.25 as shown in Fig. 2.

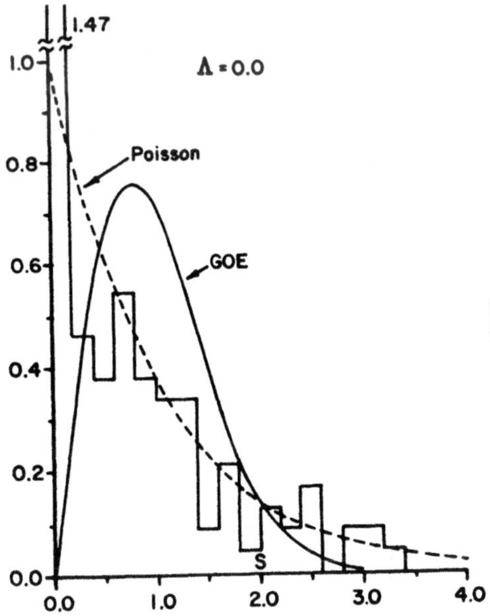

Fig. 1 Nearest-neighbour spacing distribution for Q.Q.

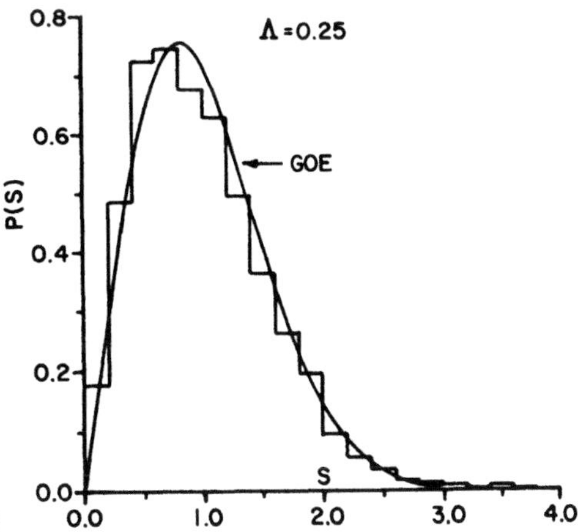

Fig. 2. Spacing distribution for H with Λ=0.25.

Next we study the number statistic n and in fig. 3 is shown the number variance[10] $\Sigma^2(\bar{n})$ for $\bar{n}=1$ as a function of Λ. We observe that the transition is fairly rapid and is essentially complete for $\Lambda=0.8$.

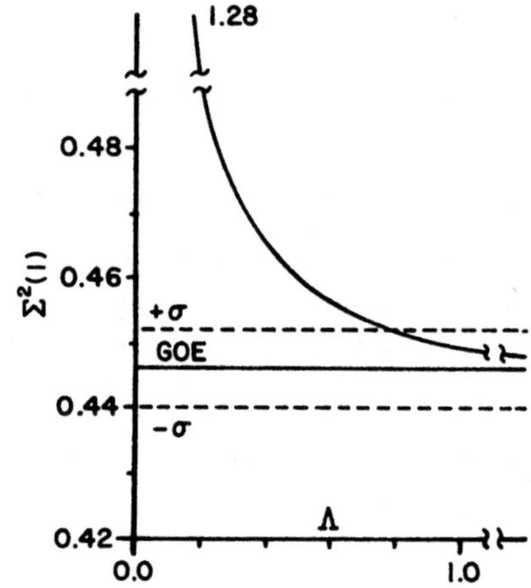

FIG. 3. The number variance $\Sigma^2(1)$ vs. Λ.

Finally in Fig. 4 we show $\bar{\Delta}_3(\bar{n})$ statistic[10] for $\bar{n}=10$. The transition is essentially complete for $\Lambda=0.6$. It should be remarked that the critical value of Λ depends on the correlation length considered and for $\bar{\Delta}_3(20)$ this value is 1.4 as shown in Fig. 5.

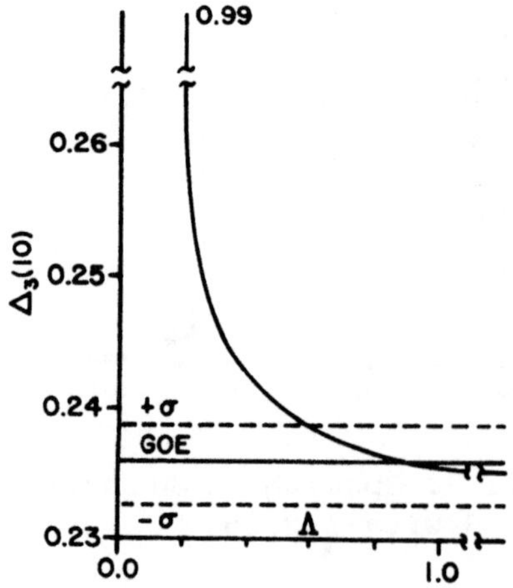

FIG. 4. $\bar{\Delta}_3(10)$ vs. Λ.

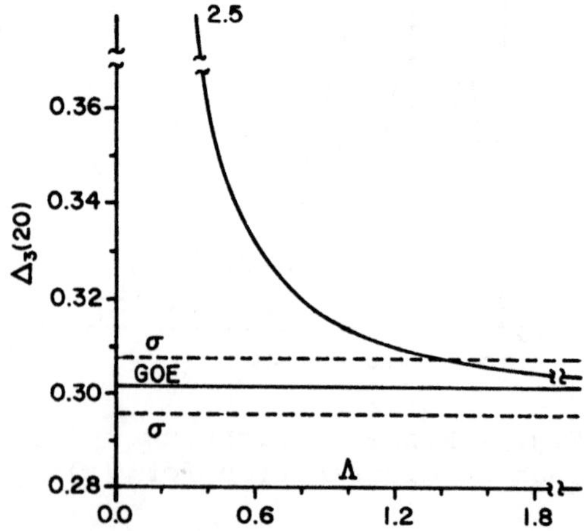

FIG.5. $\bar{\Delta}_3(20)$ vs. Λ.

We have shown that the addition of a small random part to the Q.Q.-interaction changes the fluctuation pattern dramatically and the transition to GOE fluctuations is extremely rapid. By analysing the various realistic nuclear interactions in terms of their Q.Q. component, one can fix the value of the parameter Λ (taking the remaining part as random). A comparison with the values obtained in this work would hopefully help in understanding why these interactions give GOE fluctuations. Further, by making an orthogonal decomposition of the Hamiltonian in terms of a symmetry preserving and symmetry breaking part, and taking the symmetry breaking part as random there by fixing Λ, the method may be extended to provide information on symmetry mixing.

The authors are grateful to A. Pandey for discussions regarding the transition parameter. This work has been supported by the Natural Sciences and Engineering Research Council of Canada.

REFERENCES:

1. T.A. Brody, J. Flores, J.B. French, P.A.Mello, A. Pandey, and S.S.M. Wong, Rev. Mod. Phys, 53, 385 (1981).
2. R.U. Haq, A. Pandey and O. Bohigas, Phys. Rev. Lett. 48, 1086 (1982).
3. H.S. Camarda and P.D. Georgopulos, Phys. Rev. Lett. 50, 492 (1983).
4. S. Mukamel. J. Sue and A. Pandey, Chem. Phys. Lett. 105, 134, (1984).
5. S.S.M. Wong and J.B. French, Phys. Lett. 33B, 449 (1970); O. Bohigas and J. Flores, Phys. Lett. 34B, 261 (1971).
6. S.S.M. Wong and J.B. French, Nucl. Phys. A198, 188 (1972); O. Bohigas and M.J. Giannoni, Ann. Phys. 89, 393 (1975).
7. A. Pandey, Ann. Phys. (N.Y.) 134, 110 (1981).
8. J.B. French, V.K.B. Kota, A. Pandey and S. Tomsovic, Phys. Rev. Lett. 54, 2313 (1985).
9. J.P. Elliott, Proc. Roy. Soc. 245, 128 (1958).
10. F.J. Dyson and M.L. Mehta, J. Math. Phys. 4, 701 (1963).

QUALITATIVE PROPERTIES OF EIGENFUNCTIONS OF CLASSICALLY CHAOTIC HAMILTONIAN SYSTEMS

Eric J. Heller
Departments of Chemistry and Physics
University of Washington
Seattle, Wa. 98195 USA

Abstract

We begin by discussing the properties expected of eigenfunctions of a classically chaotic Hamiltonian system, using simple Correspondence Principle arguments. The properties involve nodal surfaces, coordinate and momentum space amplitude distribution, and phase space distribution. The eigenfunctions of the stadium billiard are examined, and it is found that the periodic orbits of shortest periods and smallest stability parameter profoundly affect the eigenfunctions: "scars" of higher wavefunction density surround the special periodic orbits. Finally a theory is presented for the scars, showing that they must exist, and relating them directly to the special periodic orbits. These same periodic orbits cause level density fluctuations.

1.INTRODUCTION

1.1 *chaos and quasiperiodicity*

Suppose we know the Hamiltonian for a quantum mechanical system of physical interest, and we need to solve for its eigenfunctions and eigenvalues over a range of energies. This can be a nearly impossible task if many degrees of freedom and high energies prevail. On the other hand, we are always free to examine the classical mechanics of the same Hamiltonian; this will usually be quite easy to do. Such examination may prove extremely useful for understanding the nature of the quantum dynamics or of the eigenfunctions and eigenvalues, especially if the energy is high enough. Classical mechanics may also help *generate* accurate approximations to the quantum dynamics, eigenfunctions, and eigenvalues via one of several semiclassical approaches. It is fortunate that in the high energy regime, where quantum mechanics is so difficult, semiclassical methods are more accurate. This is so, *as long as the classical dynamics is integrable*. In the event that it is chaotic on a scale larger than or comparable to Planck's constant, we don't really have a very good idea what to do. A semiclassical method that will produce eigenvalues and especially eigenfunctions in the classically chaotic regime is not available. This is a regrettable situation, but at the same time exciting and challenging. If semiclassical methods are to fulfill their crucial role of providing an intuitive basis and a means to accurate approximation of quantum systems, then we had better understand how to implement them in the typical (i.e. classically chaotic) situations.

There has been progress concerning semiclassical methods applied to classically chaotic situations, especially for the eigenvalues. Periodic orbits, which are ubiquitous but of no measure compared to the typical chaotic orbits, have played a large role. Gutzwiller[1] has been able to come close to the explicit semiclassical construction of eigenvalues of the classically chaotic anisotropic Kepler problem. Berry and Tabor[2] give spectral density-of-states fluctuations in terms of classical information. Berry[3] has discussed the spectral rigidity using a formula from Hannay and Ozorio de Almeida[4] and other measures of the density of quantum states. In the mathematical literature, the Selberg trace formula[5] (closely related to the Green's function trace method of Gutzwiller) yeilds an *exact* spectrum in terms of periodic orbits for systems of negative curvature, such as the equation

$$y^2\nabla^2\Psi(x,y) + \kappa^2\Psi(x,y) = 0$$

on certain domains.

Berry[6], using semiclassical arguments, predicted the properties that one would expect a high energy quantum mechanical eigenstate to have in the presence of underlying classical chaos. His reasoning was qualitative and did not constitute a procedure to explicitly construct eigenvalues and eigenfunctions. We shall give the essence of his arguments below. Periodic orbits did not appear in Berry's conjectures about the eigenstates. Since they play such a dominant role in the theory of the eigenvalues, it would be odd if they did not affect the eigenfunctions. We shall see that the periodic orbits do indeed play a dramatic role.

1.2 *waves and rays*

We briefly note that none of the questions we are posing about the relationship between classical chaos and quantum mechanics are very interesting at low energy. For example, consider the lowest energy solutions of a free particle confined to a simply connected region:

$$\nabla^2\Psi(x,y) + \kappa^2\Psi(x,y) = 0 \qquad (1)$$

with the condition $\Psi = 0$ on the boundary. The ground state must be nodeless: a single "bubble" coming to zero height on the boundary. The first excited state must have one nodal line, but there are two ways of having one: either it goes from one side to the other, touching the boundary twice, or it stays on the interior, closing on itself. In every known case the node touching the boundary has the the lower of the two eigenvalues. It it conjectured but after much effort still not proven that this must be true. This seems like it ought to be a simple problem, but in fact it is very subtle.

We are primarily intersted in much higher lying states and energies, where presumably the rays have something to do with the eigenstates. At the heart of most semiclassical methods is the idea of associating wave fronts lying perpendicular to the ray trajectories, in the manner of the following picture:

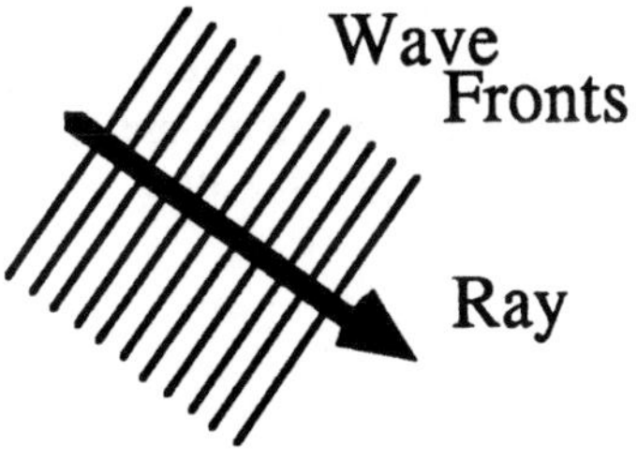

FIG.1

The mathematical statement of this relationship is

$$\Psi(x,y) = \sum_j \sqrt{P_j}\, e^{iS_j/\hbar} \tag{2}$$

where the sum is over independent ray paths through the point (x,y), P_j is the classical probability for the path, and S_j is the jth branch (corresponding to the jth independent path) of the classical action

$$\begin{aligned} S(x,y) &= \int (p_x dx + p_y dy) \\ &= \int \mathbf{p} \cdot d\mathbf{x} \end{aligned} \tag{3}$$

That the wavefronts are perpendicular to the rays is evident from

$$\nabla S = \mathbf{p} \tag{4}$$

which follows from (3).

We are glossing over certain details, such as the Maslov phase correction, which adds a phase to S at each pass according to the number of caustics touched by the trajectory since the last pass by (x,y). The whole construction, Eq.(2), is valid only if the dynamics is quasiperiodic, in which case there are a finite number of branches or ray paths through a given point. Quite typically there are four branches, which happens for the "box like" orbits exhibited by the familiar Lissajous orbits of two uncoupled incommensurate harmonic oscillators. Another case of this is the particle confined to a two dimensional rectangular box:

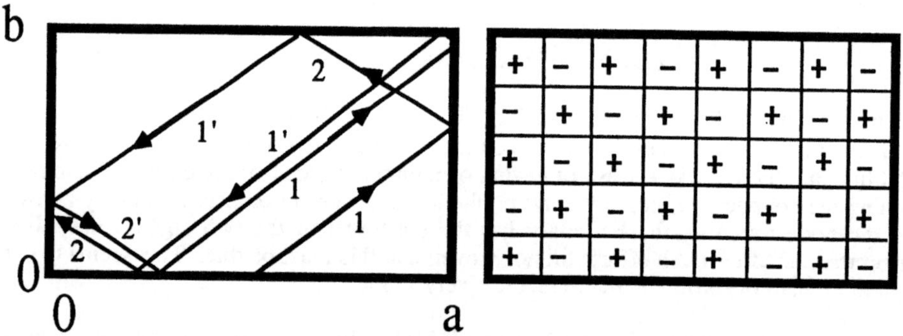

FIG. 2

Following an arbitrary ray, it is seen that after enough bounces, any location in the box will be accessed by the 1, 1', 2, and 2' ray paths. For every 1 there is a 1', and so on. This means that the complex wave $e^{iS_j/\hbar}$ can be replaced by $\cos(S+\delta)$, a real cosine wave with a phase shift, for the 1, 1' pair. This happens also for the 2, 2' pair and thus the eigensolutions are superpositions of four traveling waves or two standing cosine waves whose constant phase fronts lie perpendicular to a ray and its time reverse. To see that this is indeed the case, let us decompose the usual way of expressing the eigenfunctions, namely

$$\Psi_n(x,y) = A \sin(n\pi x/a) \sin(m\pi y/b) \tag{5a}$$

into

$\Psi_n(x,y) = A/2[\cos(n\pi x/a - m\pi y/b) - \cos(n\pi x/a + m\pi y/b)]$

$$= A/2[\cos(\mathbf{k}_1 \cdot \mathbf{x}) - \cos(\mathbf{k}_2 \cdot \mathbf{x})] \quad (5b)$$

$$= A/2[\cos(\mathbf{k}_1 \cdot \mathbf{x}) + \cos(\mathbf{k}_2 \cdot \mathbf{x} + \pi)] \quad (5c)$$

where $\mathbf{x} = (x,y)$ and $k_{1x} = n\pi/a$, $k_{1y} = -m\pi/b$, etc. We see indeed a sum of two cosine waves, with momenta chosen so that the box boundaries are also nodal lines, thus satisfying the boundary conditions. The x and y components of the wavevectors are just so as to allow an integer number of half wavelengths from end to end in the box, or (using $p\lambda = h$) what is the same thing

$$\oint p_x dx = nh$$

$$\oint p_y \, dy = mh \quad (6)$$

where the integrals are taken around a complete cycle in x and y, respectively. These integrals are simply the Bohr-Sommerfeld-Einstein quantization conditions, applied to a two dimensional particle in a box. To summarize the rectangular box procedure:

(1) Find allowed values of k_x and k_y from the Bohr-Sommerfeld-Einstein quantization conditions,
(2) Associate a cosine wave with each of the wavevectors $\mathbf{k}_1 = \mathbf{e}_x k_x + \mathbf{e}_y k_y$ and $\mathbf{k}_2 = \mathbf{e}_x k_x - \mathbf{e}_y k_y$, and
(3) Add the cosine waves $\cos(\mathbf{k}_1 \cdot \mathbf{x})$ and $\cos(\mathbf{k}_2 \cdot \mathbf{x})$ together with an appropriate phase shift for the boundary condition.

Note that the nodal lines that result from the addition of the two cosine waves line up along the directions $\mathbf{n}_1$ and $\mathbf{n}_2$, where $\mathbf{n}_1 = \mathbf{k}_1 + \mathbf{k}_2 \propto \mathbf{e}_x$ and $\mathbf{n}_2 = \mathbf{k}_1 - \mathbf{k}_2 \propto \mathbf{e}_y$

Many aspects of this simple case extend to quasiperiodic motion on smooth potential energy surfaces. Examining the figure below, we see that locally things are very much the same as the box. At the right of Fig 3, the grey arrows indicate the nodal line directions, the black arrows show the trajectory directions, and the lighter lines are the wave fronts. Note that the nodal lines form perpendicular bisectors of the "x" made by the ray (trajectory) lines. This observation is the basis of the orthogonal "nodal coordinate" system, constructed entirely out of the classical trajectories.[7,8] The idea is that the coordinates lie along the nodal lines and so form a separable system of coordinates for individual eigenfunctions.

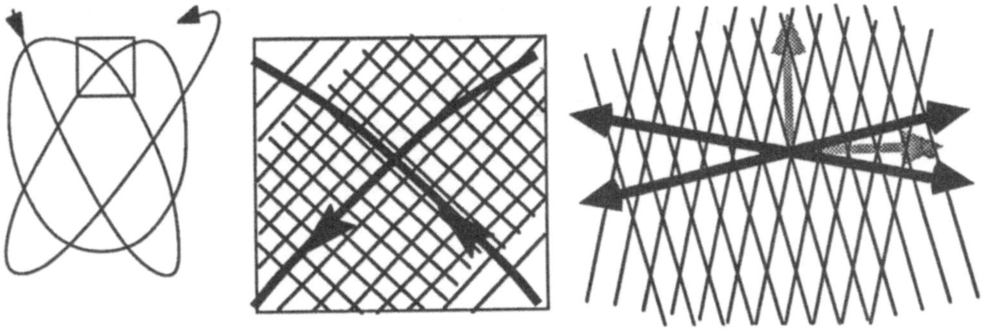

FIG. 3

The nodal coordinates stand out strongly in the eigenfunction, Fig. 4, for example (the classical trajectory is given as a probability map for comparison):

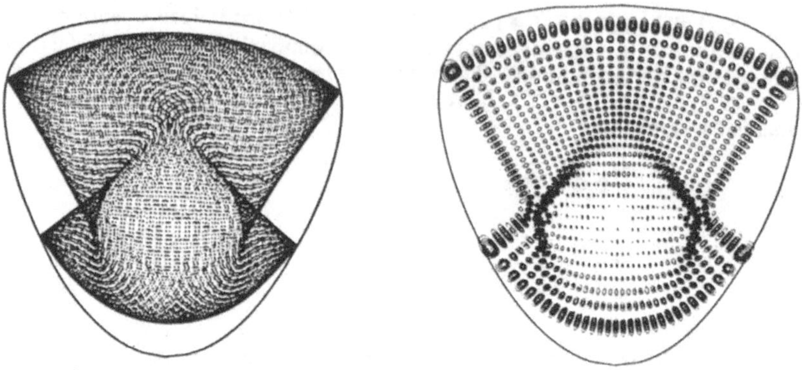

FIG. 4

This example reminds us that more than two cosine waves may contribute: note the caustic "swallowtail" region, where the trajectory folds over on itself and more waves contribute. The resulting nodal structue is quite jumbled, even though a quasiperiodic trajectory underlies this eigenfunction. Such behavior is common, especially in the vicinity of resonance zones in the classical phase space. Since multiple overlapping resonance zones lead to chaos[8], we ought to examine the qualitative effects of adding 2,4,... different cosine waves together, all with the same *amplitude*, but different direction. (The amplitude represents the square root of the classical probability , $P_j^{1/2}$, which is controlled by the kinetic energy. This is the same at a given spot on the potential energy surface no matter what the direction.) In a completely chaotic region of phase space all directions contribute equally. Perhaps the correct qualitative behavior of eigenfunctions in the classically chaotic regime and in the high energy limit can be obtained by examining the properties of random wavefront superpositions for a large number of waves. M. Berry's[7] conjecture in 1981 was that this was indeed the case.

1.3 *Qualitative study of plane cosine wave superpositions*

We thus turn to the qualitative study of superpositions of plane cosine waves of the form

$$\Phi(\mathbf{x}) = \frac{1}{\sqrt{N}} \sum_{i=1}^{N} \cos(\mathbf{k} \cdot \mathbf{x} + \delta_i) \tag{7}$$

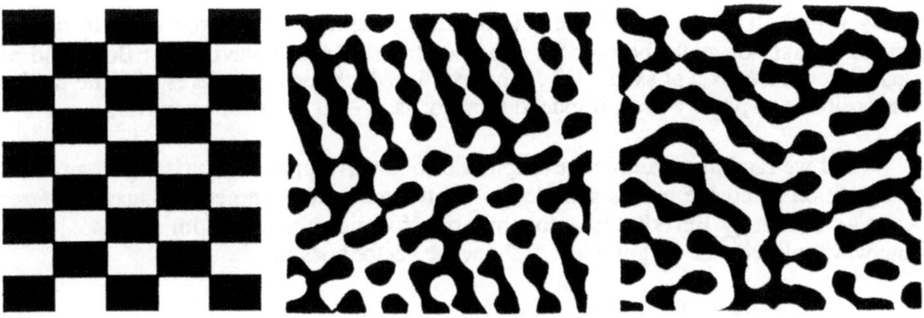

FIG. 5

1.3.1 Nodal Patterns

At the left, the simple result of superposition of two standing cosine waves is shown. In the center panel, we show the result of four waves superposed, and on the right, six are added together, both with "random" directions, but equal wavevector magnitude. The nodal pattern for a very large number of waves looks quite similiar to the six wave case. Evidently, the nodal patterns aren't telling us us very much beyond that more than a few cosine waves are contributing. This means that the nodal pattern is a misleading measure of the response of quantum mechanics to classical chaos, since six waves is far from the superposition of an infinite number of waves.

In the literature, one can find examples of nodal structure being used as evidence for "quantum ergodicity". While certainly a nodal pattern like that on the right is necessary for quantum mechanics to mimick classical chaos, it is not sufficient, since a small number of waves superposed can give wild looking patterns. Of course, if we checked a larger area or certian correlation functions one could tell the difference between say 6 and 100 waves, but one rarely gets that large an area to work with in physical applications.

1.3.2 Speckles

Superpositions of waves with random directions for $\mathbf{k}_i$ exhibit what are sometimes called "speckles": spikes of large amplitude. It is easy to see how this happens. At some values of $\mathbf{x}$, many of the cosines will be in phase, say near $\cos(\mathbf{k}_i \cdot \mathbf{x} + \delta_i) = 1$, leading to an unusually large value of $\Phi(\mathbf{x})$. At rare places they will *all* be nearly 1, leading to a huge spike if N is large.

If the quantum wavefunctions of a classically chaotic system are really to be a random superposition of an infinite number of cosine waves, we must expect spikes of *very large amplitude* at rare places. Surely this doesn't happen unless $h \to 0$, just as in the quasiperiodic case the eigenfunctions become infinite at the classical turning point caustics only as $h \rightarrow 0$. If we do not believe in infinitely high spikes in the eigenstates for *finite* h, then they must not really be random superpositions of plane waves. But Berry did not suggest they were, except as $h \rightarrow 0$. The *way* in which the eigenstates become random superpositions of plane waves, if indeed they do, as $h \rightarrow 0$ is a wide open field.

If we randomly sample the amplitudes of a wavefunction consisting of a random superposition of cosine waves sampled in coordinate space, we get a Gaussian of zero mean. As shown by Berry[6], the autocorrelation function for a random superposition of plane waves all with the magnitude k of the wave vector goes as

$$<\psi_0(\mathbf{r}')\psi_0(\mathbf{r}+\mathbf{r}')> \sim J_0(kr)$$

J_0 is a Bessel function of zero order and where the brackets imply an average over all $\mathbf{r}'$.

2. CLASSICAL CHAOTIC DYNAMICS AND QUANTUM EIGENFUNCTIONS

2.1 *The Bunimovich stadium: classical mechanics*

We are now ready to see what the eigenfunctions of a classically chaotic Hamiltonian really look like. The Bunimovich stadium plays a large role in the remainder of this paper, because it is one of the few simple classical systems to be proven to be completely chaotic. The stadium consists of two semicircular caps and finite length horozontal straight sections connecting them. A section of a typical ray path inside this enclosure is shown next:

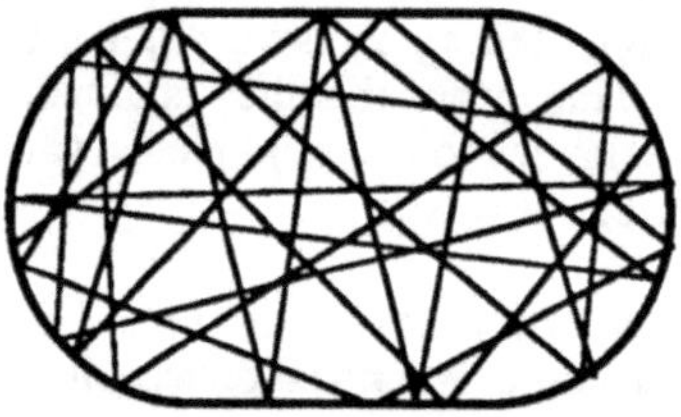

After a very large number of bounces, the ray democratically accesses every direction at every point inside the enclosure. It therefore seems reasonable to expect the properties of the eigenfunctions of the stadium to be those of a superposition of random wave fronts in the semiclassical limit.

There are however a very few special orbits, which are periodic. These are of no measure compared to the typical ones, so that one would never find a periodic orbit by chance guessing at initial conditions. Also, they are all unstable. If this were not the case, there would exist a region of stable confinement around the orbit, contradicting the claim that the dynamics is ergodic. Any trajectory starting near such a periodic orbit falls away from it at a rate which depends on the orbit, and the long time average of the trajectory becomes ergodically distributed over phase space.

A few examples of periodic orbits follow:

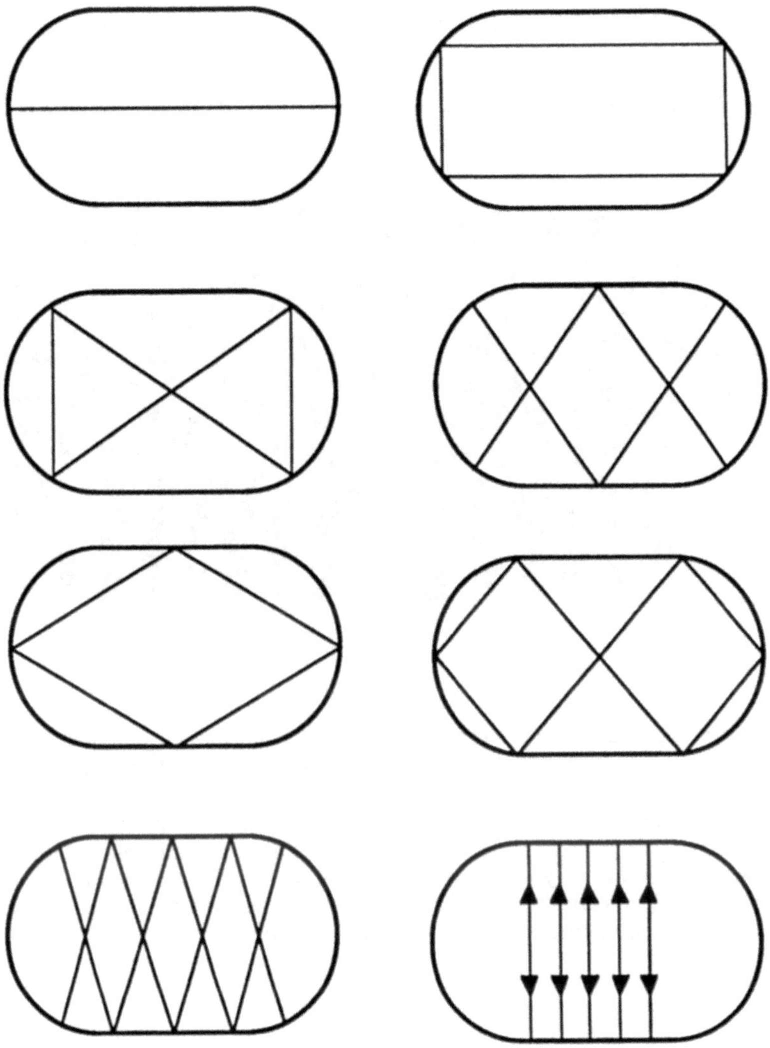

FIG.6

2.2 *The Bunimovich stadium: quantum mechanics*

Since the typical trajectory passes each point in all directions, the eigenfunctions of the stadium ought to be ideal cases for the study of Berry's conjecture: are the eigenfunctions essentially the superposition of many random plane waves?

2.2.1 Amplitude and Correlation Functions

McDonald and Kaufman[10] examined these properties with mixed results: there were definite but not spectacular deviations from the Gaussian random amplitudes and Bessel correlation function.

2.2.2 Nodal Patterns

In their publication on the quantum eigenstates of the Bunimovich stadium enclosure[9], McDonald and Kaufman[10] showed only the nodal lines of one state, although

McDonald's comprehensive and geminal Ph.D. thesis gave perspective plots of many of the eigenstates too.

We have redrawn the figure of Ref. (10b), showing it as black and white with all four sectors of the odd-odd parity state:

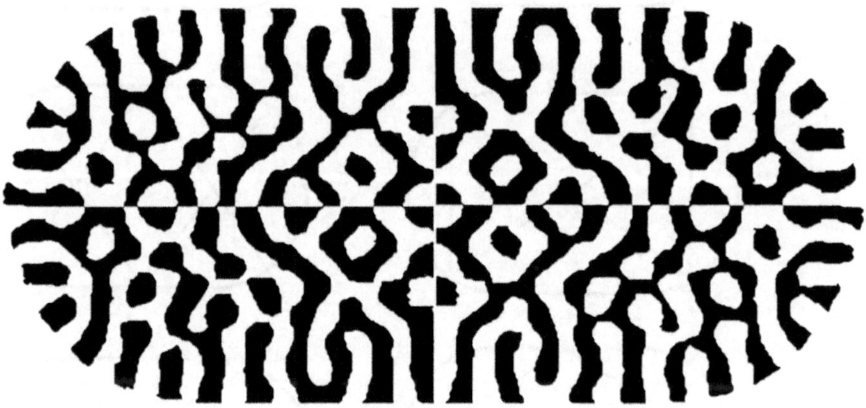

FIG. 7

References (10) show one quadrant only, which is all that is technically necessary by symmetry, but we find that for "pattern recognition" purposes it is useful to plot all four. One pattern that stands out clearly is the organizing effect of the boundaries and the symmetry reflection lines: the nodes of the eigenstates must intersect these boundaries and lines perpendicularly. This is because any nodal lines that intersect must do so at right angles, a fact which is easy to prove by inserting an appropriate Taylor series expansion about the assumed nodal line intersection into the wave equation. The boundaries of the stadium and the two symmetry lines for the odd-odd case are necessarily loci of nodal lines.

2.2.3 Amplitude Contour Maps

Nodal patterns, like energies, contain a very limited amount of information. Suppose the eigenfunction is known as a projection onto a complete set of states, say coordinate space for example. This completely specifies the eigenfunction. In Fig. 8 we display, as a $|\Psi|^2$ contour map, the same eigenfunction of the stadium for which the nodal pattern was given above. The nonuniform distribution of this state is unexpected. After all, the eigenstates of a particle in a *rectangular* box at least cover coordinate space evenly (after allowing for smoothing over the inevitable nodes). Here, we have passed to an *ergodic* system and worse, a K-system with strong chaotic properties, only to find that coordinate space is no longer evenly covered! (In an ergodic system, it is the energy hypersurface that is supposed to be evenly covered. However, in two dimensions a simple calculation shows that this implies a uniform coverage in coordinate space up to the boundary allowed by the total energy. Berry has noted that in more than two dimensions, the coordinate space probability actually *decreases* as one approaches the boundary, a phenomenon he calls *anticaustics*.) Of course, the rectangular box is far from ergodic, since momentum space is accessed in just four places, but bearing in mind the rule of thumb that the nodal surfaces are perpendicular to the corresponding classical motion it appears that the eigenfunction shown above isn't doing much better in momentum space either. *What is going on?*

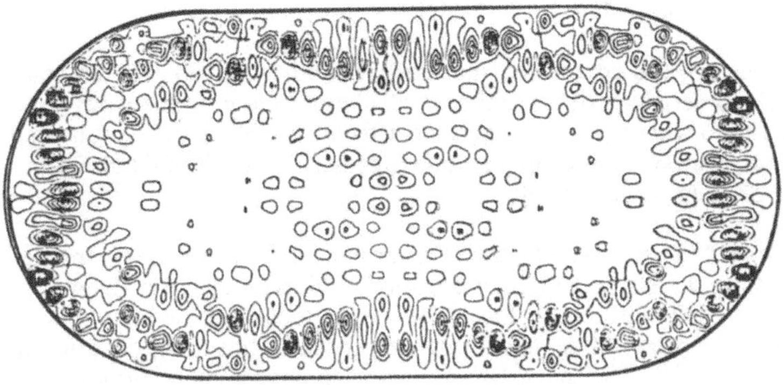

FIG. 8

There seems to be a possible relationship between the periodic orbit which is an inscribed rectangle and the high density regions in the eigenfunction just shown. To confirm this suspicion, we display a managerie of eigenfunctions, all of even higher energy than the one above:

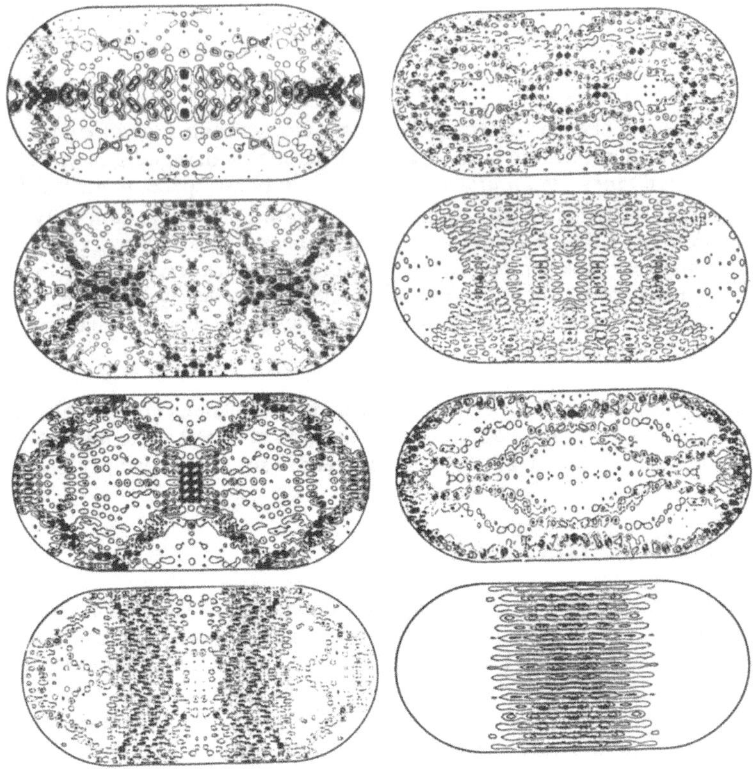

2.2.4 *Scars of the periodic orbits*

There is no avoiding the conclusion that the probability distribution of many of the eigenstates reflects certain periodic orbits . Calculations are now underway to determine what fraction and to what extent the states are affected by the periodic orbits in this way, but a casual examination in this energy regime suggests about half of the states are affected visibly by one or more periodic orbits.

McDonald[10a] was aware that such orbits were influencing the eigenstates, but his method of plotting the states (only one quadrant, and as perspective diagrams) did not reveal the extent of the influence. Later, we gave a theory of the effect which at least partially solves the mystery of the relation of the periodic orbits to the concentrations seen in the eigenfunctions[11]. The theory shows that such concentration around the periodic orbits must occur, but predicts a lower bound to the effect that is much weaker, apparently, than the actual effect. Just how strong the effect is will have to be quantified in future studies. We call the effect: **scars of the periodic orbits**. We will show below why they appear and how they may "heal" as $h \to 0$, as they ought to in order to agree with the Correspondence Principle.

Periodic orbits are the foundation of the theory of eigenvalue properties for chaotic Hamiltonian systems. Eigenvalue properties include densities of states, fluctuations in the density of states, "spectral rigidity"[3,12], nearest neighbor level spacing statistics, and the eigenvalues themselves. These aspects of the energies are the subjects of several works in this volume. Below, we will bring out the relation of the scars to fluctuations in level densities .

No randomly chosen classical orbit shows any scarring after a long time average. After a sufficiently long time, the phase space distribution of an ergodic classical trajectory converges on a *stationary and uniform distribution* in phase space, namely $\delta(E-H(p,q))$. The eigenfunctions, which are also stationary distributions, are not so uniformly distributed as the trajectories, and suffer from the scars. The scars are indeed surprising, but it is quite clear that we must now look for the proper role of the measure zero periodic orbits, which after all manage to control the eigenvalues in spite of their lowly measure. At the very least,the scars provide an interesting detour on the way to the conjectured $h \to 0$ limit of Gaussian random eigenfunctions.

In the quasiperiodic regime, the time average of a classical orbit is also a statioary distribution covering a smaller domain in phase space. In this regime however, the quantum mechanics is much better behaved, and uses the trajectory for "classical support"[13].

3. THEORY OF THE SCARS

Clearly, the evidence suggests that the eigenfunctions of the stadium enclosure are not random superposition of plane waves, even at rather large energy. The scarring is seen up to the highest energies that have been investigated, or about $k^2/2 = 5000$ for a stadium of length 4 (see Eq. (1)). We have shown[11] that this must be the case, that the scars are the necessary consequence of a certain class of periodic orbits. Here, we review and expand on that reasoning, and pose other questions about the scarring phenomenon.

3.1 *The Break Time*

The essence of the theory behind the scarring is very simple to understand. We only need to assemble a few facts to make the scarring seem more plausible. We shall then go into further detail and show that scarring must occur for certain periodic orbits embedded in the classical chaos.

(1) Eigenfunctions can be constructed from dynamics by Fourier transform:

$$\Psi_E(\mathbf{x}) = \lim_{T\to\infty} \frac{1}{T}\int_{-T}^{T} e^{iEt/\hbar}\Phi(\mathbf{x},t)\, dt \tag{8}$$

(2) The integral in Eq. (8) need not be taken to infinity, but only to some finite "break" time T such that states adjacent to $\Psi_E(\mathbf{x})$ are effectively filtered out . The state $\Psi_E(\mathbf{x})$ *can thus be determined after a finite amount of dynamics.* The break time is roughly

$$T = \frac{h}{D(E)}$$

where D(E) is the density of states. For those eigenvalues much closer together than the average, the the time needed to resolve neighboring states is longer. Fortunateley, the occurence of close eigenvalues is rare in the chaotic regime, since a Wigner repulsion of levels prevails[14]. This fact tends to strengthen the idea of a break time, since the instances of states requiring longer to filter out from their neighbors will be rare.

The break time observation pinpoints a crucial difference between classical and quantum mechanics: classical dynamics is given infinite time to develop its stationary behavior; quantum mechanics has only a finite time (in a bounded system with a finite density of states). The quantum eigenstates would be expected to be less globally distributed than the infinite time average of a classical trajectory that took longer than the break time to access the full phase space. (It is imaginable that the quantum system somehow accelerates exploration of the available space above the classical rate. It is not clear under what circumstances this is possible.)

It is easy to construct examples of "break time" quantum localization: Take a two dimensional harmonic oscillator with incomensurate frequencies, so that almost all the classical trajectories densely cover a two dimensional domain in coordinate space. Now add many very narrow (compared to a de Broglie wavelength of the corresponding quantum problem) and very tall stipples. These will occasionally scatter a classical trajectory, effectively randomising the harmonic oscillator actions over a long time, but the energy will stay constant. That is, the classical trajectory will become ergodic, but only over a very long period of time, for narrow stipples. On the other hand, the narrow stipples barely scatter waves of deBroglie wavelength much larger than the width of the stipples, and the Fourier transform of a moving wavepacket only locally affected by the stipples is an eigenfunction only locally affected by the stipples. If the stipples remain the same but the energy is increased, a point will come when the wavelength is comparable with the stipple width, and the scattering rate will become large enough to compete with the finite break time T. This time is an increasing function of the energy, since the density of states is going up. Both of these trends lead to the conclusion that the eigenstates will begin to reflect the long time classical dynamics if the energy is high enough. If the frequencies are commensurate, the situation is radically different, both classically and quantally. We will not this case here.

In the context of the stadium, the break time concept suggests that we should not expect the eigenstates to reflect a superposition of an infinite number of plane waves. The typical classical trajectory passes a given region in all possible directions, but it requires a long time before the directions are more or less densly covered. The break time may occur long before this.

The arguments given above only point a qualitative finger at the break time as the culprit in the scarring phenomenon. We shall need these arguments and several more given below to quantify the effect.

(3a) The allowed energies are given by the peak positions in the spectrum (or local density of states, or spectral density) $S_\infty(E)$ of $|\Phi(0)\rangle$:

$$S_\infty(E) = \frac{1}{2\pi} \lim_{T\to\infty} \int_{-T}^{T} e^{iEt/\hbar} \langle\Phi|\Phi(t)\rangle \, dt \tag{9a}$$

The peaks vary in magnitude according to $|\langle\Psi_E|\Phi\rangle|^2$.

(3b) A smoothed *spectral envelope*, $S_T(E)$, can be obtained by smoothing $S_\infty(E)$ over some energy bandwidth, or equivalently by cutting off the time integration, say for example as

$$S_T(E) = \frac{1}{2\pi} \int_{-T}^{T} e^{iEt/\hbar} \langle\Phi|\Phi(t)\rangle \, dt \tag{9b}$$

A seemingly trivial but important principle follows from this: whatever $|\Phi(t)\rangle$ does before time T can affect the spectrum $S_T(E)$; whatever it does after T cannot affect it.

Note that we are examining a diagonal element of the Green's function propagator, i.e.

$$S_\infty(E) = \frac{1}{2\pi} \lim_{T\to\infty} \int_{-T}^{T} e^{iEt/\hbar} \langle\Phi| e^{-iHt/\hbar} |\Phi\rangle \, dt \tag{10}$$

3.2 *Local Density of States Oscillations*

If for $|\Phi(0)\rangle$ we use coordinate states $|q\rangle$ and then we trace over all q, (or any complete set of states), we get the *total density of states* $D_\infty(E)$ form Eq.(9), or the *smoothed total density of states* $D_T(E)$ from Eq.(10). It is now known from several points of view[1,2] that certain periodic orbits play a large role, and perhaps completely determine, both $D_T(E)$ and $D_\infty(E)$. In particular, certain periodic orbits cause oscillations in total density of states $D_T(E)$. We now show that these periodic orbits cause oscillations in $S_T(E)$, the *local* density of states. Then we will relate these local density of states oscillations to the scars in the eigenfunctions.

The function $|\Phi(0)\rangle$ that is convienient for our purposes is a Gaussian wave packet $|z_0\rangle$ of controlable average initial position q_0 and momentum p_0; these evolve into q_t and p_t:

$$\langle q|z_t\rangle = e^{\frac{i}{\hbar}[(q-q_t)\cdot A_t\cdot(q-q_t) + p_t\cdot(q-q_t) + i\gamma_t]} \tag{11}$$

The γ_t term contains both phase and normalization. A_t is a matrix of N dimensions for N degrees of freedom which governs the dispersion (spread) of the wave packet. By choosing $\mathbf{q}_0$ and $\mathbf{p}_0$ to be on the periodic orbit, we can launch the wave packet in close analogy to a cluster of classical trajectories centered on the periodic orbit initial conditions. Indeed, a few classical trajectories serve to *govern* the dynamics of the quantum wave packet. This statement can be made quantitative:

(4) The classical dynamics of a trajectory in a smooth potential, together with the information on nearby orbits (or the *stability equations* for the orbit- see below) serve to fully determine the history of a Gaussian wave packet. The Gaussian remains Gaussian, the expectation values of position and momentum are those of the trajectory, and the shape of the Gaussian depends in a known way on the nearby classial orbits. Thanks to Hagedorn[15] we can say that the difference $||\psi^{\text{Gaussian}} - \psi^{\text{exact}}||^2$ is of order Bh, where B is an increasing function of time.

This means that we can quantitavely specify the evolving wave packet using only classical dynamical information up to any time as long as we please, provided h is small enough. This time may be long enough to encompass many periods of a periodic orbit, but it will generally not be as long as the break time T required to resolve the states. T increases drastically as $h \to 0$ (as h^{-N} in N degrees of freedom).

The evolution of a Gaussian launched in a smooth potential is intimately connected to both the guiding trajectory (which governs the average position and momentum of the Gaussian), and nearby trajectories, which govern the spreading of the Gaussian. A Gaussian launched on a periodic orbit will therefore return to its starting position after one classical period, but for an unstable orbit (all periodic orbits are unstable in the chaotic regime) the Gaussian will have changed shape. If we start with a minimun uncertainty wave packet, then we may be certain that it will have spread upon its return. We expect that the more unstable the orbit, the more spreading the Gaussian will do in a short time. But by assuming that h is small enough, we know that after one or even many more periods the classically determined Gaussian is quite accurately representing the evolution of the intial wave packet, thanks to the property discussed in (4) above.

Explicitly, the evolution of the Gaussian is governed by the trajectories in the following way. First, we write A_t as

$$A_t = \frac{1}{2} P_Z \cdot Z^{-1} \tag{12}$$

where

$$\begin{pmatrix} \dot{P}_Z \\ \dot{Z} \end{pmatrix} = \begin{pmatrix} 0 & -V'' \\ m^{-1} & 0 \end{pmatrix} \begin{pmatrix} P_Z \\ Z \end{pmatrix} \tag{13}$$

For N degrees of freedom, the matrix of second derivatives of the potential evaluated at the current position $\mathbf{q}_t$, V" , is N X N. $\mathbf{P_Z}$ and $\mathbf{Z}$ and their time derivatives appearing on the left are each NXN matrices. These are the classical equations one derives for the *differences* between the reference trajectory and nearby trajectory Each column of $\mathbf{P_Z}$ and $\mathbf{Z}$ describe the deviations in momentum ($\mathbf{P_Z}$) and position ($\mathbf{Z}$) of one trajectory nearby the reference trajectory ($\mathbf{q}_t$,$\mathbf{p}_t$). Taking N such nearby trajectories for N degrees of freedom completely characterizes the linearized dynamics near the reference trajectory. This information on the stability is fed to A_t through the P_Z and Z matrices, and thus the Gaussian spread is controlled by the stability equations.

The complex term γ_t carries the normalization and the semiclassical phase, which is seen to be the usual action integral:

$$\gamma_t = \gamma_0 + \frac{i\hbar}{2} \mathrm{Tr}[\ln(\mathbf{Z}\cdot\mathbf{Z}_0^{-1})] + \int_0^t \left\{ \frac{\mathbf{p}_t\cdot\mathbf{p}_t}{m} - E \right\} dt \tag{14}$$

Littlejohn has given a rigorous and very readable treatment of wave packet dynamics in terms of the symplectic transformations and metaplectic operators which may be constructed from the linearized dynamics.[16]

Two simple cases are instructive. First, for a one dimensional harmonic oscillator, the classical parameters x_t and p_t just execute the harmonic oscillator motion, while the spread parameter A_t behaves as

$$A_t = \frac{1}{2} \frac{2A_0\cos(\omega t) - m\omega\sin(\omega t)}{\frac{2A_0}{m\omega}\sin(\omega t) + \cos(\omega t)} \tag{15}$$

Note that A_t is a constant if A_0 is taken to be $im\omega/2$. The case of unstable motion is more germain, and this is illustrated by an inverted harmonic oscillator, with $V(x) = -{}^1/_2 m\omega^2$:

$$A_t = \frac{1}{2} \frac{2A_0\cosh(\omega t) + m\omega\sinh(\omega t)}{\frac{2A_0}{m\omega}\sinh(\omega t) + \cosh(\omega t)} \tag{16}$$

Suppose that we launch a wave packet along a periodic orbit. For the time being, it will not matter whether this orbit is embedded in complete chaos or is simply an unstable fixed point in a quasiperiodic domain. (The motion sufficiently close to the orbit for some finite number of round trips is the same whether it eventually leads to complete chaos or simply to quasiperiodic motion.) We can erect a Poincare surface of section perpendicular to the orbit, and it will have the appearance as in the sketch shown below, Fig.10

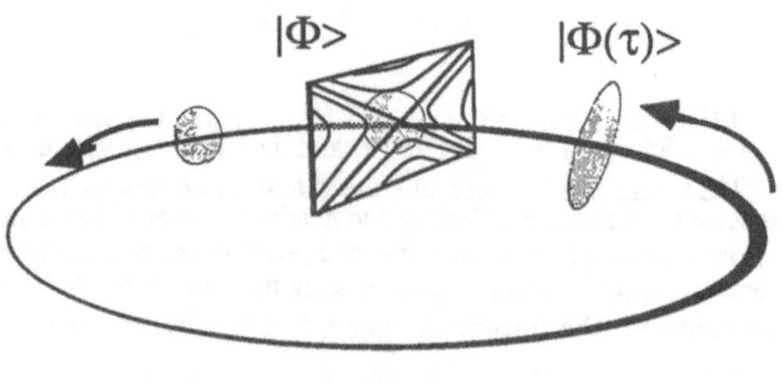

FIG. 10

Now that we know that the packet is bound to return along the orbit at every multiple of the classical period, and further that the shape is still Gaussian and the parameters are known, we can say *quantitatively* what the overlap $<\Phi|\Phi(t)>$ is that appears in Eq. (9). The wave packet spreads benignly *along* the direction of the orbit, and rapidly perpendicular to the orbit. In the phase space of the perpendicular degrees of freedom, the spreading happens along two of the four arms of the seperatrix belonging to the unstable fixed point; contraction occurs along the other two. Trajectories along these two arms seperate exponentially in phase space from the periodic orbit; the other two correspond to exponential attraction. This is exactly the situation for the one dimensional barrier. The overlap for the barrier degree of freedom goes as (using λ instead of ω)

$$<\phi|\phi(t)> = [\cosh(\lambda t) - i\,(m\lambda/2A_0 - 2A_0/m\lambda)\,\sinh(\lambda t)]^{-\frac{1}{2}} \tag{17}$$

Note that if $2A_0 = m\lambda$, this overlap simplifies to

$$<\phi|\phi(t)> = [\cosh(\lambda t)]^{-\frac{1}{2}} \tag{18}$$

Both (17) and (18) go as $e^{-\lambda t/2}$ at long times. The motion along the orbit controls the fast initial (nearly Gaussian) decay of the total overlap $<\Phi|\Phi(\tau)>$ and causes recurrances at multiples of the period $\tau = 2\pi/\omega$, and the motion perpendicuar to it controls the height of these successive recurrences. The overlap will rise and fall repeatedly at multiples of the period, with each successive rise and fall smaller than the last. This is shown in Fig. 11.

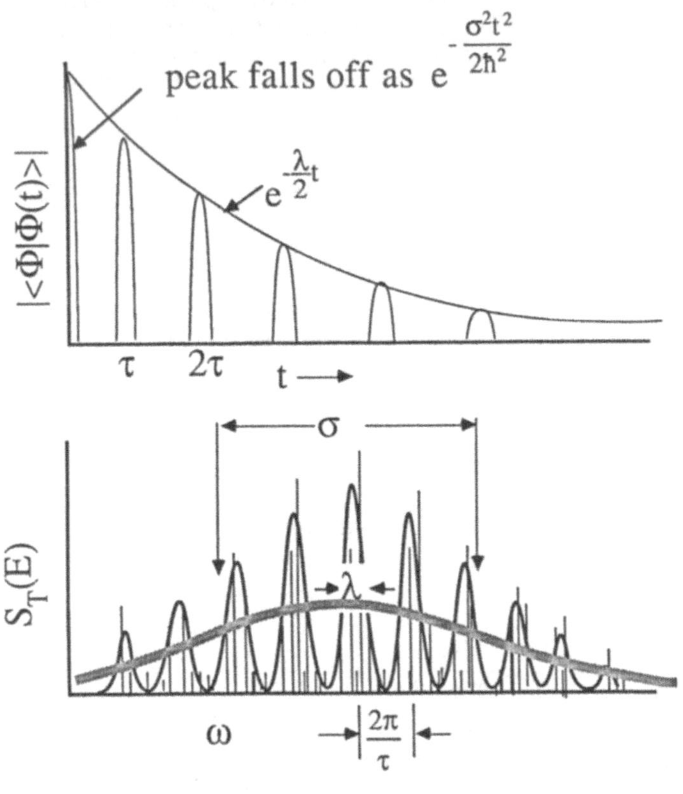

FIG.11

In the lower panel, the Fourier transform of this time dependence, $S_T(E)$, is shown, at three successively finer resolutions. It is very easy to show that at high enough energy or small enough h the initial decay of the overlap is Gaussian, with a time constant that is directly related to the initial momentum, the gradient of the potential at the initial position, and the initial A_0 matrix. This initial decay is recorded in the Fourier transform as the Gaussian envelope with standard deviation σ. Whatever the finer structure at higher resolution, it must be consistent with this envelope. The next higher resolution envelope, showing as a series of "mountains and valleys", is spaced by an energy h/τ, where τ is the classical period of the periodic orbit. This structure is a necessary consequence of the time domain recurrances, which are themselves "slaved" to the classical trajectory. Again, whatever higher resolution structure exists must be consistent with the mountains and valleys, just as the mountains and valleys are consistent with the overall Gaussian envelope. Concievably, the heirarchy of envelopes could extend many levels deep, but we shall stop here, assuming that the next deeper level resolves individual levels from one another.

3.3 *The Scars and the Local Density of States Fluctuations*

The two envelopes just discussed are obtained strictly from classical mechanics, and we can say *rigorously* that mountains and valleys exist. We can even get their height and width from trajectory data alone. This structure seen at intermediate spectral resolution is the key to the existence of the the scars. Recall that individual spectral peaks are a direct measure of the overlap between a state $|\Phi\rangle$ and the eigenstates $|\Psi_E\rangle$. Those peaks under a mountain in the local density of states at intermediate spectral resolution must have higher overlap, on the average, with $|\Phi\rangle$ than those states in the valleys. More important is the fact that this overlap is also higher than the *statistically expected* overlap based on the assumption of normalized, Gaussian random eigenfunctions. This statistically expected overlap is indicated by the smooth low resolution envelope shown in grey in the last figure. It is calculated easily by assuming that the Wigner transform of the density $|\Psi_E\rangle\langle\Psi_E|$ is uniform in phase space on the energy shell, so that we may estimate it as

$$|\Psi_E\rangle\langle\Psi_E| \rightarrow \frac{\delta(E-H)}{\mathrm{Tr}[\delta(E-H)]} = \rho_{\Psi_E}^{\mathrm{Wigner}} \tag{19}$$

which means the overlap $|\langle\Phi|\Psi_E\rangle|^2$ is

$$|\langle\Phi|\Psi_E\rangle|^2 = \mathrm{Tr}[\rho_{\Phi}^{\mathrm{Wigner}}\rho_{\Psi_E}^{\mathrm{Wigner}}] = \int_{-\infty}^{\infty}\frac{S(E')\delta(E-E')}{D(E)}\,dE'$$

$$= \frac{S(E)}{D(E)} \equiv I_{\Psi_E}^{\mathrm{statistical}} \tag{20}$$

where $S(E) = \mathrm{Tr}\,[\rho_\phi^{\mathrm{Wigner}}\,\delta(E-H)]$ is the smoothed local density of states, and $D(E) = \mathrm{Tr}[\delta(E-H)]$ is the smoothed total density of states. Neither S(E) nor D(E) has any oscillatory contributions included. The ratio $S_T(E)/S(E)$ gives the *average* local deviation of the intensities from fully statistical:

$$\frac{S_T(E)}{S(E)} = \frac{I_{\Psi_E}^{T}}{I_{\Psi_E}^{\mathrm{statistical}}} \tag{21}$$

This ratio still says nothing about the disposition of the actual intensities of individual lines; it only restricts the local average. For $S_T(E)$ we find

$$S_T(E) = \frac{\sqrt{2\pi}}{\sigma\tau} \sum_n e^{-\frac{(n\hbar\omega - E_0)^2}{2\sigma^2}} u_\lambda(\frac{E}{h} - n\omega) \qquad (22)$$

where $u_\lambda(\omega)$ is the Fourier transform of $[\cosh(\lambda t)]^{-1/2}$. The local density of states D(E) is

$$S(E) = \frac{\sqrt{2\pi}}{\sigma} e^{-\frac{(E - E_0)^2}{2\sigma^2}} \qquad (23)$$

so that the ratio is just Eq.(23) into Eq. (22). The function $u_\lambda(\omega)$ maximizes close to $4/\lambda$ at $\omega=0$, which means that for well seperated bands the local enhancement of the overlap in the center of the peaks is

$$\frac{I}{I^{statistical}} = \frac{4\,\omega}{2\pi\,\lambda} \qquad (24)$$

This ratio is automatically greater than unity if the bands are well seperated. If the bands are not so well seperated, there are still oscillations in $S_T(E)$ about the statistical average S(E), and the peaks represent local enhancement of the overlap of the eigenvectors with the initial state.

Next we note that the stability analysis of the periodic orbit is the same all along the orbit, which means that the initial state $|\Phi\rangle$ could be anywhere along the orbit and the local densities of states $S_T(E)$ and S(E) would be the same. Also, the overlap $|\langle\Phi|\Psi_E\rangle|$ equals $|\langle\Phi(t)|\Psi_E\rangle|$.*This means that the eigenvectors which have enhanced overlap with one nonstationary state centered on the periodic orbit have enhanced overlap with* ***all*** *states centered on the orbit.*

It is begining to emerge that some eigenstates have larger than statistical overlap with states which are well localized to the region of a classical periodic orbit. However, we must address the issue of the natural fluctiations in the intensities, which in any local region of the spectrum would be the square of real Gaussian random amplitudes, or a χ^2 distribution of one degree of freedom. The requirement that the average line strength be higher than statistical in the "mountain" regions simply means that some eigenstates have larger overlap with the state $|\Phi\rangle$ than would otherwise be allowed.

3.4 *Properties of the Scars*

We collect what we know so far. First, scars exist in certain eigenfunctions along the periodic orbits, as measured by localized wave packets which are probes of phase space. The enhanced density in phase space near the periodic orbits will translate into enhancements in coordinate space, as can be shown by integrating over the momentum in the phase space distributions. This is consistent with the scars that we have shown earlier, in coordinate space.

Second, since the scar density in phase space goes as ω/λ, those periodic orbits with largest ratio ω/λ are expected to have the heaviest scars. Note that this ratio is a purely classical quantity.

Third, the ratio given in Eq.(24) is only a lower bound to the effect, since the actual spectral intensities are free to be nonstatistically distributed even within the confines imposed by $S_T(E)$. That is, the correct local density of states $S_T(E)$ can be reproduced with a relatively few eigenstates having very large overlap with $|\Phi>$ and the remainder having very small overlap. Those having the very large overlap would then be more heavily scarred than the estimate, Eq. (24). One way for this to happen is for each mountain to break up at higher resolution into many narrower mountains, each with many eigenstates under it, etc. Clearly, this heirarchy can continue right down to the ultimate spectral resolution of individual eigenstates.

The numerical evidence obtained so far, which must be viewed as preliminary, suggests that the scarring is indeed much heavier than the ratio , Eq.(24). One reason for this is certainly a kind of cooperative effect, in which scarring along one periodic orbit wieghts that region more heavily, keeping other nearby (in energy) states away, making it easier for them to scar some other region more heavily, etc. (It is easy to show that a collection of all states in a given energy region must evenly cover phase space, the more so the largeer the region.) The phenomenon reminds one of phase seperation in statistical mechanics.

Fourth, the scars are expected to get narrower as $h \to 0$. This can be seen from the following argument: Suppose that we start the initial test state $|\Phi>$ just off the periodic orbit. If the deviation from the orbits is small compared to the dimension of the state in phase space, then the subsequent overlaps will hardly be affected, so the spectrum and the scar arguments all follow. But as $h \to 0$, the same absolute deviation will become far from the periodic orbit in the sense that the extent of the state $|\Phi>$ in phase space, which in any direction can be taken proportional to $\sqrt{h}$, becomes small compared to the deviation. Then, the overlap recurrances following the initial decay will not happen, since the wave packet will miss its birthplace, having been set off in a nonperiodic region. Of course, since phase space is dense with periodic orbits, the packet is always launched along some, but these are likely to be of the much more numereous variety $\omega/\lambda<<1$. Note too that the scar, as it gets narrower, does not get any denser, since the estimate, Eq.(24), is independent of h.

Since the scars are apparently so much more dense than the bound Eq.(24), it is not known whether periodic orbits with $\omega/\lambda<<1$ can contribute to scarring. One thing is certain:the Correspondence Principle must be obeyed, so most eigenstates cannot be completely dominated by one or a few scars of simple orbits like the ones shown in this paper as $h \to 0$. If they were, then narrow wave packets, which should obey nearly classical behavior for long times, would be unable to do so. This follows because the packets are composed of the eigenstates. It does seem possible that the eigenstates could become a mass of scar tissue as $h \to 0$, assuming that complex and long period orbits were "resolved" as $h \to 0$. Another possibility is that the eigenfunctions try to become as much like Berry's conjecture as possible, with a few "scratches" remaining to mar the otherwise perfectly Gaussian random surface. Whatever the answer, the future is bound to bring interesting new insights and results concerning the eigenfunctions.

Finally, we remark that the relation of the scarring to other localization phenomena in quantum systems[17-20] is of great interest. It is possible that the break time phenomenon encompases a variety of cases where quantum eigenfunctions explore less of the phase space of the system than the analogous classical mechanics.

Taylor and co-workers[21] have given an intriguing point of view on the wavefunction localization, in terms of a confining potential resulting from an adabatic seperation of certain degrees of freedom appropriate to the periodic orbits.

BIBLIOGRAPHY

1a. M.C. GUTZWILLER, Physica 5D, 183(1982); b.J. Math. Phys. 12, 343(1971).
2. M.V. BERRY and M. TABOR, Proc. Roy. Soc. A349 ,101(1976).
3. M.V. BERRY, Proc. Roy. Soc.400, 229 (1985).
4. J.H. HANNAY and A.M. ORZORIO de ALMEIDA, J. Phys A 17, 3429 (1984)
5. D.A. HEJHAL, Duke Math. Jour. 43, 441 (1976).
6a. M.V. BERRY in "Chaotic Behavior of Deterministic Systems", Les Houches Summer School Lectures, 1981) North-Holland 1983, p171;b. J. Phys. A: 10 2083 (1977).
7. P. PECHUKAS, J. Chem. Phys. 57 , 5777 (1972).
8. N. DeLEON and E.J. HELLER , Phys. Rev. A30, 5 (1984)
9. B.V. CHIRIKOV, Phys. Rep. 52, 263 (1979).
10a. S.W. McDONALD, Ph. D. Thesis, Lawence Berkeley Laboratory Report LBL 14837; b. S.W. McDONALD and A.N. KAUFMAN, Phys. Rev. Lett. 42, 1189 (1979).
11. E. J. HELLER, Phys. Rev. Lett. 53, 1515(1984).
12. T.H. SELIGMAN and J.J.H. VERBAARSCHOT, "Long Range Stiffness of Spectral Fluctuations in Integrable Scale Invarient Systems" (pre-print 1986)
13. J. W. HELTON and M. TABOR, Physica 14D , 409 (1985).
14a. T.A. BRODY J. FLORES, J.B.FRENCH, P.A. MELLO, A. PANDEY and S.S.M. WONG, Rev. Mod. Phys. 53, 385 (1981); b. O. BOHIGAS, M. J. GIANNONI and C. SCHMIT, J. Phys.Lett. 45, L1015 (1984).
15. G.A. HAGEDORN, Comm. Math. Phys. 71, 77(1980).
16. R. LITTLEJOHN, Physics Reports, to be published.
17. P.W. ANDERSON Phys. Rev. 103, 1492(1958)
18.a S. FISHMAN, D.R. GREMPEL, and R. E. PRANGE , Phys.. Rev. Lett. 49, 509(1982); b. Phys. Rev. Lett. 49, 509 (1982).
19. K. ISHII, Progr. Theor. Phys. Suppl. 53, 77(1973).
20. G. CASATI and F. VALZ-GRIS, and I. GUARNIERI, Nuovo Cimento Lett. 28 , 279(1980).
21.G.HOSE, H.S. TAYLOR, and Y.Y. BAI, J. Chem. Phys. 80,4363 (1984).

QUANTUM ERGODICITY IN A QUANTUM MEASURE ALGEBRA: INTENSITY FLUCTUATIONS IN MOLECULAR SPECTRA

E. B. Stechel
Sandia National Laboratories
Albuquerque, New Mexico 87185/USA

We pose the following question: If a system is prepared in a smooth non-stationar state, then how much of its available space is covered during the evolution of thi state? We might ask this question if we have an interest in laser directed (o assisted) chemistry. In this case we might want to know if unimolecular decompostio can occur before intramolecular relaxation (energy transfer) has been able to destro the specificity of an excitation; in other words, if one specifically places energ into a bond α of a molecule will it flow to another bond β or will it preferentiall remain in bond α? In more general terms we may be looking for an enhanced understand- ing of intramolecular energy transfer (IET). How IET occurs is perhaps obvious from classical viewpoint, but it is a more subtle problem from a quantum viewpoint Intuitively one might expect that if the system is "chaotic" and hence ergodic a non stationary state should cover all of its available space.

Unfortunately, while this question has meaning in a classical dynamical system (CDS) it does not in a quantum dynamical system (QDS). This essential difference arises because in a CDS a phase space is a measure algebra (MA) [1], which is to say that w can measure volumes and areas on surfaces (e.g. of constant energy) in phase space. The measure (area) of a subregion A of the surface of constant energy E_0 is $\mu(A) = \int_A d\underline{q}d\underline{p}\ \delta\big(H(\underline{q},\underline{p})-E_0\big)$. To make the question meaningful for a QDS we need a quantu measure algebra (QMA) [2] a mathematical construct in which the concept of volume o area (how much?) takes on meaning in quantum mechanics. We need a formulation of a QD analogous to a CDS; i.e., in a MA rather than in a Hilbert space. Given a QDS we ca then provide a definition of quantum ergodicity, one which is formulated withou reference to classical or semi-classical mechanics, yet at the same time one whic possesses a correspondence limit that is related to classical mechanics. It will tur out that in a QDS with energy states which are gaussian random in coordinate space (plausible conjecture for a "quantum chaotic" system, [3]) a smooth (on scales large ompared to $\hbar$), pure non-stationary state samples approximately $\frac{1}{3}$ of the available pace if it is real and $\frac{1}{2}$ if complex.

his paper is divided into two parts. Part I assembles three pieces of background ecessary to develop the logic behind this result. Section IA is devoted to classical lgebraic ergodic theory (AET) [4]. Section IB briefly discusses random matrix theory RMT) (for a thorough review see Ref. [5]) which should require very little development ince a large fraction of this volume is devoted to this subject. Section IC reviews he concepts in quantum "phase" space flow (P(a,b)'s and p_n^a's) which were originally ntroduced by Heller [6]. Part II develops what we will call quantum AET (the quantum nalog of classical AET) [2]. Section IIA presents the formal development with the new efinitions. Section IIB discusses intensity fluctuations in quantum ergodic (QE) pectra, leading to the somewhat surprising result that in a QE system a pure, real, mooth state samples only about $\frac{1}{3}$ of its available space.

. BACKGROUND

. Classical Algebraic Ergodic Theory

lgebraic Ergodic Theory (AET) [4] was developed in the 1930's as a branch of measure heory [1] providing the first rigorous mathematical definitions of ergodicity, weakly ixing and mixing. A classical dynamical system (CDS) [7] is a triplet (M,μ,ϕ_t) with M smooth manifold (phase space), μ a measure defined on M, and ϕ_t a one parameter group f measure preserving diffeomorphisms (Hamilton's equations of motion). The basic eneralized (or abstracted) concept is that of an abstract dynamical system (ADS) which s a triplet (M,μ,ϕ_t) with (M,μ) a MA and ϕ_t a one parameter group of automorphisms epending measurably on t (time); i.e., for any measurable sets A & B in M, $\mu(\phi_t A \cap B)$ s a measurable function of t ($\cap$ is the intersection set operation). The MA is a ime-invariant set M under an evolution operator ϕ_t with a σ-algebra X of subsets of M nd a time-invariant measure μ defined on that algebra; i.e., the measure of a time-volved set A does not change in time, $\mu(\phi_t A) = \mu(A)$. A σ-algebra has two operations, eneralized multiplication and addition. The multiplication operator is typically the ntersection ($\cap$); however, we use the symbol $\odot$. Multiplication is idempotent ($\mu(A\odot A)$ $= \mu(A)$) and commutative ($A\odot B\odot C = B\odot C\odot A$). The addition operator is typically the symetric difference (Δ). Addition has two notable properties: first $(A \,\Delta\, A) = 0$ and econd $(A \,\Delta\, (B \,\Delta\, C)) = ((A \,\Delta\, B) \,\Delta\, C)$ (associativity) . The symmetric difference is efined as the union minus the intersection, thus $\mu(A \,\Delta\, B) = \mu(A)+\mu(B)-\mu(A\odot B)-\mu(B\odot A)$.

It is along this vein that, in Part II, we define a QDS. But at this point we take the pportunity to indicate that there are a couple of hints that a measure of volume does xist in quantum mechanics, e.g., the common phrase, 'a pure state occupies a unit cell

(V=1) in phase space' of volume $(2\pi\hbar)^d$ where d is the number of degrees of freedo
However, there is no corresponding statement to tell us how much volume is occupied
a mixed state $(\hat{\rho}_A=\Sigma_i a_i|\phi_i\rangle\langle\phi_i|)$. It is fair to assume that if $\hat{\rho}_A$ is an equal mixtu
of N orthonormal states then V=N. But what about $\hat{\rho}_A=\frac{1}{4}|\phi_1\rangle\langle\phi_1|+\frac{3}{4}|\phi_2\rangle\langle\phi_2|$, clearly $1 \leq$ ≤ 2, yet the precise result (see below) $V=8/(5+3|S|^2)$ with $S=\langle\phi_1|\phi_2\rangle$ is not evide
from previously developed concepts.

The second hint comes from the Weyl correspondence (see, e.g., Refs. [8,9]), whi
states that the smoothed or average density of states can be obtained by integrati
volumes in phase space; i.e.,

$$D(E) \approx \int d\underline{q}d\underline{p} \frac{\delta\left(H^{cl}(\underline{q},\underline{p}) - E\right)}{(2\pi\hbar)^d}. \qquad [IA.1]$$

Moreover, D(E)dE counts the energy states in the incremental dE about E. Furthermor
the number of bound states supported by a potential well equals the integrated volu
of phase space below the dissociation energy divided by the volume of a unit cell.

We now give definitions for ergodicity, weakly mixing and mixing (for a thorough dis
cussion see Ref. [7]). In an ADS (M,μ,ϕ_t) with A and B subsets of M, the flow i
ergodic in M if

$$P(A,B) = \lim_{\Gamma\to\infty} \frac{1}{\Gamma}\int_0^\infty dt\, \exp\left(-\frac{t}{\Gamma}\right) \mu\left(\phi_t A \odot B\right) = \frac{\mu(A)\mu(B)}{\mu(M)} \qquad [IA.2]$$

which is to say that a time-averaged set $\{\phi_t A\}$ is stochastically independent of al
sets B of positive measure and contained in M. The definition of mixing is

$$\lim_{t\to\infty} \mu\left(\phi_t A \odot B\right) = \frac{\mu(A)\mu(B)}{\mu(M)} \qquad [IA.3a$$

which is to say that after a sufficiently long time the fractional amount of the time
evolving set $\phi_t A$ in any set B of positive measure in M is the same as the origina
fractional amount of A in M. Finally the definition of weakly mixing is

$$\lim_{\Gamma\to\infty} \frac{1}{\Gamma}\int_0^\infty dt\, \exp\left(-\frac{t}{\Gamma}\right) \left| \mu\left(\phi_t A \odot B\right) - \frac{\mu(A)\mu(B)}{\mu(M)} \right| = 0 \qquad [IA.3b$$

which states that after sufficient time the time-evolving set $\phi_t A$ only occasionally bu
of no measure deviates from equilibrium.

In terms of a stochasticity hierarchy [7], integrable systems sit at the bottom (n
stochasticity) of the totem pole (trajectories lie on d-dimensional invariant tor

embedded in a (2d-1)-dimensional energy surface). Next are the pseudo-integrable systems (dominated by invariant tori but having regions of small measure in which trajectories explore a D-dimensional space with $2d-1 \geq D > d$). The first and weakest stochastic property is ergodicity, then weakly mixing and mixing. Climbing the pole, higher (stronger) properties imply lower (weaker) ones. Strongly stochastic systems are K-systems (Kolmogorov entropy greater than zero) and C-systems (exponential divergence of trajectories).

In this paper we are concerned only with the weakest of the stochastic properties, ergodicity. In that regard we make a very important, but often overlooked distinction, which is the difference between ergodic <u>evolution</u> and ergodic <u>flow</u>. The definition of an ergodic flow was given above (see Eq. [IA.2]) and it can be seen that it is defined only with respect to the set M. In a CDS there always exists an M such that the flow is ergodic. Since this is the case we say that the <u>evolution</u> is ergodic. However, if we take M to be a constant energy hypersurface, for example, the <u>flow</u> is oftentimes non-ergodic. Since the <u>evolution</u> is ergodic the single P(A,A) (the time average that the time-evolved $\phi_t A$ intersects A), without measuring any P(A,B) for $B \neq A$, determines how many distinct sets B are accessed. If P(A,A) = 5 min/hour ($\frac{1}{12}$ of the time), then $\phi_t A$ accesses 12 cells. In other words, P(A,A) is inversely proportional to the size of the space (M_A) that $\phi_t A$ covers. Recall the question posed in the beginning. Evidently we already have a strong hint as to how to make it meaningful for a QDS.

B. <u>Random Matrix Theory</u>

It is fair to say that the first practical quantum ergodic theory is random matrix theory (RMT). RMT was largely developed with regards to statistical theories of neutron resonance reactions (see e.g. Ref. [10]). Much of the motivation arises because a nuclear system is a strongly interacting many particle system whose hamiltonian is unknown. In 1956 Wigner [11] determined the statistical distribution of eigenvalue spacings with reasonable, plausible or maybe simply guessed assumptions about the statistical nature of the hamiltonian matrix elements, H_{ij}. By considering gaussian distributed matrices of order two he arrived at the Wigner surmise for the level spacing distribution, namely

$$P(s)ds = \frac{\pi s}{2\bar{s}^2} \exp\left(-\frac{\pi s^2}{4\bar{s}^2}\right) \qquad \text{[IB.1]}$$

where s is the spacing between two neighboring (energy) eigenvalues and $\bar{s}$ is the mean energy spacing. This distribution is characterized by level repulsion ($P(s) \to 0$ for $s \ll \bar{s}$) and by a high degree of rigidity (relatively strongly peaked near its mean value, the peak of P(s) is at $s\approx .8\bar{s}$). Numerical computations with matrices of order greater than two and with matrix elements chosen from independent gaussian random distributions have qualitatively confirmed the Wigner surmise.

The purpose to the gaussian orthogonal ensemble (GOE) is to give the real eigenvalues of a system so complicated that it appears random independent of the representation. It is applicable to the internal region of the highly excited nucleus. It should also be applicable to QE systems in general, and in particular QE molecular systems, despite the fact that much of the motivation is lost in chemistry. Hamiltonians in molecular physics are well known. Bound state problems have smooth many-body potentials. Nonetheless, the fluctuation properties of "chaotic" quantum systems do appear to be indistinguishable from that of the GOE (for a review see Ref. [12]).

So far we have only referred to the fluctuation properties of the eigenvalue spacings or, in other words, the density of states. Here we want to also make connections to intensity fluctuations; i.e., fluctuations in $|\langle\phi|n\rangle|^2$ where $|n\rangle$ is an energy eigenstate and $|\phi\rangle$ is a non-statistical state, e.g. an eigenstate of a measurable which does not commute with the hamiltonian. The relevant analog in RMT is the fluctuation properties of the measurable neutron resonance strength, $\Gamma_{\lambda(c)}\propto\gamma^2_{\lambda(c)}$. Previous results indicate that the width distribution is Porter-Thomas (χ^2_1) [10,13] at least as long as no doorway states or other intermediate state structures are present. This distribution can also be obtained if $\gamma_{\lambda(c)}$ are considered to be elements of the transformation which diagonalizes the GOE hamiltonian. The Porter-Thomas distribution is

$$P(\Gamma_\lambda)d\Gamma_\lambda = \frac{1}{\sqrt{2\pi\bar{\Gamma}_\lambda\Gamma_\lambda}} \exp\left(-\frac{\Gamma_\lambda}{2\bar{\Gamma}_\lambda}\right). \qquad [IB.2]$$

C. Quantum Phase Space Flow: P(a,b)'s and p_n^a's

Heller [6], motivated by the concept that a classically ergodic system uniformly samples all of its phase space subject only to the known constants of motion, focused on a dynamic quantity which he called P(a,b). P(a,b) is the time averaged probability of starting in state ρ^a and being in state ρ^b at some later time.

$$P(a,b) = \lim_{\Gamma\to\infty} \frac{1}{\Gamma}\int_0^\infty dt \exp\left(-\frac{t}{\Gamma}\right) \mathrm{Tr}\left\{ \rho^a(t)\, \rho^b \right\}. \qquad [IC.1]$$

He reasoned that if quantum flow is ergodic, P(a,b) should equal a constant independent of b for all states ρ^a and ρ^b which are equivalent. One might imagine that two states are equivalent if they have the same mean energy and the same spread in energy. He showed the direct relation between P(a,b) and the spectral intensities, p_n^a and p_n^b where

$p_n^{a,b} = \langle n|\rho^{a,b}|n\rangle$. For a non-degenerate system $P(a,b) = \Sigma_n p_n^a p_n^b$. The p_n^a are, in turn, related to the high resolution absorption (or emission) spectrum of the state ρ^a,

$$\epsilon_a(\omega) = \omega\kappa \int dt \exp(i\omega t) \operatorname{Tr}\{\exp\left[-iHt/\hbar\right] \rho^a\} = 2\pi\omega\kappa\, \Sigma_n p_n^a \delta(\omega - E_n/\hbar), \qquad [IC.2]$$

where κ is a constant and $H|n\rangle = E_n|n\rangle$ (emission $\kappa\omega+\kappa'\omega^3$). Therefore, experimentally measuring a spectrum to infinite resolution; i.e., high enough resolution to resolve every line, informs us as to how much time ρ^a spends in ρ^a and how much time ρ^a spends in ρ^b, providing that the spectrum of ρ^b has also been experimentally measured.

As in classical mechanics and according to this definition of quantum ergodicity the quantum evolution is ergodic, since $P(a,a) = P(a,a')$ if $\rho^{a'} = \rho^a(t')$ for some time t'. The question remains can the quantum flow be ergodic and what determines the available space (M) of ρ_a? Heller noted that classically or quantum mechanically this must include all states consistent with the known constants of motion; e.g., known symmetries and spectral moments, $\langle E^k\rangle$. The spectral moments can be summarized in terms of the low resolution spectrum,

$$S_a(E;\Gamma) = \frac{1}{2\pi} \int dt \exp(i\omega t)\, \Omega(t/\Gamma) \operatorname{Tr}\left\{ \exp\left[-iHt/\hbar\right] \rho^a \right\} \qquad [IC.3a]$$

with $\Omega(x) = 1$ for $|x| < 1$ and $= 0$ $|x| > 1$. The smoothed density of states is then

$$D(E;\Gamma) = \Sigma_n \frac{1}{2\pi} \int dt \exp(i(E-E_n)t/\hbar)\, \Omega(t/\Gamma). \qquad [IC.3b]$$

This gives rise to the spectral criterion for quantum ergodicity; for all P(a,b) to be equal (equivalent a and b having the same $S(E;\Gamma)$) implies that $p_n^b = p_n^a = x_n = S_a(E_n;\Gamma)/D(E_n;\Gamma)$. In other words spectral congestion (no missing lines) would be the hallmark of ("ideal") quantum ergodicity.

In section IA we saw that the measure of the space covered by a time evolving set is inversely proportional to the time average that the set intersects itself. The measure of the accessible space is the maximum allowed by the known constants of motion. Here, since the quantum evolution is likewise ergodic and assuming we can replace set by state [15], then the quantum measurable P(a,a) (a time average) is inversely proportional to the measure of the space accessed by state ρ^a. The low resolution enevelope $\{x_n\}$ defines the minimum possible value for P(a,a). Heller [6] called this minimum $P^{sto}(a,a) = \Sigma_n x_n x_n$. The inverse of $P^{sto}(a,a)$ is the maximum measure for the space; i.e., the measure of M, hence the fraction covered is $F = P^{sto}(a,a)/P(a,a)$ (for a review see Ref. [14]).

II. QUANTUM MEASURE THEORY: THE QUANTUM GENERALIZATION OF CLASSICAL MEASURE THEORY

The goal here is to define a QDS in the same vein as a CDS or ADS. Toward this end Repeat the definition of an ADS as a triplet (M,μ,ϕ_t) [7]. An ADS is a MA which is time-invariant set M under an evolution operator ϕ_t with a σ-algebra X of subsets of and a measure defined on that algebra. A σ-algebra has two operations: (1) genera ized multiplication which is idempotent $(\mu(A\Theta A) = \mu(A))$ and commutative and (generalized addition which has the property $\mu(A \Delta A) = 0$ and is associative. T algebra is closed to both operations.

We now give the analogous definition of a QDS [2]. Differences are high-lighted underlining. A QDS, (M,μ,ϕ_t), is a QMA which is a time-invariant set M under an evol tion operator ϕ_t with a quantum σ-algebra X of subsets of M and a measure defined that algebra. A quantum σ-algebra has two operations: (1) generalized multiplicati which is idempotent $(\mu(A\Theta A) = \mu(A))$ but not commutative and (2) generalized additi which has the property $\mu(A \Delta A) = 0$ but is not associative. The algebra is closed both operations.

A. Definitions in a Quantum Measure Algebra

In a QMA (M,μ) states of the system are measurable sets. A QMA is defined by a tim invariant state ρ^M $(\rho^M_{nm} = x_n\delta_{nm})$ where $\underline{x}$ is called the envelope. If A is any state ρ of the system then the definition of measure is

$$\mu(A) = \frac{1}{\mathrm{Tr}\{(\rho^A)^2\}} . \qquad \text{[IIA.1]}$$

Basically the measure counts the number of pure states comprising a given state. Th measure of M is then $\mu(M) = 1/(\underline{x}\bullet\underline{x})$. Thus $\mu(M)$ is essentially the number o eigenstates underneath the envelope, $\underline{x}$. This measure is properly time-invariant $\mu(\phi_t A) = \mu(A)$. This answers the question: what is the measure of a mixed state? No the answer given in section IA can be verified.

The multiplication operator which is properly idempotent, $\mu(A\Theta A) = \mu(A)$ is defined fo A and B any states ρ^A and ρ^B of the system as

$$\mu(A\Theta B) = \frac{\mathrm{Tr}\{\rho^A\,\rho^B\}}{\mathrm{Tr}\{(\rho^A)^2\}\ \mathrm{Tr}\{(\rho^B)^2\}} . \qquad \text{[IIA.2a]}$$

$$\mu(A\Theta B\Theta C) = \frac{\mathrm{Tr}\{\rho^A\,\rho^B\}\ \mathrm{Tr}\{\rho^B\,\rho^C\}}{\mathrm{Tr}\{(\rho^A)^2\}\ \mathrm{Tr}\{(\rho^B)^2\}\ \mathrm{Tr}\{(\rho^C)^2\}} . \qquad \text{[IIA.2b]}$$

It is, however, easily verified that this is <u>non</u>-commutative; i.e., $\mu(A \odot B \odot C) \neq \mu(B \odot C \odot A)$. The second operation, addition or the symmetric difference is defined as in an ADS, that is $\mu(A \,\Delta\, B) = \mu(A) + \mu(B) - \mu(A \odot B) - \mu(B \odot A)$. For A and B states

$$\mu(A \,\Delta\, B) = \frac{\mathrm{Tr}\{|\rho^A - \rho^B|^2\}}{\mathrm{Tr}\{(\rho^A)^2\}\ \mathrm{Tr}\{(\rho^B)^2\}} . \qquad [IIA.3]$$

While $\mu(A \,\Delta\, A) = 0$, $\mu(A \,\Delta\, (B \,\Delta\, C)) \neq \mu((A \,\Delta\, B) \,\Delta\, C)$; i.e., addition is <u>non</u>-associative. We do not give the definition here but direct the reader to Ref. [2].

The relation between Heller's P(a,b) [6] and the QMA defined here [15] is most easily seen through a generalization which includes mixed states. Now let

$$P(A,B) = \lim_{\Gamma\to\infty} \frac{1}{\Gamma} \int_0^\infty dt\, \exp\left(-\frac{t}{\Gamma}\right) \frac{\mathrm{Tr}\{\rho^A(t)\, \rho^B\}}{\mathrm{Tr}\{(\rho^A)^2\}\ \mathrm{Tr}\{(\rho^B)^2\}}$$

$$= \lim_{\Gamma\to\infty} \frac{1}{\Gamma} \int_0^\infty dt\, \exp\left(-\frac{t}{\Gamma}\right) \mu\left(\phi_t A \odot B\right) . \qquad [IIA.4a]$$

The correspondence of this P(A,B) to the definition for its namesake in Eq. [IA.2] should be evident. Thus, since quantum mechanically the <u>evolution</u> is ergodic then $P(A,A) = \mu(A)\mu(A)/\mu(M_A)$ where M_A is the set actually covered by A during its evolution. Furthermore, if flow is classically ergodic and if the spectral criterion for "ideal" QE flow is satisfied then $P^{sto}(A,A) = \mu(A)\mu(A)/\mu(M)$ and $P^{sto}(A,B) = \mu(A)\mu(B)/\mu(M)$.

If A and B ($\rho^{A,B}_{nn} = x_n + \delta^{A,B}_n$) are subsets of M then $\mu(A \odot M) = \mu(A)$ and $\mu(B \odot M) = \mu(B)$. This then requires that $\Sigma_n \delta^{A,B}_n = \Sigma_n \delta^{A,B}_n x_n = 0$. We call $\underline{\delta}^{A(B)}$ the fluctuation of A (B) about the envelope. Hence

$$P(A,B) = \frac{\mu(A)\mu(B)}{\mu(M)} \left[1 + \frac{(\underline{\delta}^A \cdot \underline{\delta}^B)}{(\underline{x} \cdot \underline{x})} \right] . \qquad [IIA.4b]$$

Thus if flow is "ideally" ergodic then $\underline{\delta}^A \cdot \underline{\delta}^\beta$ would equal zero. Note, however, that substituting A for B in Eq. [IIA.4b] reveals that P(A,A) is strictly greater than the "ideally" ergodic value unless the fluctuation vanishes (spectral criterion). We introduce one last definition: the normalized P(A,B) or f(A,B)

$$f(A,B) = \left[\frac{\mu(M)}{\mu(A)\mu(B)}\right] P(A,B) = \left[1 + \frac{(\underline{\delta}^A \cdot \underline{\delta}^B)}{(\underline{x} \cdot \underline{x})} \right]. \qquad [IIA.5]$$

Classically ergodic flow would have $f(A,B) = 1$. Recalling the definition for the fraction, $F = P^{sto}(A,A)/P(A,A)$ or $F = \mu(M_A)/\mu(M)$, we see that F is simply $1/f(A,A)$.

Let us summarize what we have so far. We have a QMA which preserves the essence of classical MA in that (1) all sets are <u>measurable</u> with positive measure, (2) the algebr is <u>closed</u> to multiplication (intersection) and addition (symmetric difference) (se Ref. [2]), (3) $\mu(A\Theta A) = \mu(A)$, (4) $\mu(A\Theta M) = \mu(A)$ for A a subset of M, (5) $\mu(A \Delta A) =$ and (6) we assert but do not demonstrate that a disjoint class of subsets of M $\{A_j$ $\left(\mu(A_j \Theta A_k) = \mu(A_j)\, \delta_{jk}\right)$ exists such that $\mu\left(\Delta_j A_j\right) = \mu(M)$; i.e., we can <u>partition</u> an <u>cover</u> M. Therefore the classical analog of a quantum state is here a measurable set The measure decreases as the purity of the state increases. A pure state has measur one. Quantum evolution is ergodic and f(A,A) is a measurable set function which i inversely proportional to the fraction of the QMA covered. This then gives <u>meaning</u> t the question: <u>how much</u> is covered by a time evolving state?

B. Random Fluctuations

Here we attempt to motivate the existence of random fluctuations about the envelope i a QE system. Suppose $|\psi\rangle$ is a pure real state, smooth on scales large compared with $\hbar$. Suppose further that the energy states are gaussian random (conjecture Ref. [3]). If we define $y_n = |\langle n|\psi\rangle|^2$ with expected value $\bar{y}_n = x_n$, each amplitude $\langle n|\psi\rangle$ can be taken to be a gaussian random variable with variance x_n. Then the distribution for y_n is none other than Porter-Thomas χ_1^2 (see Eq. [IB.2]). This distribution is characterized by a high probability of missing lines; i.e., many $y_n \approx 0$. For more general states we consider $\rho^A = \Sigma_j \alpha_j |\phi_j\rangle\langle\phi_j|$ with $|\phi_j\rangle = \cos(\theta_j)|\psi_j\rangle + i\,\sin(\theta_j)|\chi_j\rangle$ $\left(0 \le \theta_j \le \frac{\pi}{4},\ |\psi_j\rangle\right.$ and $|\chi_j\rangle$ real, $\langle\psi_j|\psi_k\rangle = \langle\chi_j|\chi_k\rangle = \delta_{jk}$ and $\left.\langle\psi_j|\chi_k\rangle = 0\right)$. These conditions assure stochastic independence of the pure real states comprising ρ^A and do not impose any loss of generality. The distribution for $y_n^A = \langle n|\rho^A|n\rangle$ is (where the superscript A and subscript n are dropped for ease of notation)

$$P(y)dy = \frac{z^z}{\bar{y}\,\Gamma(z)} \left(y/\bar{y}\right)^{z-1} \exp\left\{-z\, y/\bar{y}\right\} \qquad [IIB.1]$$

with $z^{-1} = \Sigma_j\, \alpha_j^2 \left(1 + \cos^2(2\theta_j)\right)$. For a pure real state $z = \frac{1}{2}$ which is the χ_1^2 distribution. For a pure complex state $\left(\cos\theta = \sin\theta\right)$ $z = 1$ in which case the distribution still has a finite probability for missing lines. For the general state $\frac{1}{2}\,\mu(A) \le z \le \mu(A)$. We have verified this with GOE calculations [18].

The implications of such statistical fluctuations on the distributions for f(A,A) and f(A,B) are: the mean of $f(A,A) = 1 + \frac{1}{z}$, while the mean for $f(A,B) = 1$. But the fraction $F = 1/f(A,A)$, thus for pure states the prediction is $\frac{1}{3} \le \bar{F} \le \frac{1}{2}$. This has been tentatively verified by Heller & Sundberg [16] by numerical calculations with pure real and complex initial states in the "demonic" potential in an energy regime where the classical mechanics shows no sizeable regions of quasi-periodic behavior. However, their distributions for the fluctuations about the envelopes appear to deviate substantially from those of Eq. [IIB.1]. This can easily be traced to energy states which are substantially non-gaussian and show significant localization. For a discussion of "scars" in energy states in systems which are classically chaotic see Ref. [17] and Heller's chapter in this volume.

Nevertheless, for gaussian random energy states the variances $\sigma^2\big(f(A,A)\big)$ and $\sigma^2\big(f(A,B)\big)$ are both inversely proportional to $\mu(M)$ and are inversely proportional to z_A^2 and $z_A z_B$ respectively. However $\sigma^2\big(f(A,A)\big)$ is substantially greater than $\sigma^2\big(f(A,B)\big)$. Unfortunately we do not have the room here to illustrate the distributions which we have determined by GOE calculations [18]. Finally we note that the variances vanish in the classical limit ($\hbar \to 0$) because $\mu(M) \propto D(E)\ \Delta E \to \infty$. Furthermore, in a coarse grained $\hbar \to 0$ limit $f(A,A)$ properly goes to 1 since $z \to \infty$.

ACKNOWLEDGEMENTS

This work was performed at Sandia National Laboratories supported by U.S. Department of Energy under contract DE-AC04-76-DP00789. I am very grateful to Prof. Thomas H. Seligman for inviting me to this enlightening and enjoyable conference and for finanicial support during my stay in beautiful Cuernavaca, Mexico. I also wish to extend my appreciation to Prof. Eric J. Heller for numerous helpful conversations and to Jim Martin and Jess Wilcoxon for a critical reading of this manuscript.

REFERENCES

1. P.R. HALMOS, Measure Theory (Van Nostrand, New York, 1950).
2. E.B. STECHEL (1985) J. Chem. Phys., 82 364.
3. M.V. BERRY (1977) J. Phys. A 10, 2083.
4. P.R. HALMOS, Lectures on Ergodic Theory (Chelsea, New York, 1956).
5. T.A. BRODY, J. FLORES, J.B. FRENCH, P.A. MELLO, A. PANDEY and S.S.M. WONG (1981) Rev. of Mod. Phys., 385.
6. E.J. HELLER (1980) J. Chem. Phys. 72 1337.
7. V.I. ARNOL'D and A. AVEZ, Ergodic Problems in Classical Mechanics (Benjamin, NY, 1968).
8. M.C. GUTZWILLER (1967) J. Math. Phys. 8 1979.
9. M.V. BERRY and K.E. MOUNT (1972) Rep. Prog. Phys. 35 315.
10. J.E. LYNN, The Theory of Nuetron Resonance Reactions (Clarendon, Oxford (1968).
11. E.P. WIGNER (1956) Proc. Conf. on Neutron Phys. by Time-of-Flight, Gatlingberg, Oak Ridge Nat'l Lab. Report No. ORNL-2309.
12. O. BOHIGAS and M.J. GIANNONI, Mathematical and Computational Methods in Nuclear Physics, J.S. Dehesa, J.M. Gomez and A. Polls (eds.), Lecture Notes in Physics 209 (Springer-Verlag, 1984).
13. C.E. PORTER and R.G. THOMAS (1956) Phys. Rev. 104 483; C.E. PORTER, Statistical Theories of Spectra: Fluctuations (Academic, New York, 1965).
14. E.B. STECHEL and E.J. HELLER (1984) Ann. Rev. Phys. Chem. 35 563.
15. E.B. STECHEL (1986) submitted to J. Chem. Phys.
16. E.J. HELLER and R.L. SUNDBERG, Proc. NATO Adv. Res. Workshop on Chaotic Behavior in Quantum Systems, Como, Italy (Springer-Verlag, 1984).
17. E.J. HELLER (1984) Phys. Rev. Letts. 53 1515.
18. E.B. STECHEL, in preparation.

A BOUND ON TIME-REVERSAL NONINVARIANCE FROM TRANSITION-STRENGTH FLUCTUATIONS

Steven Tomsovic
Department of Physics and Astronomy
University of Rochester
Rochester, New York 14627

Abstract

Transition strength fluctuation measures are calculated by random-matrix methods for time-reversal non-invariant (TRNI) Hamiltonians. We find that these measures are more sensitive to a TRNI admixing than the spectral fluctuation measures, but present experimental difficulties tend to offset this advantage. The strength measures, applied to the neutron resonance data, show that the local rms multiparticle H(TRNI) matrix element $\lesssim 0.07D$ with D being the average level spacing. Spectral averaging methods reduce this to a bound on the nucleon-nucleon TRNI vs. TRI interactions, which give for ^{169}Er, a 0.3% upper bound, which is quite close to the bound from spectral analysis.

In 1967, Wigner discussed the possibility of detecting TRNI in nuclear energy-level and strength fluctuations [1]. Subsequent work has led to a good understanding of the spectral fluctuations derivable from ensembles which allow for arbitrary TRNI admixings [2,3]. Via a study [4] of neutron and proton resonance-level data, an upper bound, 0.35% [3] has been recently determined for the relative magnitude of the TRNI nucleon-nucleon interaction. Here, we turn our attention to the strength problem.

The possibility of TRNI detection arises because the character of the strength fluctuations changes with TRNI admixing; for example, the locally renormalized strength density [5] has a χ_1^2 (Porter-Thomas) form when time-reversal invariance (TRI) is good (GOE being appropriate) and χ_2^2 when completely broken (GUE being appropriate). The TRI form arises from the fact that the strength is the square of a real transition matrix element which varies from eigenstate to eigenstate as a Gaussian random variable. In contrast, the TRNI strength is the absolute square of a matrix element which has independently varying real and imaginary parts. The essential transition in fluctuations is governed by the rate at which the eigenstates become complex as TRNI increases. The same parameter Λ, as found in [2,3], governs the transition in the fluctuations, $\Lambda^{1/2}$ being defined as the local rms matrix element expressed in terms of the local level spacing D(E): $\Lambda^{1/2}=|\langle E'|H(TRNI)|E\rangle|_{rms}/D(E)$.

For the purpose of finding the Λ-dependence of strength fluctuation measures, we introduce the interpolating ensembles [3] $\{H_\alpha:v^2\}=\{H(S:v^2)+i\alpha H(A:v^2\}$; H(S) and H(A) are respectively d-dimensional real symmetric and real antisymmetric matrices. The matrix elements are chosen such that {H(S)} is the GOE and {H(S)+iH(A)} is the GUE. The diagonalization of H(S) leaves {H(A)} invariant and therefore $\Lambda=\alpha^2 v^2/(D(E))^2$.

We now briefly describe the derivation of an approximate form for the strength density in the intermediate case. As in [6], the normalized strength can be written in terms of an overlap between a real vector $|\lambda\rangle$, defined in the d-dimensional statistical space, and the eigenstates, as $K_j=d|\langle\lambda|E_{\alpha:j}\rangle|^2$; the dimensionality factor normalizes the strength to unit mean. We next decompose the eigenstates into two real orthogonal component vectors $|R_{\alpha:j}\rangle$ and $|I_{\alpha:j}\rangle$,

$$|E_{\alpha:j}\rangle = e^{i\delta_j}\{t_j|R_{\alpha:j}\rangle + i\sqrt{(1-t_j^2)}|I_{\alpha:j}\rangle\} \tag{1}$$

where the phase is chosen so that $\langle R_{\alpha:j}|I_{\alpha:j}\rangle=0$. The invariance of the $\{H_\alpha\}$ under orthogonal transformation insures that the overlaps $\langle\lambda|R_{\alpha:j}\rangle$ and $\langle\lambda|I_{\alpha:j}\rangle$ vary across the ensemble, for asymptotic d, as independent Gaussian random variables. Then, for the fixed-t_j (or equivalently fixed $\theta_j=4t_j^2(1-t_j^2)$) conditional probability density,

$$\rho_{K_j}(z|\theta_j) = \frac{1}{2\pi}\int e^{-(x^2+y^2)/2}\,\delta\left(z-\left[t_j^2x^2+(1-t_j^2)y^2\right]\right)\,dxdy$$

$$= \theta_j^{-1/2}\,e^{-(z/\theta_j)}\,I_0\left(\sqrt{(1-\theta_j)}z/\theta_j\right) \tag{2}$$

where $I_0(x)$ is the Bessel function $J_0(ix)$. The derivation of the strength density is completed by the appropriately weighted θ_j integration. However, we have found by Monte Carlo study that for the amount of data available from neutron resonances, replacing θ_j by its ensemble average $(\overline{\theta}_j)$ is adequate; note that this is the leading term of an expansion of the density in terms of the cumulants of θ_j.

We complete the derivation of an approximate density by giving the first terms in a Λ-expansion for $\overline{\theta}_j$ or equivalently the strength variance by way of the relation $\mathrm{Var}(K_j)=2-\overline{\theta}_j$. Using perturbation theory (including degenerate theory for nearest neighbors) to expand the H_α eigenstates to the relevant order, we find $|R_{\alpha:j}\rangle=|E_{0:j}\rangle$ and $t_j=\langle E_{0:j}|E_{\alpha:j}\rangle$. $\overline{\theta}_j$ then appears in terms of an inverse-square energy-weighted sum with a damping term to take account of near-degeneracies. We use the fact that the function $R_2(s;\Lambda)$ (the probability density of finding a level at a distance r away from a given one) is the sum of nearest neighbor spacing densities. Then,

$$\overline{\theta}_j = \frac{8\Lambda}{\sqrt{2\pi}} \int_0^\infty ds\ R_2(s;0) \int_{-\infty}^\infty dw\ \frac{w^2}{s^2 + 4\Lambda w^2}\ e^{-w^2/2} + \dots$$

$$= \frac{2\pi^2\Lambda}{3}\left(2 - \gamma - \ln(2\pi^2\Lambda)\right) + \dots \tag{3}$$

where we have used Dyson's result [7] for the GOE two-level function $R_2(s;0)$ in the integration; γ is Euler's constant. Our Monte Carlo calculations have verified the accuracy of (3) for small Λ and completed the transition curve for larger values of Λ. We also mention that, unlike in the GOE and GUE, there are positive correlations between nearby strengths for intermediate values of Λ.

The strength density in the small-strength region is more sensitive to TRNI than any spectral or strength two-point measure. At present, experimental uncertainties tend to offset this advantage. As a practical matter, it will turn out that most interest will be in the strengths less than a few percent of their mean value. For the distribution function F(x), the sample error from p levels is $\{F(x)(1-F(x))/p\}^{1/2}$. The strength analysis shows that, if we ignore the smallest strengths, those $\le .004$ of the mean (5% of the total levels in the GOE), the data are consistent with GOE. Adopting a 3σ confidence limit we find an upper bound on $\Lambda^{1/2}$ of 0.07; see Figure 1.

The deficiency for the smallest strengths, from the GOE prediction, could be accounted for by experimental uncertainties, resulting in missed levels (because the levels are detected via their strengths). But this implies that the spectral analysis should be adjusted for these randomly missed states (energies and strengths being uncorrelated in the GOE). For example, the number variance $\Sigma^2(r)$ acquires a term equal to fr, where f is the fraction missed. However the close agreement of the NDE [4] with the GOE predictions is not consistent with more than about 2% of the levels missing. If we

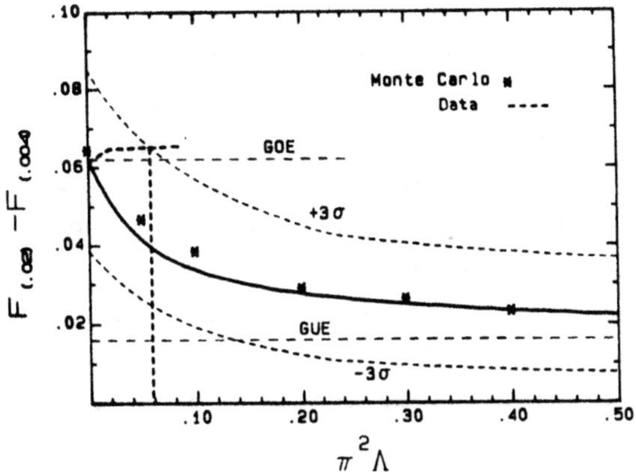

Fig. 1. The (small) strength distribution function vs. $\pi^2\Lambda$. The intersection of the data line (NDE [4], containing 21 sequences with p=1182) line with the $+3\sigma$ curve locates the upper bound. Also shown are the Monte Carlo points (*) which agree well with the theory (drawn as the solid line).

accept the lower percentage (2% as opposed to 5%), which seems natural, two interesting options immediately present themselves to explain the remaining deficiency: (i) there may be nuclear structure or non-statistical effects to be studied, and/or (ii) there may be a small TRI breaking. To examine the second possibility, we assume the first can be neglected (of course this may not be appropriate). Then for example, it is shown in figure 2 that the strength data agrees well with the calculated distribution function when the strengths less than 0.001 of the mean are excluded (2.5% of the total in the GOE) and $\Lambda^{1/2}$ is taken to be 0.04.

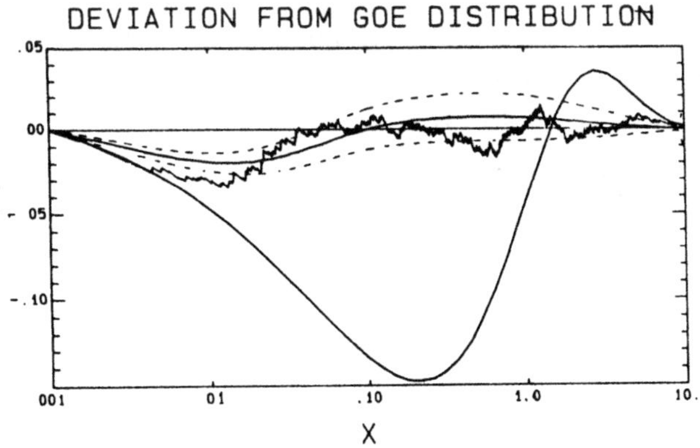

Figure 2. The deviations, from GOE, of the distribution functions for GUE, $\pi^2\Lambda$=.015 and data, each adjusted for a cutoff of the strengths below .001 of the mean. $\pm 1\sigma$ sample error curves are shown for the $\pi^2\Lambda$=.015 curve.

The spectral averaging calculation which reduces this bound on $\Lambda^{1/2}$ to a statement about the TRNI nucleon-nucleon interaction is discussed elsewhere in this conference. We just state that for ^{169}Er a bound of 0.3% on the relative norm of the TRNI interaction is implied by our analysis. Also note that the value of Λ mentioned in Figure 2 implies (if we ascribe the deficiency completely to TRNI) the value 0.15% for the relative norm of the TRNI interaction.

This work was part of a larger collaborative effort including J. B. French, V. K. B. Kota, and A. Pandey. We would also like to acknowledge R. U. Haq for helpful discussions concerning the NDE.

References:

1. E. P. Wigner, SIAM Rev. **9**, 1 (1967).
2. A. Pandey and M. L. Mehta, Commun. Math. Phys. **87**, 449 (1983); M. L. Mehta and A. Pandey, J. Phys. A **16**, 2655 (1983).
3. J. B. French, V. K. B. Kota, A. Pandey, and S. Tomsovic, Phys. Rev. Lett. **54**, 2313 (1985), and to be published.
4. R. U. Haq, A. Pandey, and O. Bohigas, Phys. Rev. Lett. **48**, 1086 (1982); O. Bohigas, R. U. Haq, and A. Pandey, in "Nuclear Data for Science and Technology", edited by K. H. Bochoff (Reidel, Dordrecht, 1983).
5. C. E. Porter and R. G. Thomas, Phys. Rev. **104**, 483 (1956); J. M. C. Scott, Philos. Mag. **45**, 1322 (1954).
6. T. A. Brody, J. Flores, J. B. French, P. A. Mello, A. Pandey, and S. S. M. Wong, Rev. Mod. Phys. **53** 385 (1981).
7. F. J. Dyson, J. Math. Phys. **3**, 166 (1962).

SEMICLASSICAL ENERGY SPECTRUM OF QUASI-INTEGRABLE SYSTEMS

A.M. Ozorio de Almeida+
Instituto de Física "Gleb Wataghin"
Universidade Estadual de Campinas
13100 - Campinas - S.P. - BRASIL

1. Introduction

The energy eigenvalues of a classically integrable system are well known in the semiclassical limit. They are governed by Bohr-Sommerfeld quantization in one dimensional motion, or its natural generalization - 'torus' quantization ('EBK' quantization) in two or more dimensions [1]. This selects out, from the continuum of invariant tori in phase space around which trajectories wind, a discrete set to represent quantum eigenstates.

Typical systems of more than one dimension are not integrable - trajectories need not lie on tori, so torus quantization cannot generally be used. Hence the importance of a substitute prescription for the energy levels E_j, or more precisely for the density of states function

$$n(E) = \sum_j \delta(E-E_j) \qquad (1)$$

due to Gutzwiller [2]. This is based on elements common to integrable and non-integrable systems, namely the infinitely numerous periodic orbits which close and retrace themselves. The prescription is that the density of states function is expressed as a 'periodic orbit sum', where each orbit gives a definite oscillatory contribution.

In integrable systems the periodic orbits come in continuous families making up 'periodic tori' (also known as resonant tori). The equivalence of the periodic orbit sum with torus quantization was shown in this case by Berry and Tabor [3,4]. The slighest perturbation of integrability destroys these families, even if it preserves most of the invariant tori, so that the periodic orbits of typical non-integrable systems are isolated loops. Isolated periodic orbits fall into

+ Research supported by FAPESP, CNPq and FINEP (Brazilian government agencies).

two classes, unstable and stable according to whether the linearized flow around then is hyperbolic or elliptic. Ergodic systems, where almost all orbits explore the whole energy surface, have only unstable periodic orbits. The original periodic orbit sum for the density of states [2] is quite explicit for unstable orbits and therefore applies directly to ergodic systems. Non-integrable, non-ergodic systems also have stable periodic orbits. These bifurcate as the energy is varied, leading to singularities in the amplitude of the orbit contribution in the original formula at the corresponding resonant energy.

Ozorio de Almeida and Hannay [5] have derived the resonant orbit amplitudes for all the generic bifurcations of stable periodic orbits of Hamiltonian systems of two freedoms. This is done by including the non-linear terms, essential for the description of the bifurcation in the Poincaré map (generated by successive travesarls by trajectories neighbouring the periodic orbit of a surface transverse to the orbit). Integrable systems are not generic in the class of Hamiltonian systems, neither are the bifurcations in which pairs of stable and unstable orbits become a 'periodic torus'. None the less the study of quasi-integrable systems (also known as KAM systems) is of fundamental importance to our understanding of general non-integrable behaviour.

This paper presents the general method for deriving transitional periodic orbit amplitudes near a bifurcation in its simplest instance - the quasi-integrable system. The main qualitative result is that, for sufficiently small perturbations, the semiclassical amplitude smooths over the details of the classical motion - the amplitude may be given correctly by the Berry-Tabor formula on a 'quasi-torus' even when the periodic torus has ceased to exist. The concluding Section analyses the equivalence of torus quantization and the periodic orbit sum. This results in criteria for the validity of applying the former method to quasi-integrable systems.

2. Amplitudes and the Non-linear Poincaré Map

Semiclassically the density of states reduces to a sum over the periodic orbits of the system and a zero time contribution [1]

$$\bar{n}(E) = (2\pi\hbar)^{-2} \int d\underset{\sim}{q}\, d\underset{\sim}{p}\; \delta\{E-H(\underset{\sim}{p},\underset{\sim}{q})\} \quad , \qquad (2)$$

where $\underset{\sim}{q} = (q_1,q_2)$ are the coordinates of the state and $\underset{\sim}{p} = (p_1,p_2)$ are the conjugate momenta. The contribution of a periodic orbit of period τ, after m repetitions (m is a positive or negative integer) is

$$(2\pi i\hbar)^{-3/2}\; \tau\, A_m \exp\,[im(\hbar^{-1}\sigma-\mu\pi/2)] \quad , \qquad (3)$$

where

$$\sigma = \int \underset{\sim}{p}\cdot d\underset{\sim}{q} \qquad (4)$$

is the action once around the orbit and μ is the number of focal points along the orbit [2]. Supposing that the conjugate coordinates of the Poincaré section are (q_1,p_1) and that the generating function for the m'th iteration of the Poincaré map is $S_m(p_1,q_{1m})$, i.e.

$$q_1 = \frac{\partial S_m}{\partial p_1} \quad , \qquad p_{1m} = \frac{\partial S_m}{\partial q_{1m}} \quad , \qquad (5)$$

the non-linear orbit amplitude A_m has the form [5]

$$A_m=(2\pi\hbar i)^{-1/2} \int dp_1 dq_1 \left|\frac{\partial^2 S_m}{\partial p_1 \partial q_{1m}}\right|^{1/2} \exp\{-i\hbar^{-1}[S_m(p_1,q_{1m})-p_1q_{1m}]\} \,. \qquad (6)$$

If polar coordinates are used in the Poincaré section or more general action-angle variables (I_1,φ_1), the corresponding Green-function [4] leads to the amplitude [5]

$$A_m=(2\pi\hbar i)^{-1/2} \sum_n \int dI_1 d\varphi_1 \left|\frac{\partial^2 S_m}{\partial I_1 \partial \varphi_1}\right|^{1/2} \exp\{-i\hbar^{-1}[S_m(I_1,\varphi_1)-I_1(\varphi_1+2m\pi)]\}. \qquad (7)$$

In this case the m'th iteration of the Poincaré map is given implicitly by

$$\varphi_1 = \frac{\partial S_m}{\partial I_1}(I_1,\varphi_{1m}) \quad , \qquad I_{1m} = \frac{\partial S_m}{\partial \varphi_{1m}}(I_1,\varphi_{1m}) \,. \qquad (8)$$

The phase functions in (6) and (7) have the wonderful property that their stationary points, satisfying

$$\frac{\partial}{\partial p_1}[S_m(p_1,q_1)-p_1q_1] = \frac{\partial}{\partial q_1}[S_m(p_1,q_1)-p_1q_1] = 0 \quad , \quad (9)$$

correspond to fixed points of the Poincaré map (i.e. periodic orbits), which is evident by comparison with (5). In action-angle variables the end point $I_1 = 0$ is also a periodic orbit. The phase function admitts to separate local approximations around each stationary point, when the periodic orbits are separated by many oscillations. Consider for instance an isolated fixed point at the origin of (p_1,q_1). The quadratic approximation of S_m, corresponding to a linear approximation of the Poincaré map, leads to the <u>stationary phase approximation</u> of (6)

$$A_m = [2\pi\hbar \left|\frac{\partial^2 S_m}{\partial p_1 \partial q_1}\right| \, |\det \underset{\sim}{\Delta}_m^{-1}|]^{1/2} \, e^{-i(\nu-1)\pi/4} \quad , \quad (10)$$

where we define the <u>Hessian matrix</u>

$$\underset{\sim}{\Delta}_m = \begin{vmatrix} \frac{\partial^2}{\partial p_1^2}(S_m-p_1q_1) & \frac{\partial^2}{\partial p_1\partial q_1}(S_m-p_1q_1) \\ \frac{\partial^2}{\partial p_1\partial q_1}(S_m-p_1q_1) & \frac{\partial^2}{\partial q_1^2}(S_m-p_1q_1) \end{vmatrix} \quad (11)$$

and ν is the signature of $\underset{\sim}{\Delta}_m$, i.e. the number of positive eigenvalues minus the number of negative ones. Introducing the <u>stability matrix</u> of the linearized Pincaré map (or the <u>monodromy matrix</u>)

$$\begin{vmatrix} p_1' \\ q_1' \end{vmatrix} = \underset{\sim}{M} \begin{vmatrix} p_1 \\ q_1 \end{vmatrix} \quad , \quad (12)$$

and using the identity

$$\det(\underset{\sim}{M}^m - \underset{\sim}{1}) = \left(\frac{\partial^2 S_m}{\partial p_1 \partial q_1}\right)^{-1} \det \underset{\sim}{\Delta}_m \quad , \quad (13)$$

the amplitude takes the simple form

$$|A_m|^2 = 2\pi\hbar \, |\det(\underset{\sim}{M}^m - \underset{\sim}{1})|^{-1} \quad . \quad (14)$$

This is just the Gutzwiller result. If $\underset{\sim}{M}$ is unstable it has eigenvalues exp $(\pm\,\beta)$, so that

$$\det\,(\underset{\sim}{M}^m - 1) = \sinh^2(m\beta/2) \quad , \tag{15}$$

whereas for a stable orbit

$$\det\,(\underset{\sim}{M}^m - \underset{\sim}{1}) = \sin^2(m\alpha/2) \quad . \tag{16}$$

Reference [5] furnishes a criterion for how close the stability angle α can be to $2\pi/m$ in the stable case, for the Gutzwiller amplitude to be used. In the present case of the breaking of integrability, both α and β are small.

3. The Reduced Hamiltonian

In general it is always possible to derive a reduced Hamiltonian, such that the Poincaré map in the neighbourhood of a fixed point is obtained by evolving the reduced system for a time τ, even if the map cannot be linearized. The situation is specially simple in the action-angle coordinates of a neighbouring integrable system (not the case treated in [5]). The full Hamiltonian is then

$$\underset{\sim}{H}(\underset{\sim}{J},\underset{\sim}{\theta}) = H_o(\underset{\sim}{J}) + \varepsilon\, H_1(\underset{\sim}{J},\underset{\sim}{\theta}) \tag{17}$$

and the coordinates of the Poincaré section of energy E are (J_1,θ_1). Suppose there is a periodic orbit passing through (J_1^*,θ_1^*), i.e. this is either a fixed point or a periodic point of the Poincaré map. To use the formalism in the preceeding section, it is necessary that they appear as fixed points. This is achieved by a canonical transformation which cancels the motion of the family of periodic orbits with $\underset{\sim}{J} = \underset{\sim}{J}^*$ for $\varepsilon = 0$. Take the family whose orbits close after n_1 circuits of θ_1 and n_2 circuits of θ_2. The requirement is then that the new velocity

$$\dot{\varphi}_1 = n_1\dot{\theta}_1 - n_2\dot{\theta}_2 = 0 \quad . \tag{18}$$

The corresponding canonical transformation is not unique. One choice is

$$\begin{aligned} \varphi_1 &= n_1\theta_1 - n_2\theta_2 \qquad & I_1 &= J_1/n_1 \\ \varphi_2 &= \theta_2 & I_2 &= J_2 + n_2 J_1/n_1 \end{aligned} \tag{19}$$

The full Hamiltonian is now

$$H'(\underset{\sim}{I},\underset{\sim}{\varphi}) = H_0'(\underset{\sim}{I}) + \varepsilon H_1'(\underset{\sim}{I},\underset{\sim}{\varphi}) \quad , \tag{20}$$

with

$$\partial H_0'(\underset{\sim}{I}^*)/\partial I_1 = 0 \quad , \tag{21}$$

where $\underset{\sim}{I}^*(\underset{\sim}{J}^*)$ is given by (19).

Since we are dealing with periodic orbits rather than points of equilibrium $\partial H_0'(\underset{\sim}{I}^*)/\partial I_2 \neq 0$ and therefore, for sufficiently small ε, $\partial H'/\partial I_2 \neq 0$ in a neighbourhood of the torus with actions $\underset{\sim}{I}$. Thus in this region we can define the function

$$I_2 = I_2(I_1,\varphi_1, -\varphi_2;E) \quad . \tag{22}$$

According to standard theory [6 § 45], I_2 is the <u>reduced Hamiltonian</u> for the flow in the (I_1,φ_1) plane, i.e.

$$\frac{d\,I_1}{d(-\varphi_2)} = -\frac{\partial I_2}{\partial \varphi_1} \quad , \quad \frac{d\varphi_1}{d(-\varphi_2)} = \frac{\partial I_2}{\partial I_1} \quad . \tag{23}$$

Note that in the $(\underset{\sim}{I},\underset{\sim}{\varphi})$ coordinates the stroboscopic 'time' $(-\varphi_2) = -2\pi$ map of the reduced flow coincides with the Poincaré map of the full flow.

The reduced flow of the unperturbed integrable system is just a horizontal shear about the fixed line $I_1 = I_1^*$. In this case the exact relation between the generating function of the Poincaré map and the reduced Hamiltonian is

$$-m2\pi I_{20} = I_1\varphi_{1m} - S_m(I_1,\varphi_{1m}) \quad . \tag{24}$$

In other words, the reduced Hamiltonian is proportional to the phase function in the integral for the amplitude. For the quasi-integrable system to first order in ε (24) generalizes to

$$\int_0^{-m2\pi} I_2(I_1,\varphi_{1m}+\omega_0\varphi_2,\varphi_2;E)d\varphi_2 = I_1\varphi_{1m} - S_m(I_1,\varphi_{1m}) \quad , \tag{25}$$

where $\omega_0 = \partial I_{20}/\partial I_1$. The stationary points of this phase function coincide with the fixed points of the Poincaré map only to first order in ε. The perturbed Hamiltonian (20) is periodic in the angles $\underset{\sim}{\varphi}$; so the phase function can be written in the form

$$I_1\varphi_{1m}-S_m(I_1,\varphi_{1m}) = -m2\pi\ I_{20}(I_1) + \varepsilon\ F(I_1,\varphi_{1m}) \quad . \quad (26)$$

4. Perturbations of the Berry-Tabor Amplitude

The origin of the (I_1,φ_1) plane corresponds to an isolated periodic orbit. For an integrable system there will be a whole family of periodic orbits with $I_1 = I_1^*$. The difference in the action σ, defined in (4), between the isolated orbit and the family, is exactly $-m2\pi\ I_{20}(I_1^*)$. In order not to count this action twice in the correct phase for the family, we have to subtract the constant reduced energy $I_{20}(I_1^*)$ from the phase function (24), to evaluate the amplitude (7) as

$$A_m=(-2\pi i/\hbar)^{1/2} \int_0^\infty dI_1\ \exp\{-i\hbar^{-1}\ m2\pi[I_{20}(I_1)-I_{20}(I_1^*)]\} \quad . \quad (27)$$

If I_1^* is sufficiently far from the origin, we can integrate (27) by stationary phase, by which

$$A_m = \left|\frac{m}{2\pi}\ \frac{\partial^2 I_{20}}{\partial I_1^2}(I_1^*)\right|^{-1/2} e^{-i(\nu-1)\pi/4} \quad , \quad (28)$$

where $\nu = \text{sign}\ (\partial^2 I_{20}/\partial I_1^2)$. The case where the periodic orbit torus cannot be separated from the isolated periodic orbit was first treated by Richens [7] and is also discussed in [5]. The full periodic torus contribution to the Berry-Tabor sum looks much simpler than the original formula [3], because of the preparatory canonical transformation (19).

The simplest perturbation to consider is the 'Chirikov resonance' [8], where only two Fourier coefficients of $F(I,\varphi_{1m})$ are non-zero and independent of I_1. With an appropriate change in the origin of φ_1, the phase function to be used in (7) becomes

$$I_1\varphi_1-S_m(I_1,\varphi_1)=-m2\pi[I_{20}(I_1)-I_{20}(I_1^*)]+\varepsilon\cos\varphi_1 \quad . \quad (29)$$

The φ_1 integral in (7) can be performed immediately, yielding

$$A_m=(-2\pi i/\hbar)J_0(\hbar^{-1}\varepsilon)\ \int dI_1\ \exp\{-i\hbar^{-1}m2\pi[I_{20}(I_1)-I_{20}(I_1^*)]\} \ , \quad (30)$$

(where $J_0(x)$ is a zero order Bessel function [9]) if we neglect the ε- dependent part of $\partial^2 S/\partial I_1 \partial\varphi_1$. Integrating by stationary phase

$$A_m = \left|\frac{m}{2\pi}\frac{\partial^2 I_{20}}{\partial I_1^2}(I_1^*)\right|^{-1/2} e^{-i(\nu-1)\pi/4} J_o(\hbar^{-1}\varepsilon) \qquad (31)$$

For small arguments

$$J_o(x) \sim 1-x^2/2 \qquad , \qquad (32)$$

so that, for $|\hbar^{-1}\varepsilon| \ll 1$, the amplitude has the Berry-Tabor form (28), with the average of I_2 along the unperturbed periodic orbit in the place of the constant I_{2o}. On the other hand, for large arguments

$$J_o(x) \sim \sqrt{\frac{2}{\pi x}} \cos (x-\pi/4) \qquad . \qquad (33)$$

Therefore, if $|\hbar^{-1}\varepsilon| \gg 1$,

$$A_m = \left|m\hbar^{-1}\varepsilon \frac{\partial^2 I_{20}}{\partial I_1^2}(I_1^*)\right|^{1/2} (e^{i(\hbar^{-1}\varepsilon-(\nu+2)\pi/4}+e^{-i(\hbar^{-1}\varepsilon+\nu\pi/4)}) \quad , \qquad (34)$$

which coincides with the Gutzwiller amplitudes discussed in section 2 for the pair of stable and unstable orbits, into which the periodic orbit family breaks up.

For general resonances, where the phase function $F(I_1,\varphi_1)$ is an arbitrary periodic function with non-degenerate maxima and minima, we first integrate (7) over I, using the method of stationary phase to obtain

$$A_m = \frac{1}{2\pi}\int_0^{2\pi} d\varphi_1 \left|\frac{m}{2\pi}\frac{\partial^2 I_{20}}{\partial I_1^2}(I_1^*(\varphi_1))\right|^{-1/2} e^{-i(\nu-1)\pi/4} e^{-i\hbar^{-1}\varepsilon F(I_1^*(\varphi_1),\varphi_1)} \ . \qquad (35)$$

The integral has the same form as (A1) in the Appendix, where a uniform approximation moddled on Bessel functions is derived, which generalizes the treatment of the Chirikov resonance.

The important parameters are the difference in the phase function $\Delta F = m2\pi\Delta I_2$ between each successive pair of stationary points (2ε in the Chirikov resonance). If they are all much greater than the wavelength $\hbar$, the result is equivalent to stationary phase integration of (35), leading to Gutzwiller amplitudes. In the opposite extreme $\varepsilon\hbar^{-1}F(I^*(\varphi_1),\varphi_1)$ will be small for all φ_1. The uniform result is then the same as that of neglecting the phase function in (35). The general criterion which establishes whether the isolated periodic orbits in the quasi-integrable system contribute separately or together as a <u>quasi-torus</u> is

$$m2\pi\ \Delta I_2 \ll \hbar \qquad \text{(quasi-torus)} \qquad (36)$$

$$m2\pi\ \Delta I_2 \gg \hbar \qquad \text{(isolated orbits)} \quad . \qquad (37)$$

In the semiclassical limit $\hbar \to 0$, all isolated orbits contribute separately. But in practice $\hbar$ is fixed though small. It then happens that, since ΔI_2 diminishes with the order of the resonance (the highest common divisor of n_1 and n_2 in section 3), there will only be a <u>finite</u> number of isolated periodic orbits in the density of states formula and an <u>infinite</u> number of quasi-tori giving Berry-Tabor contributions.

The above discussion applies only if we don't take the limit of large winding numbers $m \to \infty$. The problem is that the averaging of the Hamiltonian (35) is then no longer valid. It is clear, however, that in this limit the orbits contribute as isolated orbits - the long 'time' movement in a resonant region can in no way be approximated by a simple shear, but the amplitude of the stable orbit may have to be corrected in accordance with reference [5].

5. Conclusions

The perturbation of an integrable system generally breaks up its periodic orbit families into pairs of isolated orbits. However, the linearization around them, which leads to Gutzwiller amplitudes in the periodic sum, is valid only if the perturbation is sufficiently large. Otherwise the uniform approximations, derived in the previous section, have to be used. These are valid in the combined limit as $\varepsilon \to 0$ and $\hbar \to 0$, though the limit of $\varepsilon\hbar^{-1}$ be large or small. If $\varepsilon\hbar^{-1}$ is small enough, the isolated periodic orbits have a combined amplitude indistinguishable from the unperturbed torus - the quasi-torus. In other words, the deduction is that the semiclassical density of states is insensitive to some of the intrincate details of the classical motion; a result which had already been conjuctured [10,11]. In retrospect it justifies the use of an averaged reduced Hamiltonian in the phase function, which smooths over the internal structure of the resonant region. If the resonance is sufficiently weak, the secondary resonances will contribute as quasi-tori of very long periods. Otherwise the primary stable and unstable orbits contribute separately and we must apply perturbation theory specifically to the secondary resonances. The general scheme is that it is only for pairs of orbits that don't satisfy one of the inequalities (36) or (37) that the amplitude needs to be interpolated by a uniform approximation between the Gutzwiller and the Berry-Tabor formulae.

So far there have been few attempts to use the periodic orbit sum directly to calculate the density of states [12]. On the other hand there have been calculations of the spectrum of quasi-integrable systems using torus quantization on 'a good integrable approximation' of the system [13,14]. We are now in a position to discuss the validity of this procedure:

If all the resonances in the system satisfy (36), the periodic sum is the same as that of the unperturbed system. The equivalence of torus quantization and the periodic orbit sum derived by Berry and Tabor then implies that torus quantization is valid.

This attractive conclusion must be further scrutinized. First it must be remembered that equivalence was only proved for the quantization of a single family of tori. It the integrable approximation contains separatrices, it may be necessary to include tunnelling corrections [15].

A more serious problem is the dependence of (26) on the winding number m. After a large number of repetitions even small resonances cease to contribute as a quasi-torus. The importance of these terms in the periodic orbit sum is assessed by the following qualitative argument for the Berry-Tabor equivalence:

In general the quantized torus need not coincide with one of the periodic tori. The quantum conditions are exactly that any closed circuit on it have an action which is a multiple of 2π (in units of $\hbar$) minus the Maslov indices. If such a circuits were periodic orbits they would contribute in phase to the periodic sum. The quantized torus may be approximated arbitrarily well by periodic tori of the same energy, in the same way as any number may be approximated by a sequence of rationals. The contribution of all these infinite families of periodic orbits in the integrable system are good approximations of closed circuits on the quantized torus and are all almost in phase, leading to a peak in the density of states. In the absence of a quantized torus, neighbouring periodic tori with very different winding numbers contribute to the sum with arbitrary phases, which cancel out on the average. The role of a quantized torus is thus to organize the coherent contribution of its periodic neighbours.

Returning to the discussion of iterated windings of the periodic orbits we see that, if $\Delta\sigma$ is the difference in phase between the periodic orbit and a circuit with the same topology on the quantized torus, the difference after m windings will be $m\Delta\sigma$. The higher the winding of the periodic torus, the worse is its coherence with the circuits of the quantized torus. The peaks in the density of states are therefore brought about by the low windings of the neighbours to quantized tori. The inequality (36) with m ommitted can thus be used as a safe criterion for torus quantization of quasi-integrable systems. It is important to note that in the case where the 'Chirikov resonance' is obtained by averaging the reduced Hamiltonian I_2 over φ_2, this criterion excludes from torus quantization systems where the islands are big enough to contain quantum states of their own, in agreement with intuition.

The heuristic argument for the equivalence of torus quantization and the periodic orbit sum is local, justifying the use of the former in regions where (36) is valid, even when it does not hold everywhere. For weak violations of (36) the isolated periodic orbits will still contribute approximately in phase near a quantized torus, which makes an isolated contribution to the density of states (as observed by Heller and Sundberg [16] and by Reinhardt [17], who interprets this fact in terms of 'fragmented tori' or cantori [18]).

It is only if the quantized torus or cantorus is very close to a low order resonance (n_1 and n_2 in (18) are small)) that torus quantization completely breaks down. As argued by Berry [19] on the basis of the periodic orbit sum, this is a region of clustering in the density of states. The average position of the cluster may be close to that of the unperturbed system, but the internal distributions will be different, displaying the multiple avoided crossings studied by Marcus and co-workers in a series of papers [19].

APPENDIX

We require a uniform approximation for the integral

$$\frac{1}{2\pi}\int_0^{2\pi} d\varphi\; g(\varphi)\; e^{if(\varphi)} \quad , \tag{A1}$$

where g and f are periodic with period 2π. Suppose that $f(\varphi)$ has n maxima φ_j^+ and n minima φ_j^-. Then we split (A1) into n integrals of the form

$$\frac{1}{2\pi}\int_0^{\varphi_j^- - \varphi_j^+} d\varphi\; g(\varphi+\varphi_j^+)\; e^{if(\varphi+\varphi_j^+)}$$

and n integrals from 0 to $\varphi_{j+1}^+ - \varphi_j^-$. Each one of these can be moddled on one of the integrals [9]

$$\frac{1}{\pi}\int_0^{\pi} d\theta\; e^{ix\cos\theta} = \frac{1}{\pi}\int_{\pi}^{2\pi} d\theta\; e^{ix\cos\theta} = J_0(x) \quad . \tag{A3}$$

The way to do this is similar to that of generic 'diffraction catastrophe integrals' [21], though (A3) is not one of them. First we define the mapping $\theta \to \varphi = \varphi(\theta)$ in the interval $(0, \varphi_j^- - \varphi_j^+)$ by the implicit equation

$$f(\varphi + \varphi_j^+) = \langle f\rangle_j^+ + \Delta f_j^+ \cos\theta \quad , \tag{A4}$$

where

$$\langle f\rangle_j^+ = \frac{f(\varphi_j^+) + f(\varphi_j^-)}{2} \tag{A5}$$

and

$$\Delta f_j^+ = \frac{f(\varphi_j^+) - f(\varphi_j^-)}{2} \tag{A6}$$

The Jacobian of the mapping

$$|d\varphi/d\theta| = |\Delta f_j^+ \sin\theta / f'(\varphi+\varphi_j^+)| \tag{A7}$$

is nonsingular even at zero and $\varphi_j^- - \varphi_j^+$, since both these stationary points of $f(\varphi+\varphi_j^+)$ coincide with zeroes of sin θ. Thus (A2) is exactly equivalent to

$$\frac{e^{i\langle f\rangle_j^+}}{2\pi}\int_0^\pi d\theta\; h(\theta)\; e^{i\Delta f_j^+ \cos\theta} \quad , \tag{A8}$$

where

$$h(\theta) = g(\varphi(\theta) + \varphi_j^+)\;|d\varphi/d\theta| \quad . \tag{A9}$$

If Δf_j^+ is large, the integral will be dominated by the stationary end-points θ = 0 and θ = π. Then we only need to known h(θ) at these two points. On the other hand, for small Δf_j^+ we can neglect the exponential and the integral will be just the average value of h(θ). The calculation is simplified if

$$\int_0^\pi d\theta\; h(\theta) = \frac{h(0) + h(\pi)}{2} = \langle h\rangle_j^+ \quad . \tag{A10}$$

It is then possible to fit h(θ) to

$$\langle h\rangle_j^+ + \Delta h_j^+ \cos\theta \quad , \tag{A11}$$

where

$$\Delta h_j^+ = \frac{h(0) - h(\pi)}{2} \quad . \tag{A12}$$

The substitution of (A11) for h(θ) in the integral will result in a good approximation for Δf_j^+ large or small. At θ = 0 and π, the expansion to second order of f(φ) and cos θ determine

$$h\begin{vmatrix}0\\ \pi\end{vmatrix} = g(\varphi_j^\pm)\;|\Delta f_j/f''(\varphi_j^\pm)|^{1/2} \quad . \tag{A13}$$

So, using the generalization of (A3) [9]:

$$J_n(x) = (\pi i^n)^{-1}\int_0^\pi e^{ix\cos\theta}\cos n\theta\; d\theta \quad , \tag{A14}$$

we obtain the uniform approximation to (A2)

$$e^{i\langle f\rangle_j^+}\;\{\langle h\rangle_j^+ J_0(\Delta f_j^+) + i\Delta h_j^+(\Delta f_j^+)\}/2 \quad . \tag{A15}$$

Collecting all the n results for the intervals $(\varphi_j^+,\varphi_j^-)$ and the integrals for the n intervals $(\varphi_j^-,\varphi_{j+1}^+)$ for which the variables $\langle f\rangle_j^-$, Δf_j^-, $\langle h\rangle_j^-$, Δh_j^- are analogously defined, the full integral (A1) is then

$$\frac{1}{2\pi}\int_0^{2\pi} d\varphi g(\varphi)\, e^{if(\varphi)} = \sum_{\substack{+\,-\\ j=1}}^{n} e^{i\langle f\rangle_j^{\pm}} \{\langle h\rangle_j^{\pm} J_o(\Delta f_j^{\pm}) + \frac{i\Delta h_j^{\pm}}{2} J_1(\Delta f_j^{\pm})\, e^{i\langle f\rangle_j^{\pm}} \quad . \tag{A16}$$

It can be verified that, with the asymptotic formulae of J_o and J_1 for large argument, we recover the stationary phase evaluation of the integral.

The situation is more complicated if the two definitions (A10) of the mean don't coincide. For $\Delta f_j^{\pm} \ll 1$ we need the integral, whereas the average of $h(0)$ and $h(\pi)$ should be used for $\Delta f_j^{\pm} \gg 1$. If the difference is slight, it may be practicable to smoothly interpolate between the two. A more rigorous though cumbersome alternative is to fit $h(\theta)$ to

$$A + B\cos\theta + C\cos\nu\theta \quad , \tag{A17}$$

instead of (A11), with ν a real number in (0,1). It is then possible to fit $\varphi(0)$, $\varphi(\pi)$ and the integral in (A10) to (A17). Equation (A14) for J_n is correct for real ν to order ν, so that, using ν arbitrarily small, we obtain a uniform approximation in terms of J_o, J_1 and J_ν, valid for any $g(\varphi)$.

REFERENCES

[1] M.V. Berry, in Les Houches, Session XXXVI - Chaotic behaviour of Deterministic Systems, eds. G. Ioss, R.G.H.Helleman and R. Stora (North Holland, Amsterdam, 1983) 171.

[2] M.C. Gutzwiller, J. Math. Phys. 12 (1971) 343.

[3] M.V. Berry and M. Tabor, Proc. Roy. Soc. A349 (1976) 101.

[4] M.V. Berry and M. Tabor, J. Phys. A: 10 (1977) 371.

[5] A.M. Ozorio de Almeida and J.H. Hannay, J. Phys. A (1986) (submitted).

[6] V.I. Arnold, Mathematical Methods of Classical Dynamics (Springer, New York, 1978).

[7] P. Richens, J. Phys. A: 15 (1982) 2101.

[8] B.V. Chirikov, Phys. Reports 52 (1979) 265.

[9] H. Abramowitz and I.A. Stegun, Handbook of Mathematical Functions (Dover, New York, 1965).

[10] G.M. Zaslavsky, Phys. Reports 80 (1981) 157.
[11] M.V. Berry, N.L. Balazs, M. Tabor and A. Voros, Ann. Phys. N.Y. 122 (1979) 26.
[12] M.C. Gutzwiller, Phys. Rev. Letters 45 (1980) 150.
[13] R.T. Swimm and J.B. Delos, J. Chem. Phys. 71 (1979) 1706.
[14] W.P. Reinhardt and D. Ferelelly, J. Physique (Colloque) 43 (1982) C2-29.
[15] A.M. Ozorio de Almeida, J. Phys. Chem. 88 (1984) 6139.
[16] E.J. Heller and R.L. Sundberg, in Chaotic Behaviour in Quantum Systems, ed. G. Casati (Plenum, New York, 1985) 255.
[17] W.P. Reinhardt, in Chaotic Behaviour in Quantum Systems, ed. G. Casati (Plenum, New York, 1985) 235.
[18] R.S. Mackay, J.D. Meiss and I.C. Percival, Physics 13D (1984) 55.
[19] M.V. Berry, in Chaotic Behaviour in Quantum Systems, ed. G. Casati (Plenum, New York, 1985) 123.
[20] T. Uzer, D.W. Noid and R.A. Marcus, J. Chem. Phys. 79 (1983) 4412.
[21] M.V. Berry, Adv. Phys. 25 (1976) 1.

Localization in the Quantum Description of the Periodically Perturbed Rotor

R. Blümel

Technical University, München, FRG

S. Fishman and M. Griniasti

Dept. of Physics, the Technion, Haifa, Israel

and

U. Smilansky

Dept. of Nuclear Physics, the Weizmann Inst. Rehovot, Israel

Abstract: In this paper we present some recent results concerning localization phenomena in the quantum dynamics of the periodically perturbed rotor. We discuss the response of a planar rotor and of a diatomic molecule to a periodic train of smooth and finite field pulses and show that both cases correspond to an Anderson model on a finite grid. The second topic is the study of the localization properties for the δ-kicked rotor when the kicking strength is large, using the transfer matrix technique.

I. Introduction

Inspite of some great progress in our understanding of the quantum dynamics of the kicked rotor, [1–10] the picture we have is still not complete. To set the scene for the ensuing discussion we shall summarize in the following lines some of the established properties of this simple system. The time dependent Hamiltonian is

$$H = \frac{\hbar^2 n^2}{2I} + \hat{k} cos\theta \sum_m \delta(t - mT) \tag{1.1}$$

and the one cycle propagator is

$$U_{nm} = e^{-\frac{i}{2}n^2\tau}(-i)^{n-m} J_{n-m}(k) \tag{1.2}$$

where $\tau = \frac{\hbar T}{I}$ and $k = \hat{k}/\hbar$.

We shall denote by $| \alpha >$ and ω_α the eigenvectors and eigenvalues of the quasi-energy (q.e.) operator G where $U = e^{iG}$. These quasi-energy states have the following properties:

For τ values which are rationally related to 4π the q.e. operator has a regular continuous component in the spectrum with extended eigenstates. [4] As a consequence, the expectation value of the rotor energy $\frac{\hbar^2 <n^2>}{2I}$ increases *quadratically* in the number of applied pulses. This is a typical quantum resonance situation.

Recently, it was shown [5] that there exists a set (of measure zero) of non-resonant τ values for which the q.e. operator has a continuous component, albeit, presumably of a singular continuous nature. The effect of this component on the rotor dynamics is not yet investigated in full.

For generic values of τ, namely excluding these two sets, general analytical results were obtained so far only in the extreme quantum regime $k < \pi$. [7] By introducing a Hermitian operator W through

$$(1 + iW)(1 - iW)^{-1} = e^{-ikcos\theta} \tag{1.3}$$

and

$$W = -tan[\frac{1}{2}kcos\theta]$$

The secular equation for the q.e. operator reduces to

$$T_m^{(\alpha)} u_m^{(\alpha)} + \sum_n W_{|m-n|} u_n^{(\alpha)} = 0 \tag{1.4}$$

where

$$\begin{aligned} u_m^{(\alpha)} &= < m \mid (1 - iW)^{-1} \mid \alpha > \\ T_m^{(\alpha)} &= tan[\frac{1}{2}(-\frac{1}{2}m^2\tau + \omega_\alpha)] \\ W_{|m-n|} &= < m \mid W \mid n > \end{aligned} \tag{1.5}$$

For $k < \pi$ the matrix elements $W_{|m-n|}$ fall off exponentially in $| m - n |$ since $W(\theta)$ is an analytic function. Therefore, (1.4) is just a tight-binding model in solid state physics, with diagonal elements $T_m^{(\alpha)}$ and hopping matrix elements $W_{|m-n|}$. Its physical properties are similar to those of models where only hopping to nearest neighbors is allowed. If the sequence $T_m^{(\alpha)}$ is random, (1.4) is just the one dimensional Anderson model for localization where all the states are localized. It was argued [7] that the sequence $T_m^{(\alpha)}$ of (1.5) is effectively random. Consequently, the quasienergy states of (1.1) are exponentially localized in angular momentum space.

This explains naturally the bounded and quasi-periodic behaviour of $< n^2 >$ in time, [3,6] which contrasts with the linear growth expected in the classical theory.

For $k > \pi$ the function W(θ) is singular and the simple mapping on the Anderson model cannot be used. Numerical calculations [11] indicate, however, that the quasienergy states are localized also for $k > \pi$.

In the present paper we would like to extend the above results in two directions. In section II, we shall show that both the resonance and localization phenomena persist even when the δ-kicks are replaced by smooth time dependent pulses, and that one could actually test these effects experimentally by measuring the excitation of rotational bands of diatomic molecules in a microwave field. [10] In section III, we show that the localization mechanism is also effective in the strong interaction domain ($k > \pi$). By using a transfer matrix formulation of the q.e. problem we are able to study the behaviour of the localization length with k.

2. Continuously Driven Rotors - Localization on a Finite Grid.

In this section, we shall deal with two systems which are driven by a train of smooth pulses. The one is the planar rotor, whose discussion will serve to bridge between the schematic δ-kicked planar rotor and the realistic system of a diatomic molecule excited by a microwave field.

The smooth driving pulse can be either chosen as a periodic Gaussian pulse of width σ

$$\Delta^{(\sigma)}(t) = \frac{1}{\sqrt{2\pi}\sigma} exp[-\frac{(t-1/2)^2}{2\sigma^2}] \qquad 0 \leq t \leq 1 \tag{2.1}$$

or, by truncating the Fourier series for the periodic δ function after N harmonics

$$\Delta^{(N)}(t) = 1 + 2\sum_{m=1}^{N} cos(2m\pi(t-1/2)) \tag{2.2}$$

Both pulse forms are compared in Fig. (1) and the similarity between the functions $\Delta^{(N=7)}$ (t) and $\Delta^{(\sigma=0.3)}$ (t) is apparent. The one cycle propagator is obtained by solving numerically the time dependent Schrödinger equation. The functions Δ(t) are defined in such a way that we can still use the symbols τ and k to denote the pulse repetition time and its action over a cycle, respectively. Fig. (2) shows the absolute square of the diagonal and first off-diagonal matrix elements of the propagator with τ=2, k=2, for the driving pulse $\Delta^{(N=7)}$ (t). The dependence of $| U_{n,n} |^2$ and $| U_{n,n+1} |^2$ on n is characterized by an abrupt change which occurs at $| n | = n_L$ = 21 in this case. For $| n | < n_L$, $| U_{n,n} |^2$ and $| U_{n,n+1} |^2$ are almost constant taking the approximate values $| J_o(2) |^2$ and $| J_1(2) |^2$ respectively, which are exactly the matrix elements

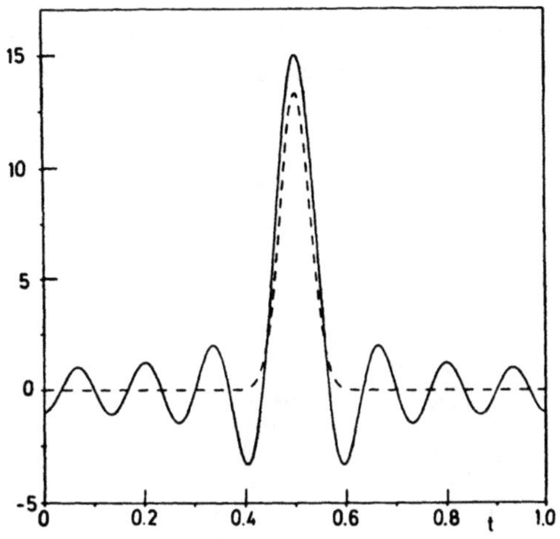

Fig. (1) Possible shapes of a microwave pulse. Broken line: Gaussian form factor $\Delta^{(\sigma=0.03)}(t)$. Full line: Truncated Fourier series for a periodic δ-function $\Delta^{(N=7)}$ (t).

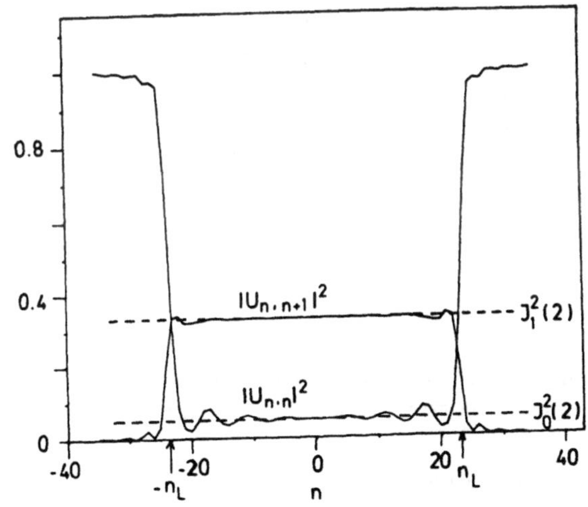

Fig. (2) Absolute squares of diagonal and first off diagonal matrix elements of the one cycle propagator U(T) for $\tau = 2$, $k = 2$ and $\Delta^{(N=7)}(t)$.

of the δ-kicked rotor. For $|n| > n_L$, $|U_{nn}| \to 1$ and $|U_{n,n+r}|^2$, $r = 1,2,\ldots$ approach zero, so that the states with $|n| > n_L$ are neither coupled to the states with $|n| < n_L$ nor do they couple to each other. This phenomenon is due to the fact that the driving field can couple only such states whose transition frequency can be matched by the field frequencies. In our case, the $n \rightarrow n+1$ transition frequency is n+1/2 and n_L satisfies

$$n_L + 1/2 = \frac{2\pi}{\tau} N \tag{2.3}$$

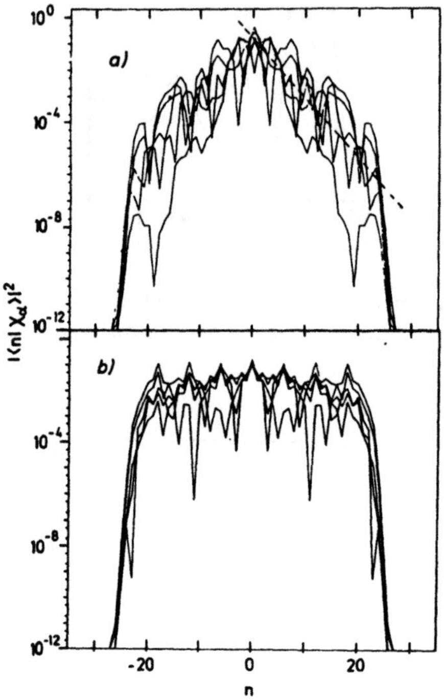

Fig. (3) Some quasi energy states characterized by a large overlap with the rotor ground state $|\,0>$ for interaction strength $k = 2$ and a) $\tau = 2$, b) $\tau = \frac{2\pi}{3}$.

The q.e. eigenstates will be correspondingly divided into two categories. The one will contain states which are (trivially) localized on rotor states with $|\,n\,|> n_L$, and they will not affect the dynamics as long as the initial state is within the coupled region. The other class will involve the ($2n_L + 1$) low n states and will couple them strongly as is apparent from the size of $|\,U_{n,n+1}\,|^2 \cong 0.33$. Now we may try to see whether there exist extended eigenstates for τ values which are rational multiples of 4π, and if the states become localized for irrational $\tau/4\pi$. Fig. (3) shows the q.e. states which have large overlap with the n=0 state for $\tau = 2\pi/3$ (resonance) and τ=2 (off resonance). We observe that the q.e. eigenvectors indeed fall off exponentially and behave as the finite length analogue of the infinite 1-dimensional, Anderson model. The localization length deduced from the mean exponential fall-off (dashed line in Fig. (3a)) is approximately 5 states, while the extended (resonance) states have almost equal probability to be in any state in the $| n |< n_L$ region.

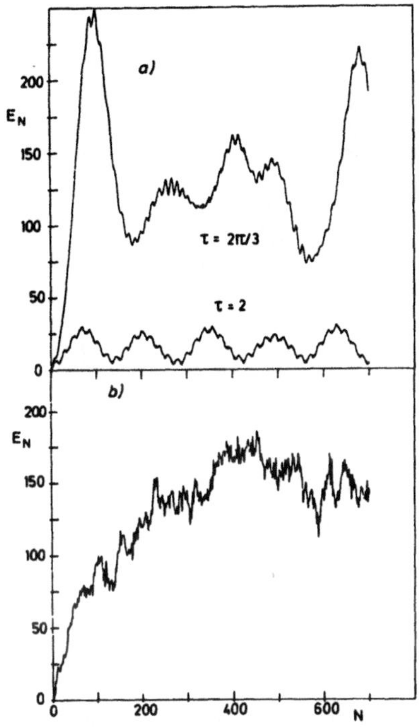

Fig. (4) Average energy of the rotor as a function of time for $k = 2$ and $\Delta(t) = \Delta^{(N=7)}(t)$. a) Quantum mechanical calculations for the localized ($\tau = 2$) and extended ($\tau = \frac{2\pi}{3}$) case. b) Classical calculation ($\tau = 2$).

In Fig. (4a), we compare the energy gained by the rotor on and off resonance. The initial quadratic rise in the mean energy on resonance is terminated after some time when all the states in the $| n |< n_L$ region are almost equally populated. The classical result for the energy is shown in Fig. (4b). Here the typical linear increase of the energy with time is terminated after the $| n |\leq n_L$ domain is populated, and there is no further diffusion to $| n |> n_L$ because the classical Chirikov criterion [1)] is only fulfilled for $| n |< n_L$. In the discussion above we have shown that the pictures which characterize the quantum dynamics of the δ-kicked rotor appear in an analogous fashion in the continuously driven problem. A similar analogy also exists for the classical dynamics, and the similarity persists independently of the detailed shape of the driving pulse. We shall now show that the above considerations can be implemented in an experiment where rotational excitation of diatomic molecules are induced by a train of microwave pulses.

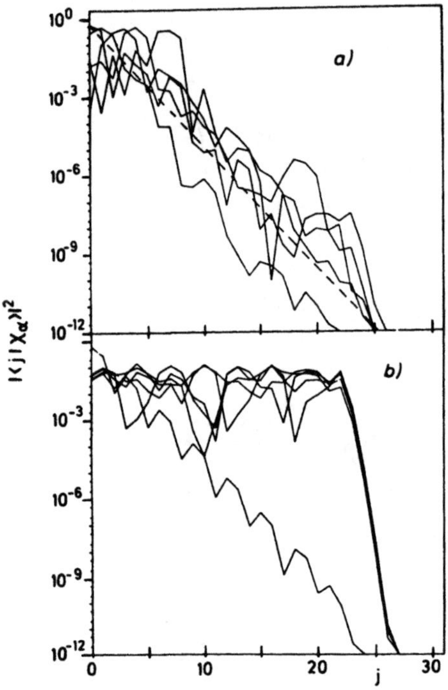

Fig. (5) Some quasi energy states characterized by a large overlap with the molecule ground state $|j=0>$ for interaction strength k = 2 and a) $\tau = 2$, b) $\tau = \frac{2\pi}{3}$.

The Hamiltonian which governs the molecule dynamics is

$$H(t) = \frac{\hbar^2 \vec{J}^2}{2I} + \mu E_\circ cos\theta\Delta(t) \tag{2.4}$$

where $\vec{J}$ is the angular momentum operator in 3 dimensions, I is the moment of inertia of the molecule along an axis perpendicular to the symmetry axis, μ its dipole moment and $E_\circ$ is the amplitude of the electric field which is applied in the z direction.

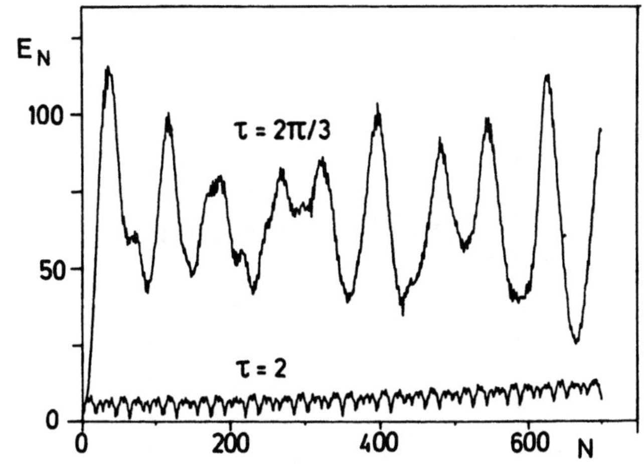

Fig. (6) Average energy of the molecule as a function of time in the localized and extended case.

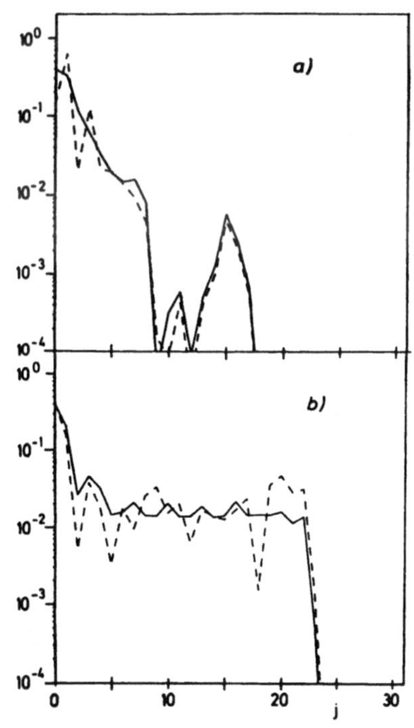

Fig. (7) Occupation probabilities of the angular momentum states of the molecule. Dashed line: "snapshot" after cycle nr. 400. Full line: average of the occupation probabilities from cycle nr. 150 to cycle nr. 200. a) Off resonance. b) On-resonance.

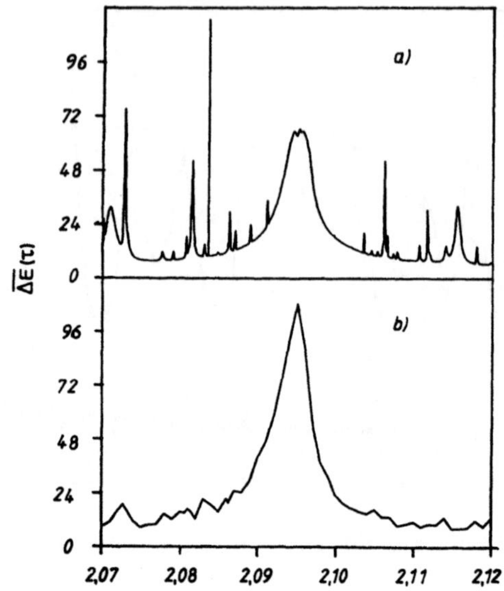

Fig. (8) Average energy gained by the molecule for τ in the vicinity of the resonance at $\tau = \frac{2\pi}{3}$, a) $T_{rot} = 0\mathrm{K}$, b) $T_{rot} = 1\mathrm{K}$.

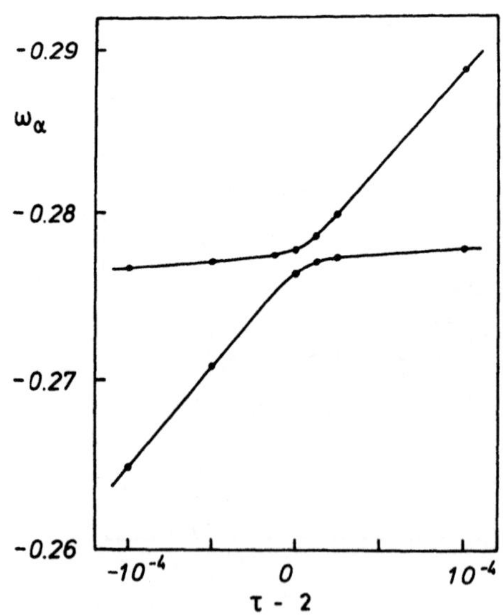

Fig. (9) Avoided crossing of two quasi energies in the vicinity of $\tau = 2$.

The eigenvalues of $\vec{J}^2$ are j(j+1), j=0,1,... and

$$< jm \mid cos\theta \mid j'm' >= \delta_{mm'}(C_j^{(m)}\delta_{j',j-1} + C_{j+1}^{(m)}\delta_{j',j+1})$$

$$C_j^{(m)} = \left[\frac{(j-m)(j+m)}{(2j-1)(2j+1)}\right]^{1/2} \tag{2.5}$$

Thus, the grid of j values is limited to the positive values only, and the rotational energies as well as the dipole matrix elements coincide with the planar rotor values only for large j. Fig. (5) shows the q.e. eigenvectors which have maximal overlap with the ground state for m=0 and for off-resonance (a) and resonance (b) conditions. It is apparent that the localization mechanism off resonance is as effective as in the planar case, and the appearance of a single localized state in the resonance case is probably due to an edge effect. The energy absorbed by the molecule is shown in Fig. (6) for the same conditions and the results are very similar to those for the planar rotor. The time averaged probabilities to excite the various j states off and on resonance are shown in Fig. (7) and they exhibit the expected behaviour.

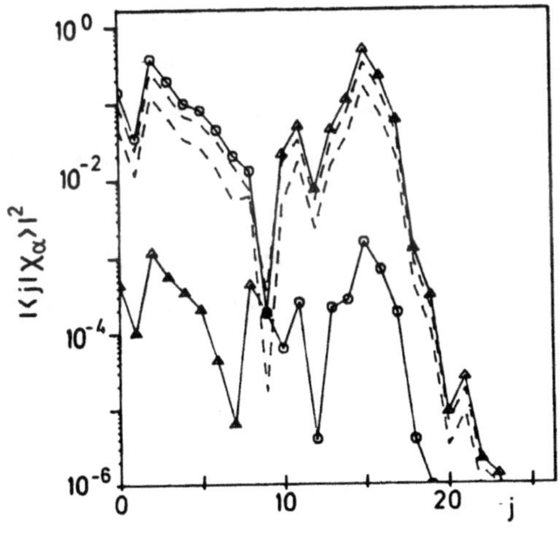

Fig. (10) Nearly degenerate quasi energy states. Full lines: Quasi energy states localized at $\mid j = 2 >$ and $\mid j = 15 >$ for $\tau = 2$ - 10^{-4}. Dashed lines: The same states for $\tau = 2$.

In order to check in more detail the behaviour of the molecule in the microwave field, we calculated the time-averaged energy transfer as a function of the repetition time τ in the vicinity of the resonance $\tau = \frac{2\pi}{3}$. Fig. (8a) shows the results where the molecule is assumed to be initially in its ground state. Apart from the resonance structure at $\tau = \frac{2\pi}{3} = 2.094$ we observe several other narrow spikes in the vicinity of the resonance. For τ values ranging from $\tau = 2.07$ to $\tau = 2.12$ none of the spikes occurs on τ values which can be assigned to a value $\frac{p}{q}\pi$ with q smaller than 43. We attribute these structures to the phenomenon of avoided level crossing which is shown in figures (9) and (10). Two q.e. eigenvalues almost cross at $\tau \approx 2$ (see Fig. (9)). The corresponding q.e. eigenstates are shown in Fig. (10) before the crossing occurs (circles and triangles) and at the point of avoided crossing (dashed lines). Before the crossing, the "circle" state is localized around j=0 and has a very small overlap with the higher j states. The "triangle" state was localized at j=15 having very low overlap with j=0. At the crossing point the states must share their structure and indeed the dashed lines show that the rotational states with j=0 and j=15 have similar overlaps with either q.e. eigenstates, reaching in probability to the 10% level. We thus have a very efficient way to transfer probability between j=0 and j=15. The region where this mechanism is effective is extremely narrow and hence the small width of the observed spikes. In Fig. (8b), we show the same quantity as in Fig. (8a), but here we assume the molecules to be produced in a beam with rotational temperature of 1K. The absorbed energy should be averaged over the ensemble of different initial j and m values. We see that the proper resonance structure is not affected by this averaging but the avoided crossing effects are smoothed out because of the high sensitivity of the avoided crossing to changes of the matrix elements (due to the presence of molecules with m≠0) and to the initial j values which involve other q.e. eigenstates. We checked [10)] that molecules like CsI or PbTe possess high dipole moment and are sufficiently heavy, so that an experiment can in principle be conducted with reasonable field strength. We checked the influence of various sources of random noise on the dynamics and found that one could carry out the proposed experiment within present day technology.

3. The Strong Coupling Domain

As was stressed before, the simple method discussed in the previous sections to show the analogy between the q.m. rotor and the Anderson model (see 1.4) is only applicable for $k < \pi$. Here we shall extend the discussion to the domain $k > \pi$ (the strong coupling domain).

The starting point is again the secular equation for the 1 cycle propagator (1.2)

$$\sum_{m=-\infty}^{\infty} J_m(k) a_{n-m}^{(\alpha)} = e^{in^2\tau/2 - i\omega_\alpha} a_n^{(\alpha)} \tag{3.1}$$

where a phase factor $(+i)^{-m}$ was absorbed into the components $a_m^{(\alpha)}$. The following properties of the Bessel functions will be relevant to the discussion

$$\text{a.} \qquad \sum_{\nu} J_\nu(k) J_{s-\nu}(k) = \delta_{s0}$$

$$\text{b.} \qquad J_\nu(k) = (-1)^\nu J_{-\nu}(k) \tag{3.2}$$

$$\text{c.} \qquad | J_\nu(k) | \sim \begin{cases} (2/\pi k)^{1/2} & |\nu| < k \\ (\frac{1}{2\pi\nu})^{1/2}(ek/2\nu)^\nu & |\nu| > k \end{cases}$$

This last property suggests to approximate eq. (3.1) by replacing the infinite sum by a finite sum over the range $| m | \leq b$ where b is an integer larger than k. Consequently the evolution operator (1.2) is a band matrix around the diagonal, with one pseudo random element in each row. Therefore, it has a form similar to the Hamiltonian of the one dimensional Anderson model for localization in a sense similar to the discussion of ref. 7, and localization might be expected to take place in this model as well. In order to establish the localization, the truncated version of eq. (3.1) is posed as a transfer matrix problem in a $2b$-dimensional vector space.

$$\vec{c}^{(n+1)} = T^{(n)} \vec{c}^{(n)} \tag{3.3}$$

where

$$\vec{c}_i^{(n)} = a_{n+b-i} \qquad i = 1, 2, \ldots 2b$$

and

$$T^{(n)} = \begin{pmatrix} t_1 & \cdots & \cdots & \cdots & t_b & \cdots & \cdots & \cdots & t_{2b} \\ 1 & 0 & \cdots & \cdots & \cdots & \cdots & \cdots & \cdots & 0 \\ 0 & 1 & \cdots & \cdots & \cdots & \cdots & \cdots & \cdots & 0 \\ \vdots & & \ddots & & & & & & \vdots \\ \vdots & & & \ddots & & & & & \vdots \\ 0 & & & & 1 & & & & 0 \\ \vdots & & & & & \ddots & & & \vdots \\ \vdots & & & & & & 1 & & \vdots \\ 0 & \cdots & \cdots & \cdots & 0 & \cdots & \cdots & 1 & 0 \end{pmatrix} \tag{3.4}$$

with

$$\begin{aligned} t_l^{(n)} &= (-J_{(-b+l)}(k) + \delta_{lb} exp(in^2\tau/2 - iw_\alpha))/J_{-b}(k) \qquad l = 1, 2, \ldots b \\ t_l^{(n)} &= (-1)^{b-l} t_{2b-l}^{(n)}, \qquad 2b > l > b \qquad t_{2b}^{(n)} = (-1)^{b+1} \end{aligned} \tag{3.5}$$

The matrices $T^{(n)}$ have the property

$$\Gamma T^{(n)}(-\Gamma^{-1}) = (T^{(n)})^{-1} \tag{3.6}$$

where Γ is the $2b$ dimensional matrix

$$\Gamma = \begin{pmatrix} 0 & \dots & \dots & 0 & 0 & \dots & \dots & 1 \\ \vdots & & & & \dots & & -1 & \\ \vdots & & & & \dots & 1 & & \\ 0 & \dots & \dots & 0 & -1 & 0 & & 0 \\ \vdots & & & 1 & 0 & & & 0 \\ \vdots & & -1 & & & & & \\ \vdots & 1 & & & & & & \\ -1 & \dots & \dots & \dots & 0 & & & 0 \end{pmatrix} (-1)^{b+1}$$

from which it follows that:

(1) $\mid det(T^{(n)}) \mid = 1$.

(2) The eigenvalues of $T^{(n)}$ appear in pairs, if λ is an eigenvalue, there is another eigenvalue λ' such that $\mid \lambda\lambda' \mid = 1$.

(3) The set of matrices having the property (2) is closed under multiplication.

Successive applications of the transfer operation (3.3) expand the volume of ν-dimensional parallelepipeds with $1 \leq \nu \leq b$. The mean expansion rate is the ν-dimensional Lyapunov exponent $\lambda^{(\nu)}$ [12)]

$$\lambda^{(\nu)} = \lim_{N \to \infty} \frac{1}{N} ln \frac{\| P^{(N)} \vec{e}_1 \Lambda P^{(N)} \vec{e}_2 \Lambda \dots \Lambda P^{(N)} \vec{e}_n \|}{\| \vec{e}_1 \Lambda \vec{e}_2 \Lambda \dots \Lambda \vec{e}_n \|} \tag{3.7}$$

where

$$P^{(N)} = \prod_{i=1}^{N} T^{(i)}$$

and $\vec{e}_j$ are any independent set of ν vectors in $2b$ dimensions. Λ stands for the exterior product.

If the limit (3.7) exists, we can use the methods of ref. (12) to calculate the $\lambda^{(\nu)}$. The one dimensional Lyapunov exponents may take at most b distinct values which we shall denote by $\{\gamma_i\}$, where $\gamma_1 \geq \gamma_2 \dots \geq \gamma_b \geq 0$;

since

$$\lambda^{(\nu)} = \max_{1 \leq l_i \leq b} \{ \sum_{i=i}^{\nu} \gamma_{l_i} \} \tag{3.8}$$

we can calculate γ_b once the set of $\lambda^{(\nu)}$ is known for ν=1, ...,b.

From the properties of the transfer matrix it follows that the one dimensional Lyapunov exponents which govern the contracting subspace are given by $\gamma_i = -\gamma_{2b-i+1} \quad i > b$. The fact that the Lyapunov exponents appear in pairs is due to the symmetry of the problem - the results should be independent of whether the transfer matrix is applied to the right or to the left.

The minimal positive Lyapunov exponent γ_b determines the localization properties of the q.e. eigenvectors. [13] This is just a consequence of the fact that it is associated with the largest length scale in the problem. If it vanishes then there exists an extended eigenvector. Otherwise, the eigenvectors are localized around some value $\overline{n}$ and the localization length $\xi = \gamma_b^{-1}$ determines the mean exponential decay of the amplitudes for n far away from $\bar{n}$. Before presenting the results, we shall briefly discuss the numerical tests which we performed to convince ourselves of the validity of our approach:

1. For any given value of k we calculated the localization length for several values of $b > k$. Convergence, namely, independence of γ_b on b, within a few percent was achieved for b$\geq k$+5, ($k \geq$1). For such values of b, $\mid \sum_{|\nu|\leq b} J_\nu^2 - 1 \mid \leq 10^{-3}$.
2. Using our numerical procedure to calculate the 2b 1-dim. Lyapunov exponents, we verified numerically that the last b exponents are negative, with $\gamma_i = -\gamma_{2b-i+1}$.
3. Setting the phase factor in $t_{l=b}^{(n)}$ to 1.0 we obtained for γ_b a very small ($\sim 10^{-6}$) number as expected for the corresponding eigenvector problem.
4. We used typically N$\sim 10^4$ iterations in (3.7). We checked ξ after every 10^3 iterations and the values show fluctuations of a few percents about the mean. Based on these observations, we shall quote our results with an uncertainty margin of $\pm$5 %.
5. We calculated the variation of γ_i in the last 10^3 iterations and found it to be of the order of 5 % as well.

A key element of the discussion in the preceding chapters was the assumption that the phase factors $e^{in^2\tau/2}$ are sufficiently uncorrelated, to warrant the applicability of the Anderson model to the q.m. rotor. In the same spirit, we shall first display and discuss the localization characteristics for a transfer problem where the phase factors in $t_{l=b}^{(n)}$ are chosen at random, with a uniform distribution on the unit circle. The resulting localization lengths are displayed in Fig. (11) for k values in the range
$1 \leq k \leq 20$. We see that for $k \geq$2, the data points can be fitted by

$$\xi_R(k) = \xi_\circ k^\mu$$

with

$$\xi_\circ = 0.517 \pm .014 \qquad (3.9)$$
$$\mu = 1.755 \pm .013$$

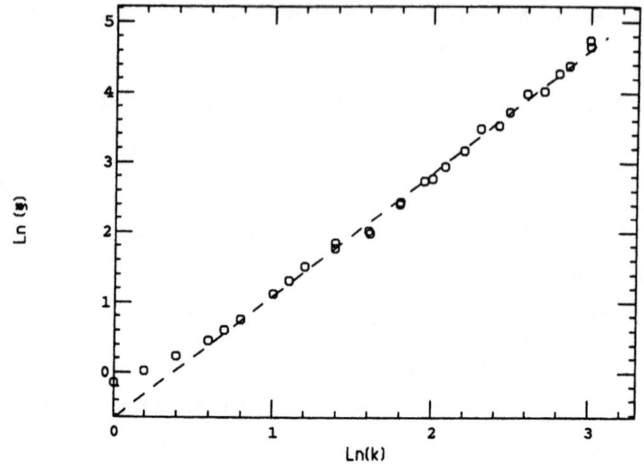

Fig. (11) Calculated $\ln \xi_R(k)$ (dots) vs. $\ln(k)$. The dashed line is the best fit whose parameters are quoted in eq. (3.9).

The parameters and the error bounds were obtained from a χ^2 fit.

In Fig. (12), we show the ratios $\xi(\tau;k)/\xi_R(k)$ for $\tau = 1$ and 2, where $\xi(\tau;k)$ is the localization length calculated with the proper rotor phase. We see that up to a change in the factor ξ_o, the localization length $\xi(\tau;k)$ seems to oscillate about a mean given by $\xi_R(k)$. The mean of the ratio $\xi(\tau;k)/\xi_R(k)$ is indicated in Fig. (12) by a dashed line. The observed oscillations are rather regular, they are certainly beyond the numerical uncertainty, and at this point we are not able to explain their origin.

We conclude that for the kicked rotor, localization of the quasienergy states persists even in the strong coupling domain. The new feature that this work reveals is that the localization length for the kicked rotor differs from the one calculated on the basis of the assumption that $e^{i\tau n^2/2}$ can be replaced by a random phase [14]. However, to a reasonable approximation the random phase calculations reproduce the overall growth of the localization length with k.

Having developed the present tool, the most natural problem to which it should be addressed is the proper semi-classical limit of the rotor, which corresponds to the limit $\tau \to 0, k \to \infty$ while $k\tau = \mathrm{K} = \text{const}$. Unfortunately, the transfer matrix approach becomes less accurate and more costly in terms of computer resources because (a) with $k \to \infty$ the dimension of the transfer matrices increase and (b) with $\tau \to 0$ correlations between the phase factors

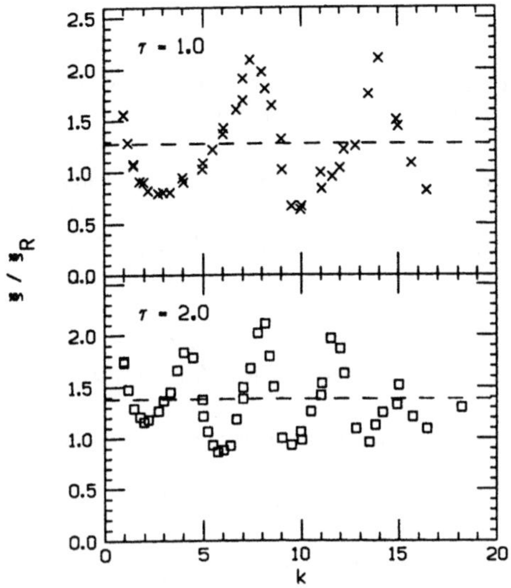

Fig. (12) The ratio $\xi(\tau;k)/\xi_R(k)$ for $\tau = 1$ and 2. The dashed lines give the mean ratio.

survive for $\sim \tau^{-1}$ iterations and therefore the total number of iterations N should be increased proportionally. Thus the approach to the semi-classical limit gets harder since $k\tau^{-1} \sim (1/\hbar)^2$.

After the work was completed, we received a preprint from D.L. Shepelyansky who used a similar method to analyze different aspects of this problem [15].

We would like to thank J. Avron, B. Shapira, R.E. Prange, and D.R. Grempel for useful discussions, and D.L. Shepelyansky for sending us the preprint of his work. The work was supported in part by the U.S.-Israel Binational Science Foundation (BSF).

References:

1. B.V. Chirikov, Phys. Rep. 52, 263 (1979). B.V. Chirikov, F.M. Izrailev and D.L. Shepelyansky, Sov. Sci. Rev. Sect. C2, 209 (1981).

2. G.M. Zaslavsky, Phys. Rep. 80, 158 (1981).

3. G. Casati, B.V. Chirikov, F.M. Izrailev and J. Ford in: "Stochastic Behaviour in Classical and Hamiltonian Systems", Lecture Notes in Physics, Vol. 93, p. 334 (Springer

N.Y. (1979)).

4. F.M. Izrailev and D.L. Shepelyansky, Teor. Mat. Fiz. 43, 417 (1980), Theor. Math. Phys. 43, 553 (80) and Sov. Phys.-Dokl. 24, 996 (1979).

5. G. Casati and I. Guarneri, Commun. Math. Phys. 95, 121 (1984).

6. T. Hogg and B.A. Huberman, Phys. Rev. Lett. 48, 711 (1982) and Phys. Rev. A28, 22 (1983).

7. S. Fishman, D.R. Grempel and R.E. Prange, Phys. Rev. Lett. 49, 509 (1982), D.R. Grempel, R.E. Prange and S. Fishman, Phys. Rev. A29, 1639 (1984).

8. J.D. Hanson, E. Ott, and M. Antonsen, Phys. Rev. A29, 819 (1984).

9. S.J. Chang and K.J. Shi Phys. Rev. Lett. 55, 269 (1985).

10. R. Blümel, S. Fishman, and U. Smilansky, J. Chem. Phys. in Press (1986).

11. M. Feingold, S. Fishman, D.R. Grempel, and R.E. Prange, Phys. Rev. B31, 6852 (1985)

12. I. Shimada and T. Nagashima, Prog. Theor. Phys. 61, 1605 (1979).

13. This was established rigorously by F. Delyon, Y. Levy, and B. Souillard (to be published) using property (2).

14. This difference was observed in preliminary calculations by M.F. Feingold and S. Fishman, to be published.

15. D.L. Shepelyansky, Phys. Rev. Lett. 56, 677 (1986).

FROM LOCALIZED TO EXTENDED STATES IN A TIME-DEPENDENT QUANTUM MODEL

Jorge V. José

Department of Physics, Northeastern University, Boston, Massachusetts 02115[*]

ABSTRACT. The problem of a particle inside a rigid box with one of the walls oscillating periodically in time is studied quantum mechanically. In the classical limit, this model was introduced by Fermi in the context of cosmic ray physics. The classical solutions can go from being quasiperiodic to chaotic, as a function of the amplitude of the wall oscillation. In the quantum case, we calculate the spectral properties of the corresponding evolution operator, i.e.: the quasi-energy eigenvalues and eigenvectors. The specific form of the wall oscillation, e.g. $\ell(t) = \sqrt{1+2\delta|t|}$, with $|t| \leq 1/2$, and $\ell(t+1) = \ell(t)$, is essential to the solutions presented here. It is found that as $\hbar$ increases with δ fixed, the nearest neighbor separation between quasi-energy eigenvalues changes from showing no energy level repulsion to energy level repulsion. This transition, from Poisson-like statistics to Gaussian-Orthogonal-Ensemble-like statistics is tested by looking at the distribution of quasi-energy level nearest neighboor separations and the $\Delta_3(L)$ statistics. These results are also correlated to a transition between localized to extended states in energy space. The possible relevance of the results presented here to experiments in quasi-one-dimensional atoms is also discussed.

INTRODUCTION. Considerable progress has been reported in this conference about defining the signatures of classical chaos in TIME-INDEPENDENT quantum problems. The essential idea of looking at the distribution of energy eigenvalues in model quantum Hamiltonians, which show chaotic behavior in the classical limit, has led to specific properties of the spectrum that distinguish between almost periodic and chaotic regimes. [1] Specifically, the nature of the spectrum of integrable Hamiltonians has been postulated to have a Poisson-like distribution for the energy level separations, and in the nonintegrable regimes, it was predicted and shown that the energy levels show the Wigner-like distribution for the energy level separations. [2,3,4] The proofs of this surmise are not yet mathematically rigorous, but the evidence is mounting in favor of this conjecture: on the one hand, we have the results from very extensive and detailed numerical calculations and on the other, a detailed WKB analysis presented in this conference by Berry. Therefore, we could say that there is a certain degree of consensus about identifying Poisson-like energy level

* Talk presented at the Second International Conference on Quantum Chaos held in Cuernavaca, Mor., Mex. January 1986

separations and Wigner-like statistics to integrable and nonintegrable time-independent quantu Hamiltonians, respectively.

The situation about the nature of the eigenfunctions is less satisfactory. Here we have both th ideas of McDonald and Kaufman about nodes of the wave functions, and the extension presente in this conference by Heller in terms of "scars" in the wave functions. One reason the study of th eigenvalue spectrum gives clearer results is that it is basis independent, whereas the eigenfunction apart from usually being harder to calculate reliably, are basis dependent.

As Jensen has pointed out [5], it is perhaps in TIME-DEPENDENT problems where answerin the question about the relevance of classical chaotic solutions in quantum mechanics may hav answers from the definitive arbiter, the experiment. Although the surface electron on heliu model proposed by Jensen, may be difficult to produce experimentally, there have been significar advances in the preparation of hydrogen atoms in highly excited Rydberg states where the necessar tests could be made.[6] These are atoms excited to very high principal quantum numbers, tha represent systems where theoretical treatments are, of necessity, nonperturbative in nature. I this conference we have heard Casati talk about the results of their studies of a one-dimension hydrogen atom in the presence of sumultaneous constant and time varying electric fields. Thes are studies intended to relate to the experiments performed by Koch et.al. and Bayfield et.al. [6 The main theoretical difficulty in these studies is to properly incorporate the continuum part c the spectrum in the calculations. Since the calculations are numeric, different tests have to b implemented to make sure that the basis chosen for the calculations is sufficiently large such tha the results are reliable.

The motivation for the study presented in this paper is twofold; first, to try to answer th question as to the general statistical properties of the Quasi-Energy-Spectrum (QES) for time dependent Hamiltonians in the nonperturbative regime, to see if the classification of the time independent problem holds also for the time-dependent case. Second, to search for universa properties of time-dependent Hamiltonians that may have a bearing on the experimental result on quasi-one-dimensional atoms. With regard to the first point, there have been studies of th extensively studied "periodically-kicked-quantum-rigid-rotor-model" (PKQRRM), introduced b Casati et.al.[7] Of interest in this paper is the analysis of the distribution of QES level separation carried out by Feingold et.al. [8] There they show that the QES is Poisson-like for the range o parameters they studied. The Poisson result was, in a sense, expected. Fishman et.al., or a they are perhaps better known, the Maryland group, showed that under certain conditions th PKQRRM maps into the one-dimensional tight-binding random Anderson model. [9] This model is known, rigorosly, to have localized states and also to have an energy-level separation distribution o the Poisson type. [10] Almost simultaneously with the work to be discussed in this paper, Izrailev considered a finite-dimensional Hilbert space representation of the PKQRRM in which he finds a Wigner-like distribution for the separation of QES eigenvalues. In the limit where the dimension of the Hilbert space goes to infinity, however, the statistics are expected to become Poisson.[11]

In this paper, I introduce another model that could conceivably have a Wigner-like distribution, for certain regions in parameter space, in the limit of an infinite-dimensional Hilbert space. The results discussed here are mostly numerical, and therefore one should subject the above conjecture to different tests to ascertain if, indeed, the QES is Wigner-like in the infinite-dimensional case. Part of the results discussed here have appeared in print [12] and others will be forthcoming. [13]

THE QUANTUM FERMI MODEL. Consider a free particle moving inside a box with rigid walls. One of the walls is fixed and the other one is oscillating periodically in time. The collisions between the particle and the walls are perfectly elastic. Classically, on the average, as time increases the particle will gain energy from the oscillating wall, and one would expect that the energy gained should increase with time without bound. The way in which the particle will gain this energy will depend on the specific funtional dependence of the wall oscillation. If the functional form of the wall oscillation is the same for all times, then there will be a value of the energy for which the period of oscillation of the wall and the round trip time for the particle will be on resonance. At these resonance points the particle will be trapped in a region of momentum space, and thus, the energy will be bounded as a function of time. An appropriate phase space to describe the motion of the particle is formed from the value of the particle velocity before the nth collision with the wall, normalized to the maximum wall velocity, let us call it $\mathcal{V}$, and the coordinate that describes the location of the wall before the nth collision. This last coordinate can be taken to be an angle, ϕ, ranging from $[0, 2\pi]$. For a given initial particle velocity $\mathcal{V}_0$, one can plot the different points in this phase space visited by the particle for different times. It is found that, for small $\mathcal{V} \leq 1$,the orbits are essentially chaotic. [14] As $\mathcal{V}$ increases, there are regions of phase space that start to be filled by small adiabatic, or high order resonant, regions until the wall period is equal to the round trip transit time of the particle, and the velocity is bounded by a Kolmogorov-Arnold-Moser (KAM) trajectory. These are the basic qualitative features of the classical motion of a particle inside a rigid box with one wall oscillationg periodically in time. This model was first conceived by Fermi when he was trying to explain the existence of high energy cosmic rays. [15] He proposed the model, with the role of the oscillations produced by the effect of interstellar oscillating magnetic fields, to explain why the energy of the charged particles was so large. Unfortunately, as mentioned above, there is a clear upper bound in the amount of energy that the particle can gain and thus the Fermi assumption is not valid. Further extensions and simplifications of the model by Ulam and others helped to clarify the properties of this model. [14] Notice, however, that the upper bound would disappear if the wall oscillations, instead of being periodic in time, are random. In this case, the energy gained by the particle will not have a generic upper bound.

The question of interest in this talk is then, to which extent can we analyze a quantum version of the Fermi-acceleration model. After all, a particle in a rigid box with fixed walls is one of the simplest and most informative models one can study in quantum mechanics. The model is defin by the Schrödinger equation,

$$i\frac{\partial\Psi}{\partial t} = -\frac{\hbar}{2m}\frac{\partial^2\Psi}{\partial x^2}$$

Here, $\hbar$ is Planck's constant divided by 2π, and m is the mass of the particle. The spatial coordina $x\epsilon[0,\ell(t)]$. The length of the box, $\ell(t)$, satisfies the symmetry relations $\ell(t) = \ell(-t)$, and $\ell(t+T)$ $\ell(t)$, with T the period of oscillation of the moving wall. Since the walls are rigid, the bounda conditions are,

$$\Psi(x=0,t) = \Psi(x=\ell(t),t) = 0, \quad \int_0^{\ell(t)} |\Psi(x,t)|^2\, dx = 1$$

These time-dependent boundary conditions (b.c.) are all important in this problem. There a ways in which these b.c. can be incorporated in a partial differential equation with fixed b.c., b the price to be paid is that the equation would no longer be Hermitian. Instead, what we can is to propose a solution to Eq(1) as an expansion in terms of the natural complete time depende basis,

$$| n,t >= (-1)^n\sqrt{\frac{2}{\ell(t)}}\sin\frac{n\pi x}{\ell(t)}$$

Therefore, a Fourier expansion of Ψ in terms of this basis is,

$$\Psi(t) = \sum_n B_n(t)|n,t>$$

The resulting equation for the expansion coefficients B_n is given by,

$$i\dot{B}_n = \frac{e_n(t)}{\hbar}B_n - \sum_{n\neq m}[i\frac{\dot{\ell}}{\ell}\frac{2nm}{(n^2-m^2)}B_m(t)]$$

Here, $e_n(t) = \frac{\pi^2\hbar^2 n^2}{2m\ell(t)^2}$. For a general $\ell(t)$,the solving of the simultaneous equations for B_n is n easy. However, if we take the particular functional dependence for $\ell(t)$ as

$$\ell(t) = \ell_0\sqrt{1+2\delta|t|}$$

with ℓ_0 the minimum box size, and $|t| \leq 1/2$, we see that Eq(5) simplifies considerably,

$$i\dot{B}_n = \frac{\ell_0^2}{\ell^2}[\frac{n^2}{T_0}B_n +^- i\delta\sum_{m\neq n}\frac{2nm}{(n^2-m^2)}B_m(t)]$$

where the + sign is for $t > 0$, the – sign is for $t < 0$, and

$$T_0 = \frac{4m}{h\pi}$$

T_0 and δ are the two important parameters in this problem; δ is related to the wall deviation fro the minimum value ℓ_0, and T_0 is the Bohr period for the ground state of the particle when the bo has both walls fixed in time. T_0 gives the intrinsic time of the particle in the problem. We see that Eq(7) has the form $i\dot{B}_n = \sum_{m\neq n} H_{nm}B_m(t)$, with H_{nm} given by the coefficient of B_n in Eq(7). To find $B_n = B_n(t)$ we have to calculate the unitary operator $U(t,s) = \mathcal{T}\ exp\frac{-i}{\hbar}\int_t^s \hat{H}(\tau)\,d\tau$, with $\mathcal{T}$ the time ordering operator. Because $\ell(t)$ is periodic in time, U satisfies the periodicity condition U(t+T,s+T)=U(t,s). We can then invoke Floquet's theorem to evaluate U only over one period to obtain U for an arbitrary time. In this case we can write U as $U(0,t) = S(t)\exp\frac{-i}{\hbar}\,F\,t$, with S(t+T)=S(t). Since $\ell(-t) = \ell(t)$,it is easy to show that $H_{nm}(-t) = H_{mn}(t)$, which implies that F is a real symmetric matrix. For arbitrary time we have that $\Psi(MT) = U(MT,0)\Psi(0)$, with $U(MT,0) = U(T,0)^M$. Therefore, the solution to Eq(1) can be written in terms of the spectral properties of U as,

$$\Psi_n(x,t) = \sum_m (\lambda_m|n>|\lambda_m)\,e^{-i\lambda_m t} \tag{9}$$

Here $|\lambda_m)$ are the eigenvectors of U with $w_m = e^{-i\lambda_m T}$ its eigenvalues. The set $\{\lambda_m\}$ are the quasi-energy eigenvalues that satisfy the symmetry $\lambda_m \to \lambda_m + 2\pi N$ with N an integer. We see from Eq(9) that once we obtain the eigenvectors and eigenvalues associated to U, we have the complete solution of Eq(1).

RESULTS. It is clear that the treatment described above is valid for arbitrary $\ell(t)$. In order to calculate U we need to keep track of the time ordered product in the calculation. A numerical nonperturbative calculation of U, for arbitrary $\ell(t)$, would require dividing the time interval in slices of size Δt and the resulting U would be obtained from the matrix product of N matrices, such that $N\Delta t = t$. This could be a very time consuming and computer memory intensive process, if matrices of resonable sizes have to be diagonalized to reliable accuracy. Here is where the usefulness of the form of $\ell(t)$,given in Eq(6), becomes evident. As seen from Eq(7), the corresponding U for $t\epsilon(-1/2,0)$ and $t\epsilon(0,1/2)$ commute for this particular $\ell(t)$. Thus, we can look at relatively large matrices since the problem reduces to the diagonalization of only one matrix, although complex and unitary.

We have calculated the spectrum of U for different size matrices. We checked our numerical results in different ways. In the weak coupling regime, the coupling constant parameter in the theory is given by δT_0, and one can easily calculate the first order corrections to the eigenvectors and eigenvalues of U. Our analytic calculations correlate rather well with our numerical calculations in the weak coupling regime. In the non-perturbative regime, we made sure that the unitarity was satisfied up to the length of the computer word. Since we are truncating the matrix U to a finite-dimensional Hilbert space, more checks are necessary to ascertain if the results obtained are not size dependent. We looked at different matrix sizes, from 50x50 to 600x600, and after sizes of about 100x100 we did not see significant changes, mainly in the distribution of the QES. The eigenfunctions can be more significantly affected by the finiteness of the Hilbert space considered.

However, we did not see significant differences when changing the size of the matrices, but of course more tests should be impossed to make sure that the eigenfunctions results represent reliably the infinite matrix results. Here we will concentrate on the QES, which we feel is better defined in our calculations and will come back to the detailed discussion of $|\lambda_m)$ elsewhere.

There are essentially two quantities that I will discuss here. The distribution of energy level separations P(S), where $S = \lambda_{m+1} - \lambda_m$, and the Dyson-Metha statistics, $\Delta_3(L)$, that has been used successfully to separate between different types of P(S)'s . [3] In figure 1 we see different results for P(S) as a function of T_0 for δ fixed to 1/2. The matrices used to obtain these results are of 300x300 in size.

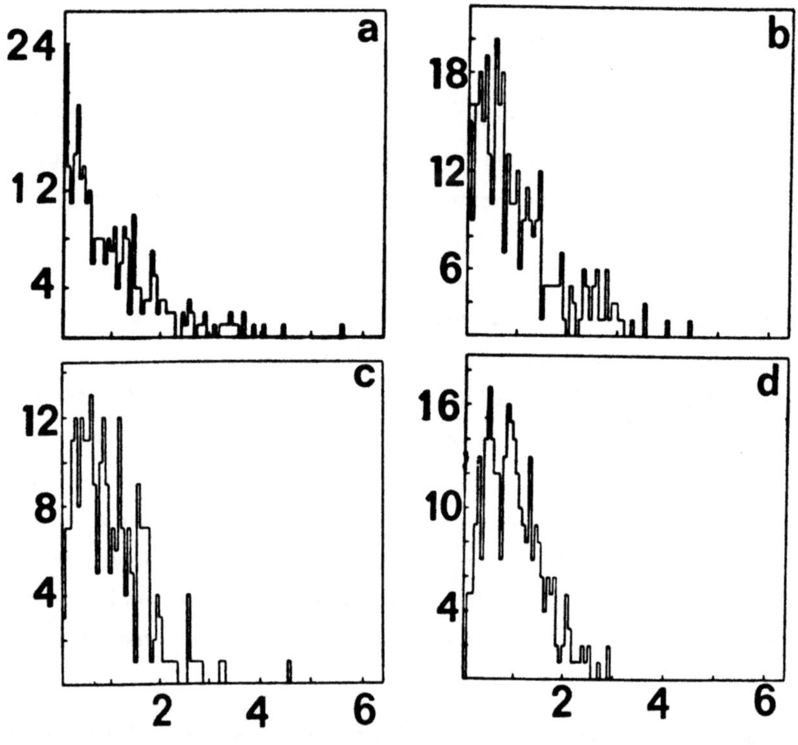

Figure 1. Results for P(S) vs S for; (a) $T_0 = 0.5, \delta = 0.2$, (b),(c) and (d), $\delta = 1/2$ and $T_0 = 8, 10, 100$ respectively.

The distribution P(S) goes from having a finite value at the origin, for small values of T_0 and δ, to a zero value at the origin for larger values of T_0 and δ. This transition seems to be smooth. We see that as the strength of the "coupling" increases, we go from a Poisson-like distribution to a Wigner distribution with energy level repulsion. To further test the reliability of these results, in Figure 2 we show also the $\Delta_3(L)$ results. The asymptotic result for a Poisson distribution, $\Delta_3(L) = L/15$, is followed in the small T_0, δ regime. As T_0 increases, with δ fixed, $\Delta_3(L)$ tends asymptotically to the Gaussian-Orthogonal-ensemble result, $\Delta_3(L) = \frac{1}{\pi^2}\log L - 0.007$. Notice that we show here results obtained **from only one** matrix and not from an ensemble of matrices as done by Bohigas et.al. [3] and by Izrailev. [11]

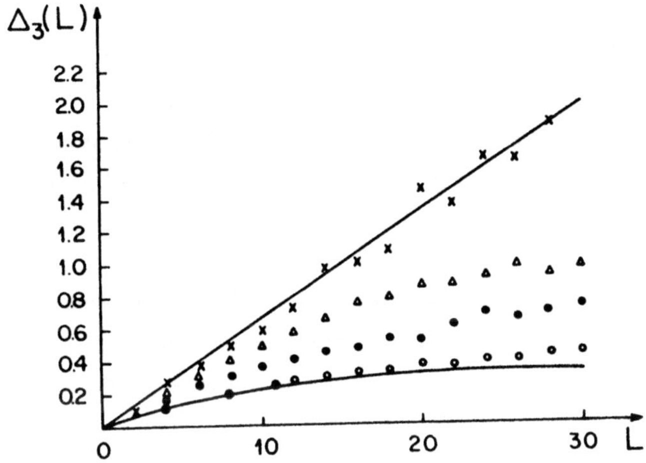

Figure 2. $\Delta_3(L)$ as defined in reference 3, for the same parameter values used in Figure 1. The solid lines represent the asymptotic analytic values for Poisson and GOE as discussed in the text.

Further tests of the statistical properties of the QES discussed here have been obtained in terms of higher order correlation functions [16] . The conclusion is that the QES shows a transition from a Poisson-like to a GOE-like distribution of nearest neighboor separations.

CONCLUSIONS. In this talk we discussed a quantum version of a Fermi-acceleration model. We were able to study in some detail the quasi-enery-spectrum because of the particular functional form of the wall oscillation, Eq(6), that allowed us to diagonalize the evolution operator exactly within one period. This particular form is somewhat close to the ones studied in the classical context.[14] In fact, when studying the classical problem using the form for $\ell(t)$ given in Eq(6) the problem simplifies as well.[13,17]

Since we have calculated the spectrum of U we can in principle calulate any quantum mechanical physical property of this model. Here we have concentrated in the analysis of the quasi-energy-spectrum and have found that as we vary the coupling constant in the model, which is inversely proportional to Planck's constant, there is a transition from a regime without energy level repulsion to one where the QES eigenvalues repel each other. The physical meaning of this result is somewhat clear. Lets take as initial condition that the particle is in the nth energy level. As the wall oscillat there will be a probability of having the particle jumping to other energy states. When the coupli is weak, the probability of having jumps to states different from n-1 and n+1 is exponentially sma Thus the states are localized and there is no energy level repulsion. As the coupling increases, t probability of jumping to states far away from n increases and the probability of "resonating" aw to a very distant state incrases as well. Terefore, there will be states separated by large energi that can overlap and, since we expect no degeneracies in this problem the levels will not cross.

A transition from localized to extended states of a similar nature to the one discussed in th talk has been found in the study of a one-dimensional hydrogen atom by Casati et.al. [18] V have to answer to which extent are these studies representatives of the infinite dimensional Hilbe space. In the hydrogen problem the question of including the continuum properly is of paramou importance. In this conference Casati has presented results that seem to agree with their previo studies. The caveat in these discussions is that, if the localization length is finite but larger tha the size of the matrices we are considering here, the results will not be representative of the re problem. Therefore, the results should be subjected to more stringent tests to answer this questic definitively. I personally think that in the present model there is a chance that indeed the infini dimensional result will not differ from our finite-dimensional numerical results. On the one han this model differs from the PKQRRM in some essential ways. First, the PKQRRM is describe in terms of a translationally invariant short-ranged, tight-binding type model, whereas the Fern model is described by a non-translationally invariant, long-ranged tight binding model. From soli state physics studies it is known that potentials with long range interactions in the infinite si limit can have extended states.

Further studies of these issues will clarify the question as to the existence of extended stat in the model considered in this talk in the infinite-dimensional case, and thus of quantum-chaos

I wish to thank R. Cordery, my collaborator in the initial stages of many of the aspects of th work reported here. Helpful conversations with S. Fishman and T. Seligman and corresponden with J. Ford are aknowledged. This work was supported in part by NSF Grant No.8500035.

REFERENCES

1. See for example, M.V.Berry, *Chaotic behavior in deterministic systems.* Les Houches summ school XXXVI, Ed. R. Helleman and G. Joos. (North Holland 1981). G. Zaslasvskii Phys. Rep.**8** 157.(1981)

2. M.V.Berry and M.Tabor Proc.Roy.Soc.London,Ser **A356**, C375(1977); S.W.McDonald an A.N.Kaufman Phys.Rev.Lett. **42**,1189 (1979);G.Casati et.al. Nuovo Cimiento Lett. **28** ,27 (1980); M.V.Berry Ann.Phys.(NY)**131** , 163 (1981). 3. O.Bohigas, M.J. Giannoni, and C. Schm Phys.Rev.Lett. **52**, 1.(1984); T.H. Seligman, J.J. Versbaarschot, and M.R. Zirnbauer ibid. **5**

215.(1984); E.Heller,H. Koppel and L.S. Cerderbaum ibid. **52** 1665.(1984); T.H.Seligman and J.J.Versbaarshot Phys.Lett. **A108** 183. (1985) and J.Phys. A18 2751. (1985).

4. See the recent updated review by T.A. Brody et.al. Rev.Mod.Phys. **53**,385.(1981)

5. R.V.Jensen Phys. Rev.Lett. **49** 1365(1982) and Phys.Rev. A30 386(1984).

6. E.Bayfield and P.M Koch Phys.Rev.Lett. **33**,258(1974). P.M.Koch and D.R.Mariani ibid. **46**, 1275 (1981); K.A.H. van Leeuwen et.al. ibid **55**, 2231 (1985), and references therein. J.Bayfield, "Fundamental Aspects of Quantum Theory", (Notes in Physics, Springer Verlag, 1986). J.E.Bayfield and L.A.Pinnaduwage, Phys. Rev Lett. **54** 313 (1985). J.N. Bardsley at.al. ibid. **56**, 1007 (1986)

7. G.Casati,B.V.Chirikov,F.M.Izrailev, and J.Ford in, Stochastic behavior in classical and quantum Hamiltonian Systems, ed G.Casati and J.Ford (Lecture Notes in physics **Vol 93**, 334.(1979). Springer, NY).

6. F.M.Izrailev and D.L.Shepelyanski, Dokl.Akad.Nauk **SSSR 249**,1103 (1979). [Sov.Phys.Dokl. **24**, 996 (1979)]; T.A. Hogg and B.A.Huberman Phys.Rev.Lett. **48**,711 (1982); Phys.Rev. **A28**, 22.(1983); S.J.Chang and K.J.Shi Phys.Rev.Lett **55**,(1985).

8. M.Feingold,S.Fishman, D.R.Grempel and R.E.Prange hys.Rev. **B31**, 6852(1985)

9. S.Fishman, D.R.Grempel and R.E.Prange Phys.Rev.Lett. **49**, 509 (1982). D.R.Grempel and R.E.Prange and S.Fishman Phys. Rev. **A 29**, 1639 (1984)

10. A.Molcanov Comm.Math.Phys. **78**, 429(1981)

11. F.M.Izrailev Pys. Rev. Lett. **56**, 541 (1986)

12. J.V. José and R.Cordery ibid. **56**, 290 (1986)

13. J.V. José, (to be publhised)

14. G.M.Zaslavskii and B.Chirikov Dokl. Akad. Nauk. **SSSR 159**, 306 (1964); [Sov.Phys.Doklady **9**, 760 (1965)]; A.Brahic Astrophys. **12** 98 (1971); M.A.Lieberman and A.J.Lichtenberg Phys.Rev. **A5** 1852 (1972). For an excelent review see "Regular and stochastic motion" by A.J. Lichtenberg and A.M. Lieberman, Publ. Springer-Verlag ,(1983)

15. E.Fermi Phys. Rev. **75**, 1169.(194

16. C. Roman and T. Seligman (in this proceedings).

17. M. Berry, has also reconized in the past that the type of functional forms given by Eq(6), do have the property of simplifying highly nonlinear problems (private communication).

18. G.Casati, B.Chirikov and D.L. Shepelyanskii, Phys. Rev. Lett. **53**, 2525 (1984)

QUANTUM LIMITATIONS OF CHAOS AND SUBTHRESHOLD IONIZATION IN HYDROGEN ATOM

Giulio Casati
Dipartimento di Fisica dell'Universita,Via Celoria,16-20133 Milano, Italy

Italo Guarneri
Dipartimento di Fisica Teorica e Nucleare dell'Universita, 27100Pavia, Italy

1. Introduction

Self-generated stochasticity is a very common occurrence in Hamiltonian systems subject to time-periodic perturbations. In macroscopic systems of this type, chaos is directly responsible for easily observable, (and often undesirable) effects.

The problem, whether any such effects survive also in Quantum Mechanics is an important one, expecially in connection with studies on microwave ionization of Rydberg atoms. Previous works [1-8] have shown, that strong ionization and excitation can take place even for frequencies well below the one photon ionization threshold. There are strong indications that this quantum phenomenon is connected with the appearance of chaotic motion in the corresponding classical system; indeed, in the classical model of a Hydrogen atom under an external periodic field, a stochastic transition takes place, leading to unlimited diffusion in phase space and eventually to ionization [2,3].

On account of such results, we are led to suspect some quantum process simulating classical chaotic diffusion. On the other hand, previous studies [9-12] on different models have shown that quantization places severe limitations on classical chaos, that may lead to a complete suppression of chaotic diffusion, even in the semiclassical region. We are then faced with the problem of understanding why in Hydrogen atom this quantum limitation fails to produce its full paralyzing effect, leaving room for "diffusive" excitation and ionization.

In this paper we summarize some recent results [5,6,13,14] of ours that we hope may contribute to the clarification of this problem. First, we shall discuss some general features of quantum systems under periodic perturbations, and we shall illustrate the quantum limitation of chaos on the kicked rotator model according to [15]. Then we shall show how some relevant ideas can be fruitfully borrowed from the rotator model in order to analyze the effect of the quantum limitation of chaos in the Hydrogen atom problem.

The result of this analysis will be that the onset of diffusive excitation in Hydrogen atoms is determined by a quantum delocalization border for field intensity. Since this border is in principle distinct from the classical chaotic border, the picture of diffusive excitation in Hydrogen atoms is more complex than one might guess on purely classical grounds. Indeed, both a delocalized and a localized quantum regime can occur above the classical threshold. In the former regime, quantum motion follows to some extent the classical diffusion; this regime was explored by recent experiments [8]. In the latter regime, however, no quantum diffusion should be expected, in contrast to classical predictions. The present level of experimental art makes it quite possible to test these predictions of a transition between these two different regimes of quantum motion.

2. The Classical Analysis

An integrable Hamiltonian system subject to an external periodic perturbation is described by an Hamiltonian

(1) $$H(\mathbf{I}, \boldsymbol{\varphi}, t) = H_0(\mathbf{I}) + \epsilon\, V(\mathbf{I}, \boldsymbol{\varphi}, t)$$

where $V(\mathbf{I}, \boldsymbol{\varphi}, t)$ is periodic in $\boldsymbol{\varphi}$ and t with periods 2π and T respectively, and $\mathbf{I}$, $\boldsymbol{\varphi}$ are action-angle variables for the unperturbed system described by the Hamiltonian H_0.

As specific examples, we will consider the "δ-kicked rotator"[9]:

(2) $$H(I, \varphi, t) = I^2/2 + \epsilon \cos\varphi \sum_n \delta\,(t-nT)$$

and the one-dimensional model for a hydrogen atom in a monochromatic electric field[4]:

(3) $$H(I, \varphi, t) = -\,1/2I^2 + \epsilon\, x(I,\varphi) \cos \omega t; \qquad \omega = 2\pi/T;\ x > 0.$$

Here, $x(I, \varphi)$ is the x coordinate of the electron, expressed as a function of action-angle variables of the unperturbed atom.

Model (3) has a direct physical relevance, in that it can be used to study the effect of a microwave field on a hydrogen atom initially prepared in an "extended" state. The use of the one-dimensional approximation (3) proves theoretically and experimentally justified for such initial states [4].

Instead, model (2) has no such immediate physical meaning. Nevertheless, it is a particularly convenient one in order to analyze the transition to chaos; indeed, by studying this model, an insight on the effect of quantization on dynamical chaos was gotten, which proved very useful for model (3) also.

Both classical systems (2) and (3) exhibit a transition to chaos when the coupling parameter ϵ exceeds some "stochasticity threshold", that can be estimated by means of Chirikov's resonance overlapping criterion [16] .

For model (2), this transition occurs for

$$\epsilon T > 1 \tag{4}$$

Above this threshold, the motion is more conveniently described in statistical terms. Specifically, considering an ensemble of trajectories leaving with a fixed value I_0 of the action and with randomly distributed phases φ, one finds that the distribution function $f(I, t)$ in action space is well approximated by the solution of the Fokker- Planck equation:

$$\partial f(I, t)/\partial t = 1/2 \;\; D \, \partial^2 f(I,t)/\partial I^2$$

with $D \approx \epsilon^2/2T$ and with the initial condition $f(I,0) = \delta(I-I_0)$.

In particular, the rotator kinetic energy increases linearly with time, according to the law $\overline{\Delta I^2} \approx Dt$.

For model (3), trajectories leaving from initial states with $I=n_0$ undergo a stochastic transition for[2,3]

$$(5) \qquad \epsilon_0 > \epsilon_s = 1/(50\omega_0^{1/3}); \quad \epsilon_0 = \epsilon n_0^4 \, ; \quad \omega_0 = \omega n_0^3 \, .$$

for $\omega_0 > 1$. When $\omega_0 < 1$, the analysis of the chaotic transition is more involved [3].

The motion in the stochastic regime can again be described by a Fokker-Planck equation:

$$(6) \qquad \partial f/\partial t = 1/2 \, \partial/\partial I \, (D(I)\partial f/\partial I) \, .$$

Here the diffusion coefficient D(I) depends on the action I: $D(I) \approx \pi^{-1}\epsilon^2 I^3 \omega^{-4/3}$. A quite remarkable feature of the diffusion ruled by eq.(6) is that the moment $\langle I \rangle = \int I f(I, t) dI$ grows to infinity - i.e., the atom ionizes - in a finite time $t_I \approx 2\omega^{4/3}/(\epsilon^2 n_0)$.

3. <u>The Quantum Localization Phenomenon.</u>

Let's now quantize models (2) and (3). In both cases we get Schroedinger equations of type

$$(7) \qquad i \, \partial\psi/\partial t = H_0 \, \Psi + \epsilon \, V(t)\Psi$$

with V(t) an operator depending periodically on time.

A particularly convenient way to study equations of this type is introducing the Floquet operator $S = U(0,T)$ where $U(s,s+t)$ is the unitary operator which gives the evolution of states ψ over the time t according to $\psi(t+s) = U(s,s+t)\psi(s)$. Indeed, in order to analyze the long-time behaviour of the solutions of eq.(7), it is sufficient to study the iterates S^n of this Floquet operator[17].

A first qualitative classification of various types of behaviour that solutions of (7) can show is provided by the nature of the quasi-energy spectrum, which is by definition the spectrum of the self-adjoint operator G such that $S = \exp(iG)$. As a matter of fact, in both models (2) and (3) the unperturbed Hamiltonian H_0 possesses a discrete spectral component, and one is interested in the time evolution of states initially coinciding with some unperturbed eigenstate. Then, it can be shown[17] that a continuous quasi-energy spectrum would enforce an indefinite spreading of such wave packets over the unperturbed spectrum. Instead, a pure point quasi-energy spectrum would be associated with a recurrent behaviour of the wave packets.

In the rotator case (model (2)) it was found that the nature of the quasi-energy spectrum depends in a sensitive way on the value of the period T. If T is a rational multiple of 4π, then the spectrum of the Floquet operator S possesses an (absolutely) continuous component, which causes the spread of the wave packet over the unperturbed eigenstates $\langle n^2 \rangle$ (which in this case is proportional to the kinetic energy of the rotator) to increase with time according to an asymptotic t^2 law. This phenomenon is known as quantum resonance and cannot be understood in classical terms; its appearance in model (2) is due to the particular nature of the unperturbed spectrum of this model[18].

When $T/4\pi$ is irrational, numerical experiments provide evidence that, in most cases, the quasi-energy spectrum is pure point. Far from spreading indefinitely, the wave packet remains <u>localized</u> in momentum space. Thus, even in the semiclassical regime ($\epsilon \gg 1, T \ll 1$) and for such parameter values that the classical rotator is chaotic ($\epsilon T \gg 1$) the quantum rotator does not reproduce the classical diffusive increase of $\langle n^2 \rangle$, except for a more or less extended, but anyway finite, time scale t_B, after which $\langle n^2 \rangle$ enters a seemingly steady-state oscillatory regime.

However, even though localization seems to be a fairly typical occurrence for irrational $T/4\pi$, we were able to prove [19] that a dense set of irrational $T/4\pi$ values exists, yielding a <u>continuous</u> quasi-energy spectrum. It is not yet clear whether this continuous nonresonant spectrum has any physical relevance, nor it is clear whether it is possible to observe it also in different models where the unperturbed spectrum is pure point. On account of numerical and theoretical analysis, these values of the period look somewhat exceptional. Therefore, despite the intriguing character of the spectral problem, our basic question about the persistence of chaos in quantum mechanics must be answered in the negative for model (2). The typical picture for the quantum rotator in the classically chaotic regime is the following: wave packets initially concentrated on a single unperturbed eigenstate start spreading in momentum space, but after a certain "*break-time*" t_B, quantum interference effects become dominant, which prevent further spreading. Then, the average number of unperturbed eigenstates significantly excited by the wave packet evolution, settles to a stationary value, which is called localization length and provides also a measure for the number of unperturbed eigenstates significantly excited by one single quasi-energy eigenstate.

In the semiclassical region a simple estimate can be given for both the localization length and the break time. We now sketch this estimation since the idea behind it is basic also to our present understanding of the quantum model (3).

Before the time t_B, the spread Δn of the wave packet over the unperturbed levels (quantized actions $I = n$) will grow in time, following to some extent the law $\Delta n=\Delta n(t)$ that is predicted by the Fokker Planck equation that rules classical diffusion. However, this diffusive growth will stop, because the discrete character of the quasi-energy spectrum will eventually become manifest; the time t_B necessary for this can be assumed to be $t_B \approx \alpha N$, with α a constant and N the number of q.e. eigenstates significantly excited by the single initial unperturbed eigenstate. In fact, $2\pi/N$ is just the average spacing of q.e. eigenvalues significantly contributing in the wave packet evolution. Then we can reasonably assume that N also measures the number of unpertubed eigenstates spanned by a single q.e. eigenstate, i.e., $N \sim l$ (the localization length). On the other hand, the latter coincides in order of magnitude with $\Delta n(t_B)$; thus we get an equation for t_B:

$$\alpha \Delta n(t_B) \approx t_B$$

In the rotator case, classical diffusion obeys $(\Delta n(t))^2 \approx Dt$, whence it follows that $t_B \approx \alpha^2 D$. Numerical experiments support very well this estimate, provided that the value $\alpha \approx 1$ is chosen [15].

4. The Quantum Delocalization Border

Now we come to the hydrogen atom model (3). Here the question about the nature of the quasi-energy spectrum is clearer; indeed, it is known that the q.e. spectrum is absolutely continuous[20]. This means that, unlike the classical atom, the quantum atom will eventually ionize, no matter how small ϵ. The corresponding ionization mechanism, however, at least for small ϵ has nothing to do with the classical chaotic phenomenon, and involves very long time scales. From the mathematical point of view, this entails that identifying the nature of the q.e. spectrum is not enough for answering the question of the existence of "quantum chaos". Indeed, this question calls for the analysis of time evolution of wave packets over the time scale in which quantum and classical evolution may be expected to agree to some extent.

In the present case, the short time behaviour of the atom is dominated by resonances (in the sense of scattering theory). In other words, the q.e. spectrum is absolutely continuous, but there are poles of the resolvent operator lying very close to the real axis[20]. For practical purposes, it can still be assumed that the q.e. spectrum has a pure point component, in which each level has a small width.

Therefore, we can expect that the same mechanism which leads to quantum localization of the classical dynamical chaos in the rotator case will be working here, too. There is an essential difference, however; indeed, if it is possible to choose such parameter values that the quantum $\Delta n(t)$ follows the classical diffusion for a time larger than the classical chaotic ionization time, then we may expect "diffusive" ionization also in the quantum case. Conditions for this <u>delocalization</u> phenomenon can be easily derived; in fact, the break time can be estimated by the same procedure outlined above for the rotator case - the difference being that here the appropriate Fokker-Planck equation (6) must be used. In this way we find that, for $\omega_0 > 1$ delocalization will occur for[5,13]

$$(8) \qquad \epsilon_0 \gtrsim \epsilon_s \qquad\qquad \epsilon_0 \gtrsim \epsilon_q = \omega_0^{\,7/6}/\sqrt{(6n_0)}$$

A remarkable fact is that in order to obtain the estimate (8) for the *delocalization border,* the parameter α was taken ≈ 1 just like in the rotator case. The good agreement of (8) with numerical results implies that, however particular, the rotator model still retains features that have some generality in problems about quantization of chaotic systems.

Like in the rotator case, the model (3) was the object of extensive numerical investigations, which fully confirm the main result of the above discussion[9]. Namely, three distinct quantum regimes can be observed according to the particular choice of ϵ_0, ω_0 and of the initial state n_0 . If ϵ_0 lies below both the classical stochasticity border ϵ_s and the delocalization border ϵ_q , then both the classical and the quantum model will exhibit localization in action space; this is the case of fig. 1, where $\omega_0 = 1.5$, $n_0 = 100$, $\epsilon_0 = .01$ and where the quantum and classical

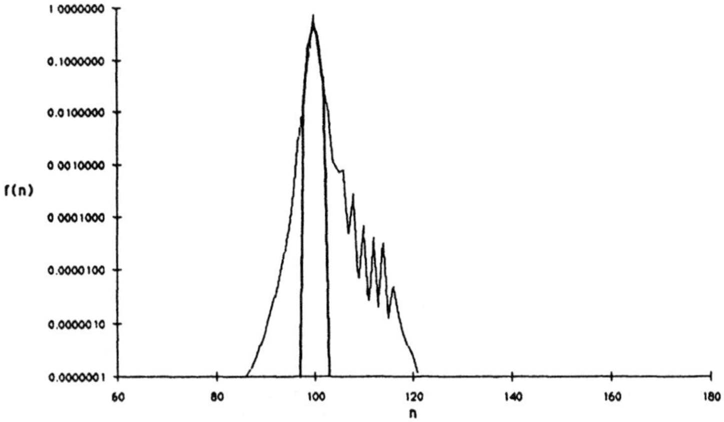

Fig.1 Classical(—)and quantum(-) distribution function $\overline{f}_n$ averaged over 40 values of $\tau=\omega t/(2\pi)$ within the interval 80 $<\tau<$ 120. Here $\omega_0=1.5$, $\epsilon_0=.01$, $n_0=100$. For these parameters value, both classical and quantum packets are localized. Notice the small quantum tunneling through the classical Kolmogorov invariant curves.

distributions over unperturbed actions are shown after 120 periods of the external field.

If ϵ_0 lies in between the thresholds ϵ_s and ϵ_q , we observe classical chaotic motion and quantum localization; here the classical atom would ionize diffusively, but the quantum one does not. (Fig. 2: $n_0=100$, $\omega_0 = 1.5$, $\epsilon_0 = .03$, distributions after 480 periods). In Fig. 3 we show the comparison of classical and quantum probability of excitation above $n=1.5\ n_0$, as functions of time. Quantum localization of classical chaos is here quite evident.

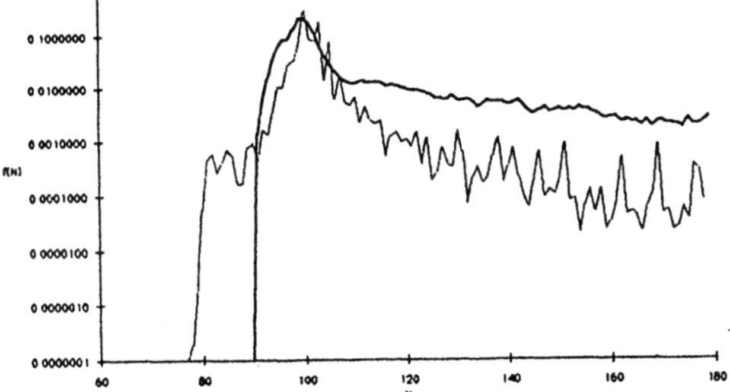

Fig.2 Same as Fig. 1 but with $\epsilon_0 = .03$. The classical (—) and quantum (-) distribution functions are averaged over 40 values of τ within the interval $440<\tau<480$. In this case $\epsilon_s < \epsilon_0 < \epsilon_q$, and the quantum packet is localized as expected (besides a small resonant plateau). On the contrary, the classical packet is strongly diffusing as it is also shown in fig. 3.

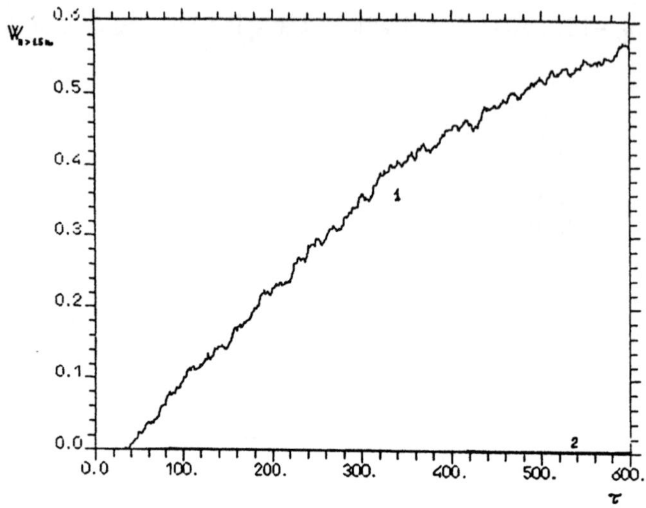

Fig. 3a

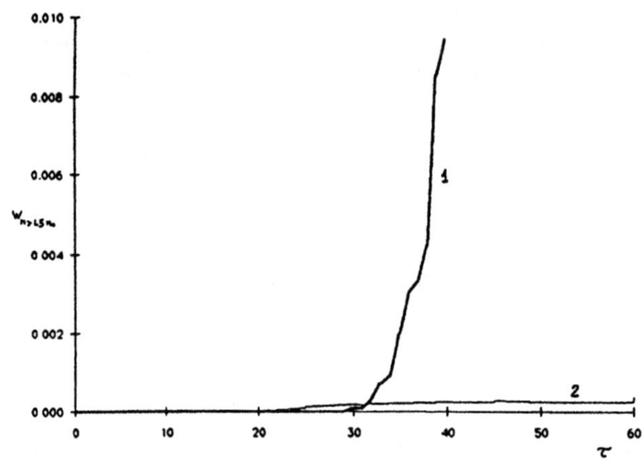

Fig. 3b

Fig.3 Classical(1) and quantum(2) total probability $W_{n>1.5n_0}$ above level $n=1.5n_0$ for the case of fig.2. The classical ionization probability is order of magnitudes higher than the quantum one(fig.3a). The comparison of the two ionization probabilities for short times is shown in fig. 3b.

Finally, if ϵ_0 exceeds both thresholds, we observe a qualitative agreement between classical and quantum results (Fig. 4, n_0 =100, ω_0= 1.5, ϵ_0 =.15). Here a quantum ionization mechanism is at work, that cannot be understood in terms of standard perturbation theory. Indeed, this mechanism is effective at much lower frequencies than those ($\omega_0 > n_0/2$) predicted by the pertubative theory of the photoelectric effect.

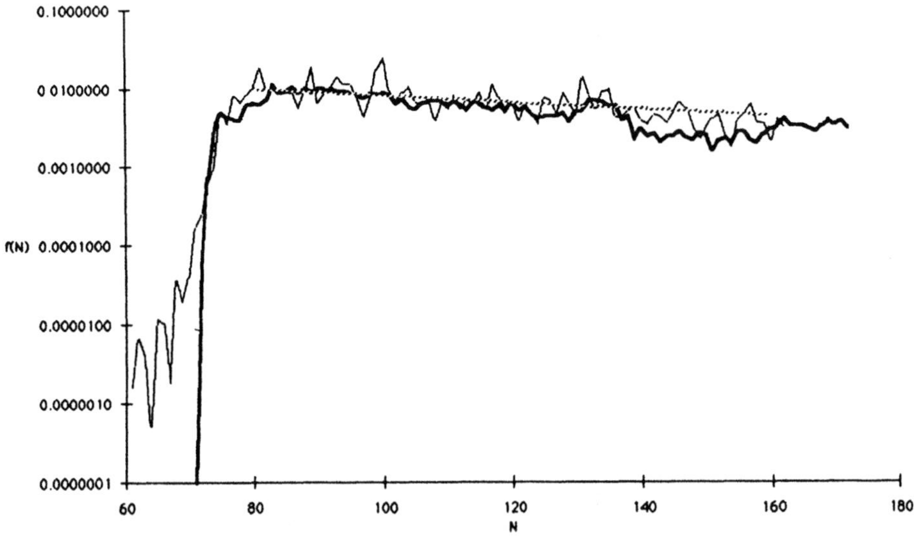

Fig.4 Same as fig.1 but with $\epsilon_0 = .15$. Here, the classical (–) and quantum (-) distribution functions are averaged over 40 values of τ within the interval $0<\tau<40$. In this case $\epsilon_0 > \epsilon_q$ and both classical and quantum motion obey a diffusion law given by Fokker-Planck equation. The dotted line gives the solution of the Fokker Planck equation.

5. Underthreshold Ionization

Due to the delocalization phenomenon, the dependence of the ionization probability W_I on frequency at fixed field intensity displays a number of unexpected features. In Fig. (5), we plot W_I against ω_0 for $n_0 = 66$, $\epsilon_0 = .05$. First of all, since the classical diffusion is due to resonances between the external field and the harmonics of the electron motion, it takes place only for frequencies ω_0 above some critical value ω_c of the order of 1, when $\epsilon_0 \geq 50^{-1}$. (for $\epsilon_0 \leq 50^{-1}$, the value ω_c has to be determined from eq. (5)). Notice that, since the diffusion coefficient D that determines the rate of diffusive excitation is decreasing with ω_0 at fixed ϵ the critical value ω_c for frequency has classically a marked threshold character for diffusive excitation.

In Fig. (5) we see that for the given field the quantum atom is delocalized at $\omega \geq \omega_c$, and quantum and classical excitation go in approximately the same way. However, since $\epsilon_q > \epsilon_s$, at $\omega_l \approx (6n_0\epsilon_0^2)^{3/7}$ one falls below the delocalization border. Then localization occurs, and the ionization probability drops as compared to the classical value. Upon further increasing ω_0, the one photon threshold is reached; beyond that, the numerically computed ionization probability is in

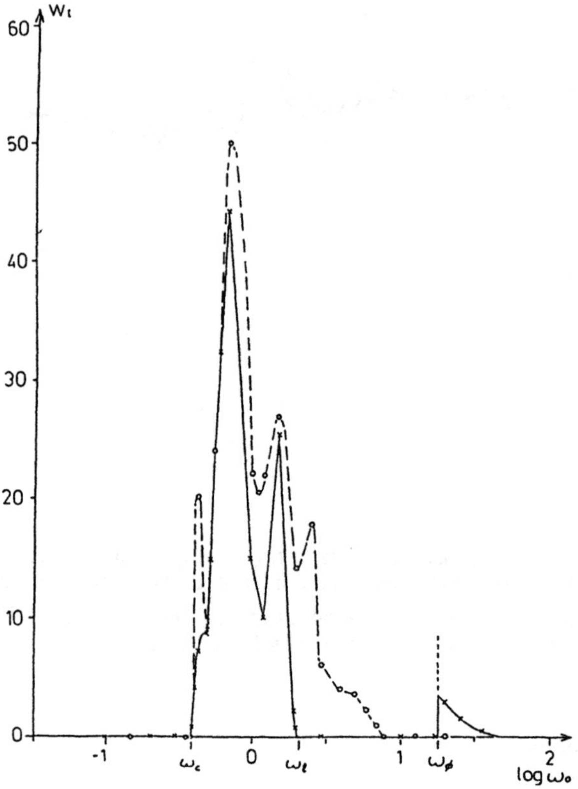

Fig.5 Ionization probability $W_I = \sum_{n>\bar{n}} |c_n|^2$ versus field frequency ω_0 after a number of periods $\tau=40\omega_0$ which corresponds to the same real physical time t for all frequencies. Here $n_0= 66$, $\epsilon_0 =.05$, $\bar{n}=99$. (×) Quantum case, (o) Classical case. The position of the threshold ω_Φ is due to the definition of W_I adopted here.

perfect agreement with standard theoretical predictions.

As is apparent from Fig. (5), diffusive ionization occurs at very low frequencies as compared to the one photon threshold. Even more strikingly, its rate is much higher.

REFERENCES

[1] J. E. Bayfield and P.M. Koch, Phys. Rev. Lett. 33 (1974) 298.

[2] N.B. Delone, V.P. Krainov, D.L. Shepelyansky, Usp. Fiz. Nauk 140 (1983), 355 (Sov. Phys. Uspeky, 26 (1983), 551).

[3] R.V. Jensen, Phys. Rev. A,30 (1984), 386.

[4] D.L. Shepelyansky, Proc. Int. Conf. on Quantum Chaos, Como 1983 (Plenum ed. 1984, p. 184).

[5] G. Casati, B.V. Chirikov, D.L. Shepelyansky, Phys. Rev. Lett. 53 1984) 2525.

[6] G. Casati, B.V. Chirikov, I. Guarneri, D. L. Shepelyansky: "A New threshold for Photoelectric Effect in Hydrogen Atom" Submitted to Phys. Rev. Lett. December 1985;

[7] R. Blumel and V. Smilansky: "Subthreshold Ionization of Rydberg Atoms in a Radiation Field". Preprint.

[8] K.A.H. Van Leeuwen, G.V. Oppen, S. renwick, J.B. Bowlin P.M. koch, R.V. Jensen, O. Rath, D. Richards, J.G. Leopold, Phys. Rev. Lett. 55 (1985), 2231.

[9] G. Casati, B.V. Chirikov, J. Ford, F.M. Izrailev, Lecture Notes in Physics 93 (1979) 344.

[10] B.V. Chirikov, F.M. Izrailev, D.L. Shepelyansky, Soviet Scientific Reviews, 2C (1981), 209

[11] D.L. Shepelyansky, Physica, 8D (1983), 208.

[12] S. Fishman, D.R. Grempel, R.E. Prange, Phys. Rev. Lett. 49 (1982), 509; Phys. Rev. A, 29 (1984), 1639.

[13] G. Casati, B.V. Chirikov, I. Guarneri, D.L. Shepelyansky. Preprint.

[14] G. Casati, J. Ford, I. Guarneri, F. Vivaldi: "On the search for randomness in the kicked rotator" Phys. Rev. A, (to appear).

[15] B.V. Chirikov, D.L. Shepelyansky, Preprint 85-29, INP, Novosibirsk 1985.

[16] B.V. Chirikov, Phys. Rep. 52 (1979) 263.

[17] J. Bellissard, in 'Trends in the Eighties", Lect. Notes in Math, Blanchard 1985.

[18] F.M. Izrailev, D.L. Shepelyansky, Teor. Mat. Fiz. 43 (1980) 417.

[19] G. Casati, I. Guarneri, Comm. Math. Phys. 95 (1984) 121.

[20] S. Graffi, V. Grecchi, H. J. Silverstone, Ann. Inst. Poincare 42 (1985) 215.

RESONANCE OVERLAP IN QUANTUM SYSTEMS

L. E. Reichl and W. A. Lin
Center for Studies in Statistical Mechanics
University of Texas
Austin, TX 78712

Recently much attention has been focused on the behavior of quantum systems such as the kicked rotor and billiard systems whose classical counterparts are known to exhibit fully developed chaos. The goal is to identify and quantify the existence of "quantum chaos" in these systems. In classical systems, chaos results from resonance overlap. This has been shown for both time independent[1] and time dependent Hamiltonian systems.[2-5] Resonance overlap is most easily studied in time dependent systems because the size and position (in phase space) of nonlinear resonance zones can be varied by varying external field parameters. Even the simplest nonlinear one degree of freedom mechanical system has an infinite number of natural frequencies which vary as the energy of the system varies. A monochromatic external field will resonate with the system at each of these natural frequencies but for different energies of the underlying system. A strobe plot of the phase space shows clearly the existence of these resonance zones.[2-5] For small amplitude external fields, resonance zones are separated by KAM tori. However, as the external field amplitude increases, these zones overlap, allowing chaotic diffusion by the state of the system throughout the region of phase space occupied by the overlapping resonance zones.

One of the simplest nonlinear one degree of freedom systems consists of a particle in an infinite square well potential. The classical Hamiltonian for this system in action-angle space is $H_0 = \pi^2 J^2/8ma^2$ where J is the action, m is the mass of the particle and a is the half width of the square well potential. If we now drive the particle by a monochromatic external force of the form $F = \lambda\cos\omega_0 t$ (λ is the amplitude of the force) the effect in action angle space is to perturb the system by an infinite number of traveling cosine potential wells. In terms of momentum p, and position, x, of the particle the perturbed Hamiltonian can be written

$$H_0 = \frac{p^2}{2m} + V_0[\eta(x-a) + \eta(-x-a)] + \lambda x \cos\omega_0 t \tag{1}$$

($V_0 \to \infty$), where $\eta(x) = 0$ for $x < 0$ and $\eta(x) = 1$ for $x > 0$. In action angle space the Hamiltonian takes the form

$$H = \frac{\pi^2 J^2}{8ma^2} - \sum_{\substack{n=-\infty \\ (\text{odd})}}^{\infty} \frac{4\lambda a}{n^2\pi^2} \cos(n\theta - \omega_0 t) \tag{2}$$

where θ is the angle variable. In reference 5, we have shown that this system exhibits chaos due to resonance overlap and yet it has such a simple structure that renormalization group techniques can give excellent predictions for the values of λ and ω_0 at which overlap occurs.

The quantum Hamiltonian for this system may be written $\hat{H} = \hat{H}_0 + \lambda\hat{x}\cos\omega_0 t$, where $\hat{H}_0$ is the Hamiltonian operator for a particle in an infinite square well potential and $\hat{x}$ is the position operator for the particle. Eigenstates of $\hat{H}_0$ are denoted $|n\rangle$ and corresponding eigenvalues can be written $E_n^0 = \hbar\Omega n^2$ where $\Omega = \hbar\pi^2/8ma^2$. If we let $|\psi(t)\rangle$ denote the exact state of the system at time t and let $\langle n|\psi(t)\rangle \equiv \psi_n(t)$ denote the probability amplitude to find the system in the nth energy level at time t, then the Schrödinger equation for $\psi_n(t)$ can be written

$$i\hbar \frac{\partial \psi_n(t)}{\partial t} = \hbar\Omega n^2 \psi_n(t) - \frac{4a\lambda}{\pi^2} \cos\omega_0 t \sum_{\substack{N=1 \\ (\text{odd})}}^{\infty} \frac{1}{N^2} \{\psi_{n-N}(t) + \psi_{n+N}(t)\} \tag{3}$$

with the condition $\psi_{-n}(t) = -\psi_n(t)$. This difference equation can also be written in differential form if we introduce the generating function $F(\theta,t) = \sum_{n=-\infty}^{\infty} e^{in\theta}\psi_n(t)$. Then the Schrödinger equation in the angle picture takes the form

$$i\hbar \frac{\partial F}{\partial t} = -\hbar\Omega \frac{\partial^2 F}{\partial \theta^2} - \frac{4a\lambda}{\pi^2} \sum_{\substack{N=1 \\ (odd)}}^{\infty} \frac{1}{N^2} \{\cos(N\theta-\omega_0 t) + \cos(N\theta+\omega_0 t)\}F \tag{4}$$

Eqs.(3) and (4) are exact. From Eq.(4) we see that for the quantum system the effect of the external field is to induce an infinite number of cosine traveling potential wells in the angle representation of the system. For each N, the cosine potentials given rise to a resonance zone which affects a specific set of energy levels, $|n\rangle$. In reference 6, we have shown that the center of the Nth resonance zone lies at energy level $\bar{n}_N = \omega_0/2\Omega N$. The halfwidth of the Nth resonance zone is $\Delta n_N = (64\lambda ma^3/\hbar^2 N^2 \pi^4)^{1/2}$. That is, the Nth resonance zone contains $2\Delta n_N$ energy levels. In the absence of the external field (λ=0), n is a good quantum number. If the system is initially in the state $|n\rangle$, it will stay there. However, the external field induces transitions among the energy levels. When $\lambda \neq 0$ if we initially populate a single level lying within one of these zones the probability will spread throughout the region of the zone but remain confined to that region. The number of levels inside a given zone increases with $\sqrt{\lambda}$. At a critical value, $\lambda_c = ma\omega_0^2/4(N+1)^2$, resonance zones N and N + 2 will overlap. When that occurs, probability can spread freely between the region of influence of the two resonance zones. In Figures (1)-(3) we illustrate this behavior. Let us first note that since, for small λ, the resonance zones are isolated from one another we need not keep all cosine potentials in Eq.(4) in describing any two resonance zones as long as the other zones do not overlap with these two. The Schrödinger equation which includes only resonance zones N = 1 and N = 3 can be obtained from Eqs.(3) and (4). It is written

$$i\hbar \frac{\partial \psi_n(t)}{\partial t} = \hbar\Omega n^2 \psi_n(t) - \frac{2a\lambda}{\pi^2} [e^{-i\omega_0 t}(\psi_{n-1}(t) + \frac{1}{9}\psi_{n-3}(t)) + e^{+i\omega_0 t}(\psi_{n+1}(t) + \frac{1}{9}\psi_{n+3}(t))] \tag{5}$$

We have solved Eq.(5) numerically for $a = 10a_B$, $\omega_0 \approx (2\pi)^2/1.6920\tau_B$, $\lambda = \Lambda e^2/a_B^2$ (with Λ = 4 and 9), and $m = m_e$ where a_B is the Bohr radius, τ_B is the period of the first Bohr orbit, e is electron charge and m_e is the mass of the electron. Our theory predicts that the resonance N = 1 is located at $\bar{n}_1 \approx 150$ and has halfwidth $\Delta n_1 \approx 51$ when $\Lambda = 4$ while resonance zone $N = 3$ is located at $\bar{n}_3 \approx 50$ with halfwidth $\Delta n_3 \approx 17$. In Figures (1) and (2) we show solutions to Eq.(5) for $\Lambda = 4$.

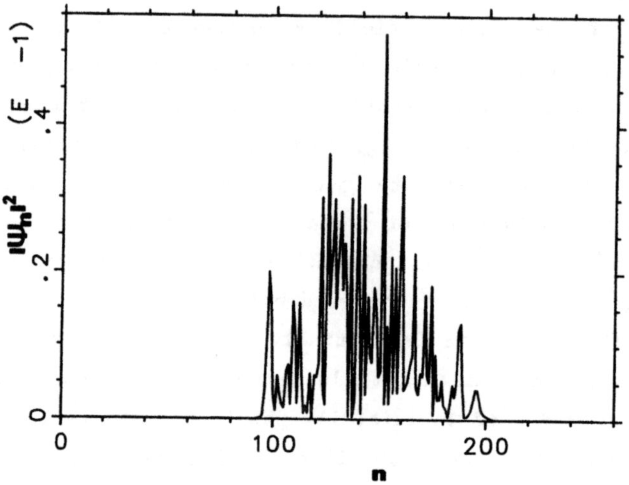

Figure 1: Solution to Eq.(5) after time $t \approx (45)(2\pi/\omega_0)$ for $\Lambda = 4$, $\psi_{130}(0) = 1$ and $\psi_n = 0$ otherwise.

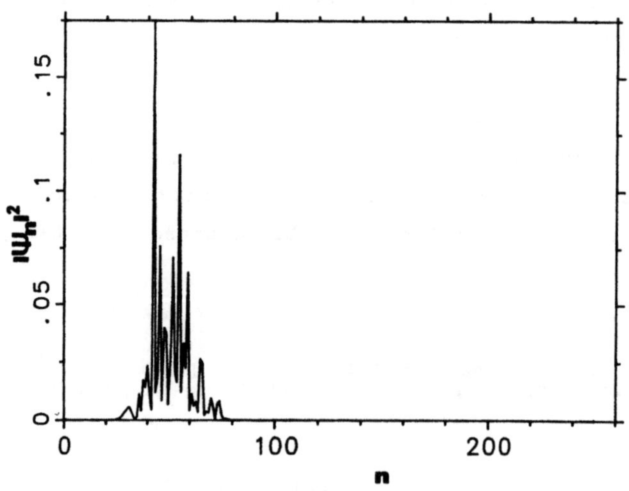

Figure 2: Solution to Eq.(5) after time $t \approx 45(2\pi/\omega_0)$ for $\Lambda = 4$, $\psi_{55}(0) = 1$ and $\psi_n(0) = 0$ otherwise.

In Figure 1 our initial condition was $\psi_{130}(0) = 1$ and $\psi_n(0) = 0$ otherwise. While in Figure 2 our initial condition was $\psi_{55}(0) = 1$ and $\psi_n(0) = 0$ otherwise. In these figures we see the spread of probability after 45 external field periods. The probability has reached its maximum spread. We see that the resonance zones remain well separated. Our theory predicts that overlap can occur at $\Lambda = 8.6$. In figure (3) we show the solution to Eq.(5), for $\Lambda = 9$ and initial conditions $\psi_{130}(0) = 1$ and $\psi_n(0) = 0$, after about 21 periods of the external field. In all cases 250 amplitudes ψ_n were used in the integration and error in normalization remained below 6×10^{-5}. It is clear that overlap has occurred and the probability is free to spread throughout the region of influence of the two resonance zones. These results show that for nonlinear quantum systems a KAM type mechanism for breakdown of invariants ("good" quantum numbers) appears to exist.

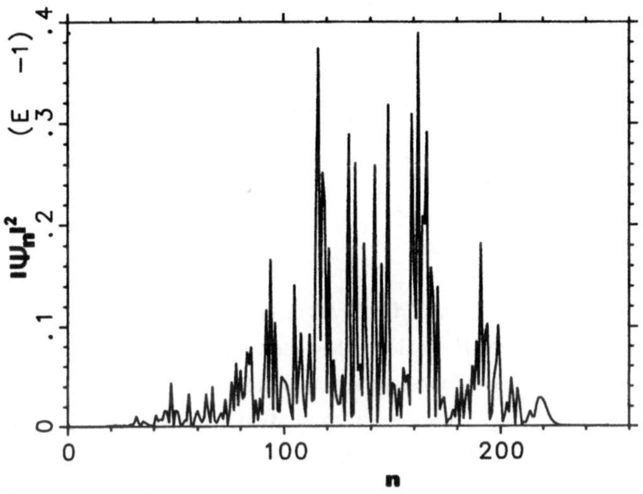

Figure 3: Solution to Eq.(5) after time $t \approx 21(2\pi/\omega_0)$ for $\Lambda = 9$, $\psi_{130}(0) = 1$ and $\psi_n(0) = 0$ otherwise.

ACKNOWLEDGMENTS. The authors wish to thank U.S. Navy Air Systems Command (Contract MDA903-85-C-0029) for support of this work.

References

1. C. H. Walker and J. Ford, Phys. Rev. 188 416 (1969).
2. B. V. Chirikov, Phys. Reports 52 263 (1979).
3. L. E. Reichl and W. M. Zheng, Phys. Rev. A29 2186 (1984); A30 1068 (1984).
4. W. A. Lin and L. E. Reichl, Phys. Rev. A31 1136 (1985).
5. W. A. Lin and L. E. Reichl, Physica 17D, (1986).
6. L. E. Reichl and W. A. Lin, to appear in Phys. Rev. A Rapid Comm., 1986.

ON THE UNIVERSALITY OF FLUCTUATIONS OF ENERGY AND QUASI-ENERGY SPECTRA*

C.E. Román, T.H. Seligman
and J.J.M. Verbaarschot**
Instituto de Física, Laboratorio de Cuernavaca
University of Mexico (UNAM)
México, D.F.

In a number of energy and quasi-energy spectra a smooth transition from Poisson- to GOE-like fluctuations has been observed as a function of the relevant parameters of the problem. The purpose of this paper is to analyse to what extent the character of this transition is universal.

To improve our understanding of this subject we study three and four point correlation functions in the form of the statistics γ_1 and γ_2 given by

$$\gamma_1 = \langle(n(L)-L)^3\rangle/[\langle(n(L)-L)^2\rangle]^{3/2}$$

$$\gamma_2 = \langle(n(L)-L)^4\rangle/[\langle(n(L)-L)^2\rangle]^2 - 3.0$$

where n(L) is the number of levels in the interval L and the average can be an energy or ensemble average or a mixture of both. They contain information not contained in the Δ_3 statistic or the number variance which depend exclusively on the two point correlation function. The nearest neighbour spacing distribution contains only little information on the three and four point functions.

We analyse the energy spectra of homogeneous polynomial Hamiltonians (1) and the eigenphases of the time evolution operators of the fermi-oscillator (2). As both systems are described elsewhere in this volume (3,4), we shall not go into any detail; let us just rmphasize that the spectra of refs (3,4) were used, and in the case of energy spectra the same unfolding procedure was applied. In refs (1), a single parameter random matrix model was developed that described the transition measured by the Δ_3 statistic and the nearest neighbour spacing distribution. Of course, the model does not include the well-understood non-universal behaviour associated with the "kink" of Δ_3.

With the availability of quasi-energy spectra that do not display the "kink", we are in the pleasant circumstances of performing a fit without any arbitrary cut-off. furthermore, the higher order statistics γ_1 and γ_2 provide a crucial test for the quality of the parametrization. The random matrix model is defined by matrix elements

$$M_{ij} = G_{ij}\, e^{-\left(\frac{|i-j|}{\sigma}\right)^{\kappa}}$$

---

* Work supported by Consejo Nacional de Ciencia y Tecnología, México.

** Permanent address: Dept. of Physics, Univ. of Illinois, Urbana/Illinois.

At first sight we have 2 parameters σ and κ but as mentioned in (1) κ = 1,2 and 4 yielded after adjusting σ the same results for the Δ_3- statistic and the nearest neighbour spacing distribution.

We ran extensive tests including γ_1 and γ_2 we found that for $\kappa>1$ we always have the same results, for $\kappa<1$ we find differences principally for Δ_3 and no agreement with experiment is found; details will be reported elsewhere (5). Choosing κ = 2 we have compared the available spectra and the ramdom matrix model. We will present some typical examples.

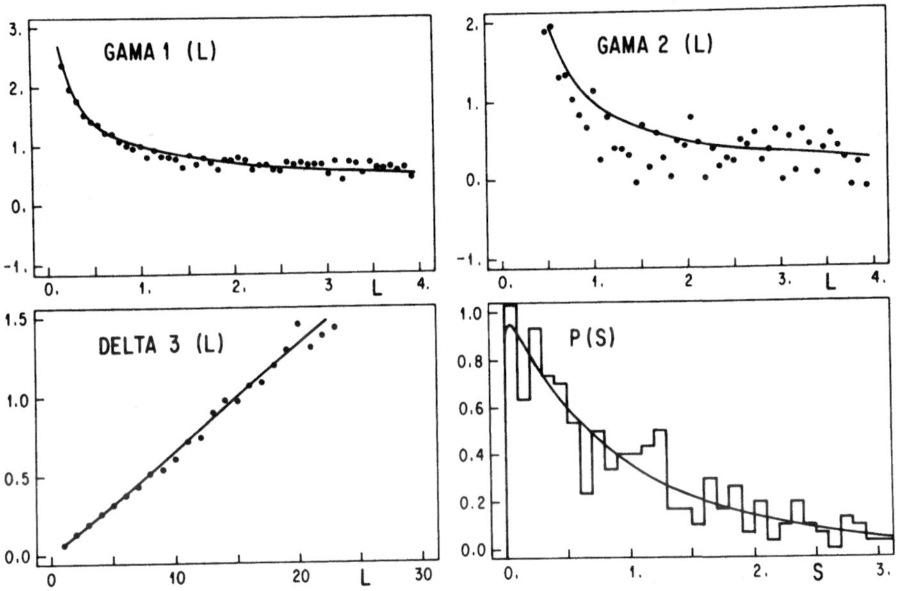

Fig. 1a. The statistics γ_1, γ_2, Δ_3 and nearest neighbour spacing for 300 quasi-energies of the fermi oscillator with parameters that yield a near Poisson distribution, the full lines show the Poisson result.

Figs. 1 shows the results for three quasi-energy spectra, a case close to the Poisson limit, an intermediate case and one close to the GOE limit. Note the absence of a kink in Δ_3. The statistical errors are fairly large which is indicated by the scattering of the points. Within this error the agreement with the model is excellent. Figs. 2 shows a similar analysis for the fourth order polynomial Hamiltonian. The statistical error is considerably smaller a more levels are available. Below the kink the agreement with the model is still very satisfactory. Yet unfortunately this agreement is fallacious in the integrable limit; in this case we expect the Poisson result. Fif. 2a shows that while the nearest neighbour spacing distribution and Δ_3 are also consistent with Poisson, this is not the case for γ_1 and γ_2. The question arises whether this discrepancy has the same origin as the "kink". A this point it may be worth noting that the kink in the in the number variance appears at much

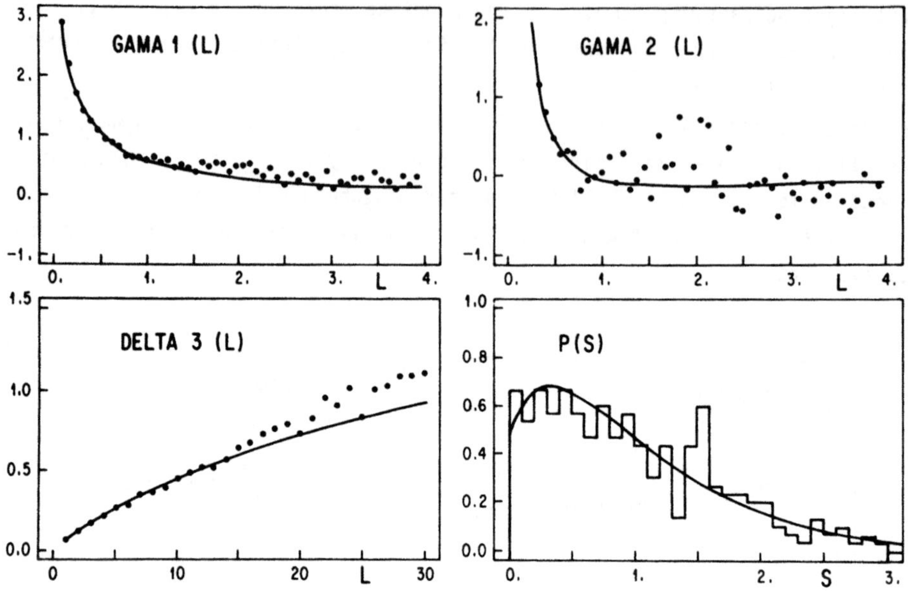

Fig. 1b. As in 1a with parameters that yield typical intermediate statistics. The full line represents a fit with the random matrix model.

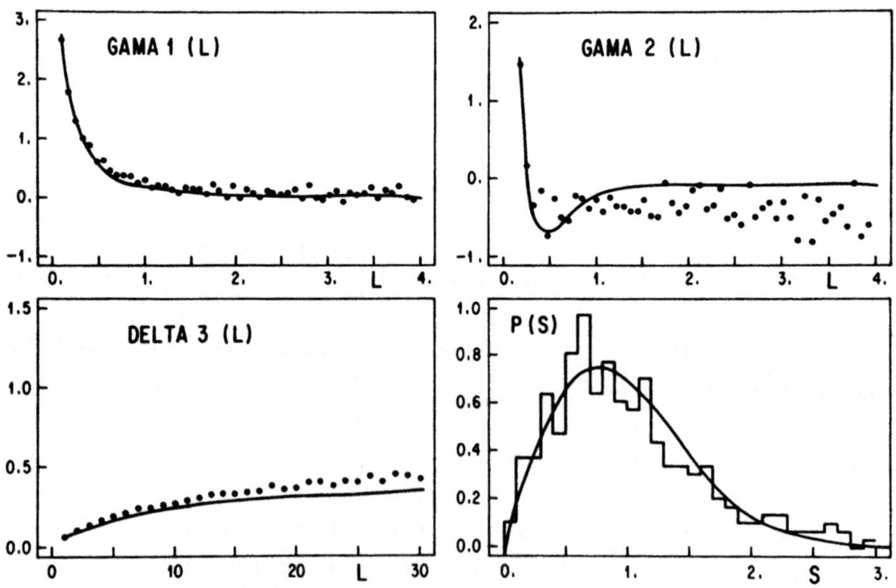

Fig. 1c. As in 1a with parameters that yield statistics a near GOE distribution. The full line shows the GOE result.

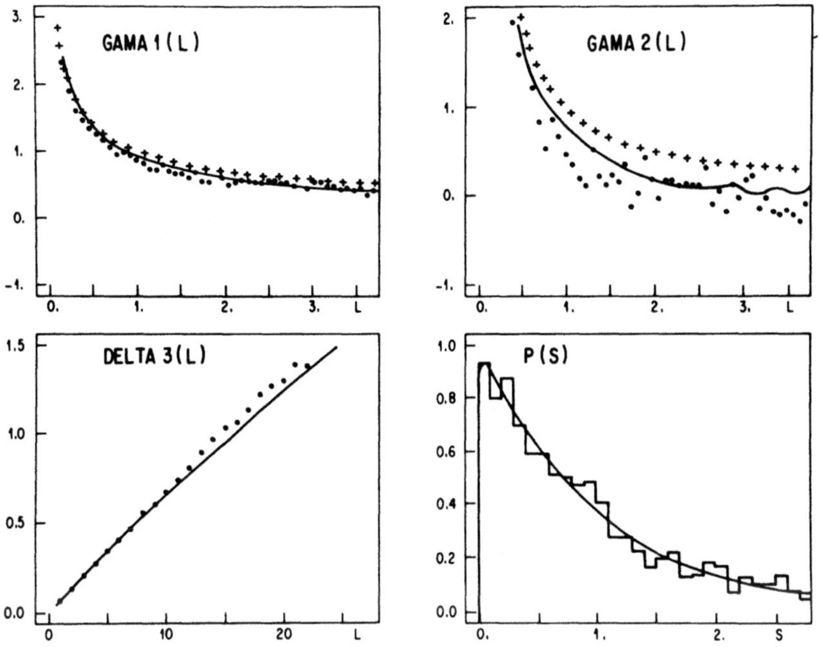

Fig. 2a. The spectral statistics γ_1, γ_2, Δ_3 and nearest neighbour spacing for an integrable fourth order polynomial Hamiltonian using $\sim$ 800 levels. The full line is a best fit with the random matrix model ($\sigma \neq 0$). The crosses indicate the Poisson result.

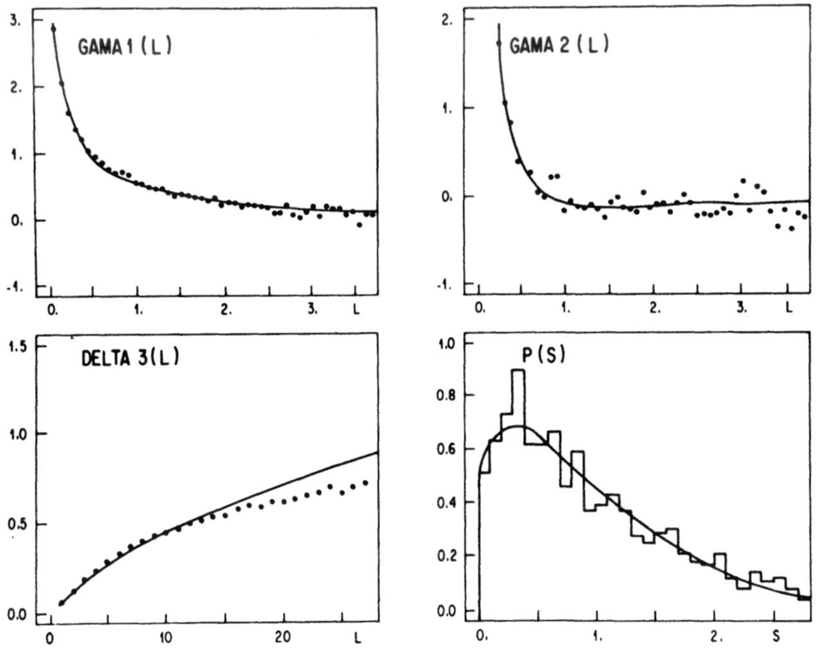

Fig. 2b. As in 2a for a non-integrable fourth order Hamiltonian that gives classically riseto ordred and chaotic motion. The full line is a best fit with the random matrix model.

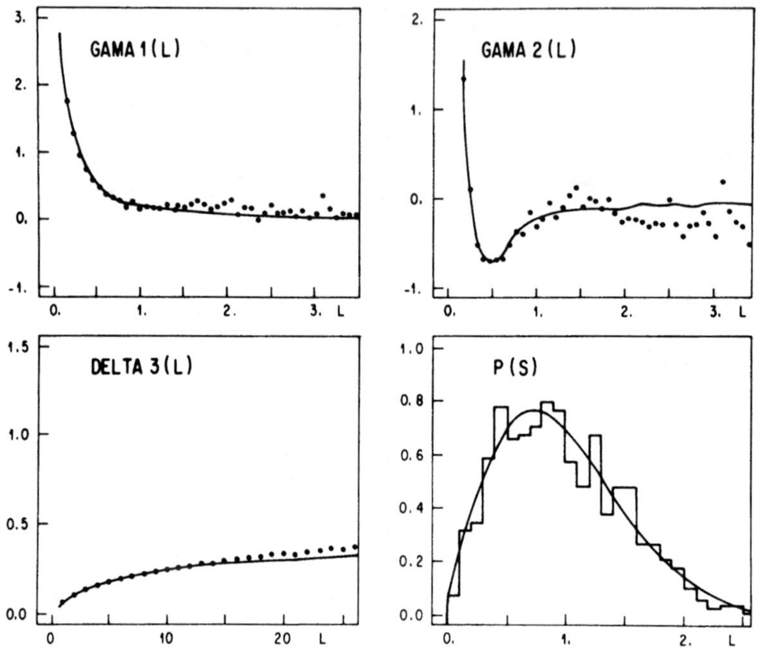

Fig. 2c. As in 2a for a non-integrable fourth order Hamiltonian that gives classically rise to chaotic motion. The full line represents the GOE result.

smaller distances than the kink in the Δ_3 .
Therefore, effects at small distances in γ_1 and γ_2 are not altogether unexpected. A semiclassical calculation for a limit is desirable, but meanwhile we may use the sim-ple criterion of checking whether the discrepancy disappears higher up in the spectrum Fig. 3 shows γ_1 and γ_2 for 32'000 levels somewhat above level 150'000 of the x^4 poten-tial and we no longer find any discrepancies.
Analogous results have been found for the other spectra discussed in refs (1,2). We therefore find that for a wide range of problems including energy and quasi-energy spectra, the transition from Poisson to GOE spectral fluctuations may be parametrized

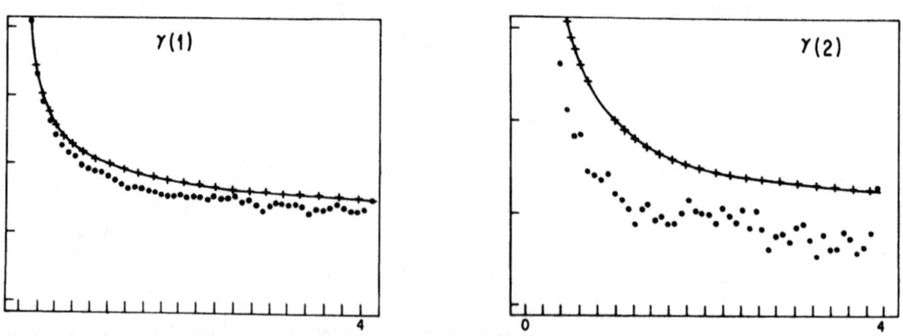

Fig. 3. Shows γ_1 and γ_2 as in 2a a but the full line here represents the results for ~ 32'000 levels somewhat above level 150'000.

by a single parameter. This strongly suggests that this transition is universal for a wide class of problems yet to be delimited.

A final comment may be in order: At this point the banded model has no adequate justification (we refer to it as a parametrization); nevertheless, its success is remarkable and urges us to look for a better understanding. Actually it may simply belong to the same universality class as the other problems, and the structure of this model suggests that localization may be the underlying connection.

ACKNOWLEDGEMENTS

The authors are grateful to J. José for generously giving us access to the quasi-energy spectra, and to him and O. Bohigas for many useful discussions.

REFERENCES

1) T.H. Seligman, J.J.M. Verbaarschot and M.R. Zirnbauer, Phys. Rev. Lett. 53 (1984) 215, J. Phys. A 18 (1985).
2) J. José, R. Cordery, Phys. Rev. Lett. 56 (1986) 290.
3) T.H. Seligman and J.J.M. Verbaarschot, This volume.
4) J. José, This volume.
5) C.E. Román, T.H. Seligman, J.J.M. Verbaarschot (to be published).

PROBABILITY FLOW IN MULTI-STEP REACTIONS

William A. Friedman
Department of Physics, University of Wisconsin
Madison Wisconsin 53705 USA

I wish to discuss a simple and general approach to multi-step processes. These are processes which progress from condition c to condition c' passing through a set of states grouped by classes[1].

A concrete example of this type of process is the multi-step compound nuclear reaction. These proceed from a channel (continuum) state c, through a number of classes of states of the compound nucleus (bound), and finally reemerge back in another channel state c'. A variety of formalisms based on different microscopic, and statistical assumptions for the classes of states and their couplings have been applied to this problem[2,3,4]. We will see here that the stucture of the different formulae which result from these is governed, to a large extent, by the features imposed by the more general type of process examined here which has applicablity beyond that of the compound nucleus problem.

Our approach will be to examine the probabilities for taking various routes from c to c', while visiting a succession of classes along the way. The primary feature of the approach is that at each stage of the process the next step is governed by probability-conserving branching ratios which are independent of previous history (Markovian). We thus consider the flow of probability through the process subject only to the constraints of unitarity. The probability for a given process is the sum of the probabilities for the individual routes. We begin by considering an illuminating and efficient procedure for carrying out this summation.

Let us consider N classes of states. In the context of the the compound nucleus problem, for example, these may be classes of bound nuclear states. We characterize each class, i, by the following quantities: its coupling to external (channel) states, characterized by transmission coefficient, τ_i^c, which gives the probability for entrance from a given continuum states; and its interclass branching ratios, μ_{ij}. Other convenient properties which we consider are: the average density of states, ρ_i; the partial width for decay from the class states to the channel states, Γ_i^c; and the class width, Γ_i. The conservation of probability is given by:

$$1 = \sum_j \mu_{ij} + \sum_c \Gamma_i^c/\Gamma_i.$$

We use these quantities to express the probability to pass from channel state c to channel state c' over each route, and we designate the total total (summed) probability as, $\sigma_{cc'}$. This quantity is to be assocated with the reaction cross section. (We here suppress the geometric factor, $\pi/k^2(2\ell+1)$.) This cross section gives, for example, the multi-step compound contribution to the reaction cross section.

Let us label the N classes in ascending order 1,2,...N. We then define the partial cross section, $\sigma_{cc'}^n$, as the probability for going from c to c' by all routes which visit class n at least once but which reach no class ν $(\nu > n)$. The cross section, is then simply, $\sigma_{cc'} = \sum \sigma_{cc'}^n$.

For the particular situation in which the classes are chosen with $\Gamma_1 \gg \Gamma_2 \gg \Gamma_3$... the partial cross sections, $\sigma_{cc'}^n$, have the following physical interpretation. Each can be associated with a life time which is imposed by its class with the smallest width, and hence with a life time of the order of $\hbar/\Gamma_n$. In general, however, no simple physical property need be associated with the partial cross section.

We next construct the partial cross sections taking care to include all appropriate routes. This procedure is systematized as follows: We consider the entrance into any of the classes i $(i < n)$ from the channel c. This probability is provided by τ_i^c. Next we consider all the routes beginning at i, the entrance class, and ending upon first reaching n. Then, we consider all routes which begin and end in class n, visiting classes k $(k < n)$ in between. Next, we treat all routes leaving n and reaching an exit channel j . Finally we consider the probability for going from class j to the channel c'.

First let us consider the probability for leaving i and arriving at n taking all possible routes which involve the classes $j < n$. This summed probability, designated M_{ij}, is provided by

$$M_{ij} = \sum_{j<n} A_{ij}^{-1} \mu_{jn} \tag{1}$$

where $A_{ij} = \delta_{ij} - \mu_{ij}$, (with $\mu_{ii} = 0$). For convenience we define the diagonal elements $M_{nn} = 1$. The probability for going from channel c to the class n, T_n^c, is provided by: $T_n^c = \sum_{i,n} \tau_i^c M_{in}$. Next, consider the probability for leaving n, going to the classes k $(k < n)$, and returning to n any number of times. This probability is provided by, $(1-\sum_{k<n} \mu_{nk} M_{kn})^{-1}$, when all such routes are summed. To determine the probability for going from class n to an exit class j, M_{nj}^T, we consider the time reversed routes for going from j to n,

$$M_{nj}^T = \sum \mu_{nk}\ A_{kj}^{-1} = M_{jn}(\Gamma_j \rho_j / \Gamma_n \rho_n)$$

where A_{ij} is defined following above The probability to pass from exit class j to final channel c' is given by, $\Gamma_j^{c'}/\Gamma_j$, so that the probability to pass from n to c' is given by

$$\sum_{j<n} M_{nj}^T (\Gamma_j^i / \Gamma_j) = \sum_{j<n} M_{nj}^T \tau_j^{c'} / 2\pi \Gamma_j \rho_j$$

From the relations above we have $\sum_{j<n} M_{nj}^T \Gamma_j^{c'} / \Gamma_j = T_n^{c'} / (2\pi \Gamma_n \rho_n)$. Combining the three stages (c-n,n-n,n-c'),we obtain

$$\sigma_{cc'}^n = T_n^c [2\pi \Gamma_n \rho_n (1 - \sum_{k<n} \mu_{nk} M_{kn})]^{-1} T_n^{c'} \tag{2}$$

Next let us consider the sum of the terms T_n^c over all channel states. This represents the probability for passing to class n from any of the continuum state. Using the relations above it can be shown that

$$\sum_{c''} T_n^{c''} = 2\pi \Gamma_n \rho_n (1 - \sum_{i<n} \mu_{ni} M_{in})(1 - \sum_{\nu>n} \eta_{n\nu})$$

where $\eta_{n\nu} = (\mu_{n\nu} + \sum_{i<n} M_{ni}^T \mu_{i\nu})/(1 - \sum_i \mu_{ni} M_{in})$. The quantity, $\eta_{n\nu}$, has only values for $\nu > n$, and represents the probability for going from class n to class ν $(\nu > n)$, visiting classes k $(k < n)$ along the way. In terms of this symbol, the partial cross section assumes an especially simple form.

$$\sigma_{cc'}^n = \frac{T_n^c T_n^{c'}}{\sum_{c''} T_n^{c''}} \left(1 - \sum_{\nu>n} \eta_{n\nu}\right) \tag{3}$$

Since the probability factors M_{in} may be obtained from the following: $M_{in} = \eta_{in} + \sum_{s=i+1}^{n-1} M_{is} \eta_{sn}$ as well as form Eq.(1), $\sigma_{cc'}$ may be expressed either as a function of (μ_{ij} and τ_{ij}^c), or as a function of (η_{ij} and τ_{ij}^c).

To confirm that all of the routes have been accounted for and that probability has been conserved, we sum $\sigma_{cc'}$ over all the final channels, c', to obtain the probability for leaving channel c and entering the classes (compound nucleus). $\sigma_c = \sum_{c'} \sigma_{cc'}$:

$$\sigma_c \equiv \sum_{c'} \sigma_{cc'} = \sum_i \tau_i^c$$

which does indeed represent the probability sought.

We next consider an equivalent expression for the cross section $\sigma_{cc'}$,

Summing over the entrance and exit channels last, this expression takes the following form,

$$\sigma_{cc'} = \sum_{i,j} \tau_i^c \Pi_{ij} \tau_j^{c'} \tag{4}$$

where

$$\Pi_{ij} = \sum_n M_{in} \left(2\pi\Gamma_n\rho_n(1 - \sum_i \mu_{ni}M_{in}) \right)^{-1} M_{jn}$$

$$\Pi_{ij}^{-1} = 2\pi\rho_i\Gamma_i(\delta_{ij} - \mu_{ij})$$

To recapitulate, we have found three equivalent expressions for the probability sum by following Eq. (2), Eq. (3), and Eq. (4). At this point let us associate these with the three major multi-step reaction formalisms.

The formalism of Feshbach-Kerman-Koonin[3] has the same structure as the form given in Eq. (2), with additional constraints. The doorway concept is used so the entrance class is restricted to class $i = 1$. The class properties are so specified that the interaction only links neighboring classes. Finally it is assumed that the probability flow between classes is in one direction. The latter is consistant with the assumption that $\Gamma_{i+1}\rho_{i+1}/\Gamma_i\rho_i$ is very large. When these constraints are included the following form is achieved from Eq. (2).

$$\sigma_{cc'} = \tau_1^c \Pi_{j=1}^{n-1} \mu_{j,j+1} \Gamma_n^{c'} / \Gamma_n$$

In the nested-average formulation[4] the classes are chosen such that $\Gamma_1 \gg \Gamma_2 \gg \Gamma_3$... This allows for the generalization of the Hauser-Feshbach formalism. The various optical S-matrices are obtained through energy averages of the full S-matrix over energy intervals I_n, such that $\Gamma_{n-1} > I_n > \Gamma_n$. The resulting cross section takes the form of Eq. (3).

Finally the formulation of AWM[2] starts with a model Hamiltonian and unrestricted definitions of classes. With the assumption of phase randomness for the coupling matrix elements this approach leads to the Π-matrix below Eq. (4) and the cross secion of the form given in Eq. (4).

While the starting points and subsequent assumptions for each of these froms are different the resulting forms may all be related to different aspects of the probability flow summations. The latter contains only the constraint of unitarity and the loss of memory.

This work was supported in part by the U.S.National Science Foundation.

REFERENCES

1. W.A. Friedman, Phys. Rev. C22, 1(1980).
2. D. Agassi, H.A. Weidenmüller, and G. Mantzouranis, Phys. Rep. 22C,145 (1975).
3. H. Feshbach, A.K.Kerman, S.E. Koonin, Ann. Phys. (N.Y.) 125,429 (1980).
4. W.A. Friedman, M.S. Hussein, K.W. McVoy, and P.A. Mello, Phys. Rep. 77C,49 (1981).

A MACROSCOPIC APPROACH TO STATISTICAL PHENOMENA

P.A. Mello*
Instituto de Física, U.N.A.M.
Apdo. Postal 20-364
01000 México, D.F., MEXICO

Abstract. The statistical distribution of quantities that involve many microscopic variables is frequently independent of the assumptions made at the microscopic level and depends only upon very general macroscopic properties of the problem. In two examples where this behaviour occurs, we show that the distribution of the quantities of interest can be found by working entirely at a macroscopic level. The first example is the statistical theory of Nuclear Reactions and the second one, the Theory of One-Dimensional Disordered Conductors.

I. INTRODUCTION

"Essentia non sunt multiplicanda praeter necessitatem" (William of Occam, 1300-1349).

William of Occam's famous dictum, known as the "Occam razor", will be the guiding philosophy behind the topics to be presented in this talk. Occam's statement, which, translated into English, means: "Entities do not have to be multiplied without necessity", was rephrased by Bertrand Russell[1] as: "If in a certain science everything can be interpreted without a certain hypothesis, there is no reason to use it".

In the statistical description of physical phenomena one is sometimes interested in the statistical distribution of quantities that involve very many *microscopic* variables. The attempt is then made to find that distribution on the basis of a microscopic theory. It frequently happens, though,

---

* Also at Departamento de Física, U.A.M. - Iztapalapa, and fellow of the Secretaría de Educación Pública, México.

that the results are to a large extent independent of the assumptions made at the microscopic level and depend only upon very general *macroscopic* properties of the problem. A central-limit theorem is normally responsible for such a behaviour. When this happens, it appears to be conceptually simpler to construct directly the statistical distribution of the quantities of interest, by imposing on it the macroscopic requirements that we have evidence of. This procedure may determine the distribution uniquely; if not, we have found it advantageous to make a selection on the basis of a maximum-entropy criterion. Such a criterion allows one to pick the "most probable" distribution among those that satisfy the given requirements; alternatively, one could say that it treats the statistical variables of the problem "as randomly as it is allowed" by the constraints.

We shall illustrate the above ideas with two examples, taken from very different fields of Physics: the first one is the Statistical Theory of Nuclear Reactions to be discussed in the next Section; the second one is the Theory of One-Dimensional Disordered Conductors, which is studied in Section III.

II. THE STATISTICAL THEORY OF NUCLEAR REACTIONS

We show in Fig.2.1 the differential cross section for the process[2] $^{35}Cl(p,\alpha_0)^{32}S$. The energy of the incident proton is about 10 MeV, meaning that the excitation energy in the compound system ^{36}A is about 18.5 MeV. In that energy region, the width-to-spacing ratio Γ/D of nuclear resonances is much larger than unity, so that the peaks shown in the figure cannot be ascribed to individual resonances. Indeed, a striking feature of the cross section of Fig.2.1 is that it varies *erratically as a function of energy.*

There is no theory of the nucleus capable to reproduce each individual peak of Fig.2.1, and even if there were one, such a prediction would be completely uninteresting. A statistical analysis of the data is indeed much more interesting and significative. For instance, one is interested in the *energy-averaged* cross section (a,b indicate a pair of

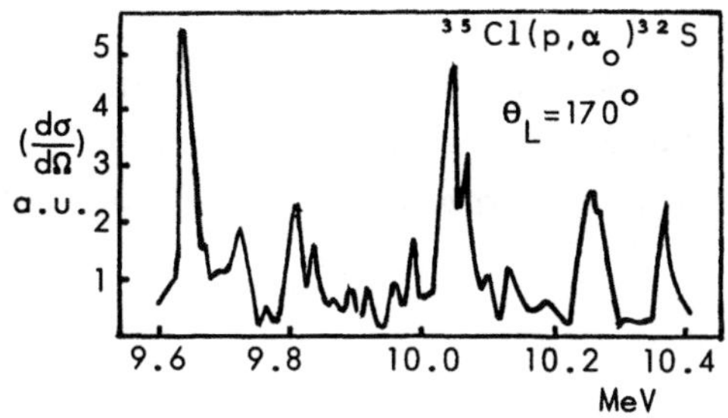

Fig. 2.1

channels, like (p,α) in the above example)

$$\overline{\sigma}_{ab} = \frac{1}{\Delta E}\int_{\Delta E} \sigma_{ab}(E)\,dE\,, \tag{2.1}$$

the *fluctuations* around it

$$\mathrm{var}\,\sigma_{ab} = \overline{(\sigma_{ab} - \overline{\sigma}_{ab})^2} \tag{2.2}$$

and eventually the full *distribution*.

We recall that, up to geometrical factors, the cross section can be written as the square of an amplitude, i.e.

$$\sigma_{ab} = |\delta_{ab} - S_{ab}|^2\;, \tag{2.3}$$

where S is the *scattering matrix*. In (2.3) we are assuming a fixed partial wave J^π .

The most significative properties of S are:

From flux conservation $\Longrightarrow$ unitarity: $SS^\dagger = I$

From time-reversal invariance $\Longrightarrow$ symmetry: $S = S^T$

From causality $\Longrightarrow$ analyticity properties.

In a problem like that of Fig. 2.1, $S(E)$ varies *erratically as a function of energy*. We are thus interested in its *energy average* $\overline{S(E)}$ (the <u>optical</u> S-matrix[3]), the *fluctuations* around it and eventually the full *distribution* dP(S). Find ing that distribution will be our concern in what follows.

We shall restrict ourselves, in the analysis to be presented below, to those processes that can be described in terms of a *direct* and a *compound-nucleus* (or fully equilibrated) contribution.

The one-channel case

Unitarity implies that S can be written as $S = e^{i\vartheta}$, ϑ being twice the phase shift, so that S is restricted to move *on* the unit circle (Fig. 2.2).

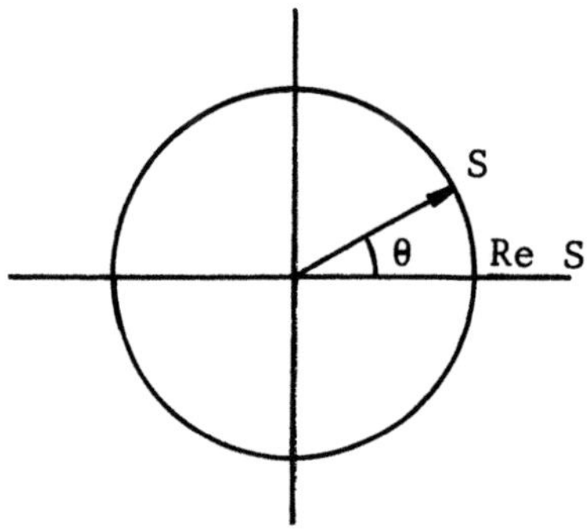

Fig. 2.2

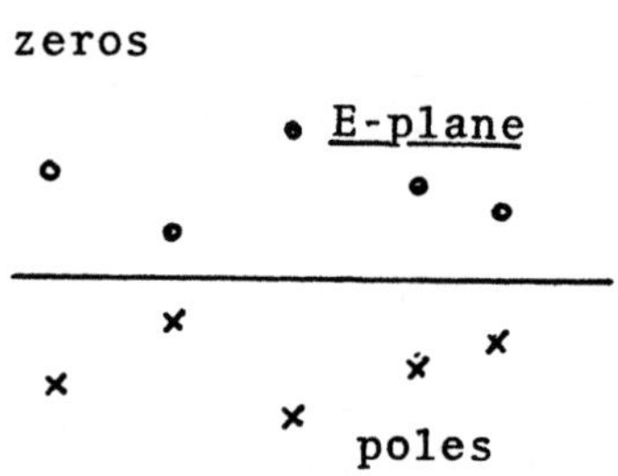

Fig. 2.3

From *analyticity*, S has poles only in the lower-half of the complex-energy plane (Fig. 2.3).

Suppose now that we perform an average of the k-th power of S, $S^k(E)$, with a *Lorentzian weighting function* centered at the energy E_0 and having a half-width I. From Cauchy's theorem one then has

$$\left[\overline{S^k}\right]_{E_0,I} \equiv \int S^k(E) \frac{I/\pi}{(E-E_0)^2+I^2} dE = S^k(E_0+iI) , \tag{2.4}$$

and hence

$$\left[\overline{S^k}\right]_{E_0,I} = \left[\bar{S}\right]^k_{E_0,I} , \tag{2.5}$$

i.e. the average of the k-th power of S coincides with the k-th power of the average of S.

As the energy E changes, S of Fig. 2.2 moves on the unit circle and ϑ changes accordingly. The integral in (2.4) is over E; we ask if it can be converted in an integral over ϑ, in such a way that

$$[\overline{S^k}]_{E_0,I} = \int_0^{2\pi} e^{ik\vartheta} p(\vartheta)\, d\vartheta = \int S^k\, dP(S) \ , \tag{2.6}$$

with

$$dP(S) = p(\theta)\, d\vartheta \tag{2.7}$$

a (non-negative) measure independent of k. The analysis of this question is so simple that we can afford presenting it here in a few lines[4]. If we expand $p(\vartheta)$ in a Fourier series

$$p(\vartheta) = \sum_k a_k e^{ik\vartheta} \ , \quad (a_k = a^*_{-k}) \ , \tag{2.8}$$

we can calculate the coefficient a_{-k} as

$$a_{-k} = \frac{1}{2\pi}\int p(\vartheta) e^{ik\vartheta} d\vartheta = \overline{S^k}/2\pi = (\overline{S})^k/2\pi \ , \tag{2.9}$$

and summing up the series we find

$$p(\vartheta) = \frac{1}{2\pi}\, \frac{1-|\overline{S}|^2}{|S-\overline{S}|^2} \ , \qquad S = e^{i\vartheta} \ . \tag{2.10}$$

Thus the answer to the question posed above is that $p(\vartheta)$ *exists, is unique* and is given by Eq. (2.10). In Eq. (2.10) only the parameter $\overline{S}$ enters. For a given S(E), which could consist of one single pole, a regular pattern or a statistical distribution of poles, $\overline{S}$ has a numerical value that depends on the centroid E_0 and the half-width I of the Lorentzian of Eq. (2.4).

For a given S(E), we can thus replace energy averages by S-space averages using the measure (2.10).

Suppose now that S(E) has resonances in the full interval $(-\infty, \infty)$ *and* that, as the averaging interval $2I \to \infty$, $\overline{S}$ becomes independent of E_0 and I. In that case, the S-space measure (2.10) depends parametrically only upon the single complex number $\overline{S}$, *any other information about S(E) being irrelevant!* In many cases of interest it also occurs that as $I \to \infty$, Lorentzian averages become indistinguishable from the more physical "box averages" (i.e. with a weight equal to $1/2I$ within $\Delta E = 2I$, and zero outside). If that happens, $dP(S) = p_{\overline{S}}(\theta)\, d\vartheta$ of (2.10) has the appealing interpretation of the *fraction of "time" spent within* $d\vartheta$ by S(E), as E runs from $-\infty$ to $+\infty$.

In order to use probability concepts in this problem, it is useful to introduce a collection or *ensemble* of S-matrices, that must possess the property of *ergodicity*: i.e., an energy average (for almost all members of the ensemble) should coincide with an ensemble average evaluated at a fixed energy. From the above we conclude that such an ensemble *must* have, at a fixed energy, the measure $dP_{\bar{S}}(\vartheta)$ of Eq. (2.10). Denoting with brackets an ensemble average, our ensemble must have the property

$$\langle S^k \rangle = \langle S \rangle^k \,, \tag{2.11}$$

which has often been called the *analyticity-ergodicity* (AE) requirement[8], because it is a consequence of the analytic structure of the S-matrix and the ergodicity of the ensemble. Eq. (2.11) is equivalent to the so called *reproducing property*[5]

$$f(\bar{S}) = \int f(S)\, dP_{\bar{S}}(S) \,, \tag{2.12}$$

that must be fulfilled for any function $f(S)$ which is analytic in its argument (and hence can be expanded in a power series of S). Eq. (2.12) also arises in the problem of finding the two-dimensional electrostatic potential inside a circle, knowing its value on the boundary; this problem gives rise to the so called *Poisson's kernel*, which is precisely our measure (2.10)!

We now want to compare the results obtained from the above *macroscopic* approach with those of *microscopic* calculations.

Suppose that we express the (1-dimensional) S-matrix in terms of a K-matrix and construct the latter in terms of resonances (the microscopic variables), i.e.

$$S = \frac{1 + iK}{1 - iK} \,, \qquad K = \sum_{\lambda} \frac{\gamma_\lambda^2}{E_\lambda - E} \,. \tag{2.13a,b}$$

One stretch of resonances (each determined by its position E_λ and its strength γ_λ^2) determines one S-matrix. A collection of stretches of resonances following given statistical laws is then used to construct an ensemble of S-matrices. Two ensembles, for which there is numerical evidence of ergodicity, were constructed[4]: in the first one the E_λ's were chosen at random, the spacing x between successive resonances thus following Poisson's law e^{-x}; in the second one, the spacing x was sampled from a Wigner distribution $(\pi x/2)e^{-\pi x^2/4}$ and the various spacings were taken as statistically independent. In both cases, the γ_λ's were taken as statistically independent Gaussian variables. In Fig. 2.4 we show the probability density for the variable ϑ (at a fixed energy E) obtained numerically from Monte Carlo calculations for the above two ensembles (*microscopic* approach), compared with Poisson's measure (2.10) resulting from a *macroscopic* approach. The agreement is excellent.

A central-limit theorem of a new kind is responsible for the stability of the resulting distribution. Poisson's result (2.10) for S implies a Lorentzian distribution for the quantity K of (2.13a). Thus viewing K as the sum over resonances (2.13b), our result implies that, in the limit when the number of such resonances grows, the resulting K becomes eventually distributed as a Lorentzian; this occurs no matter what the distribution of the resonances is, as long as one has ergodicity.

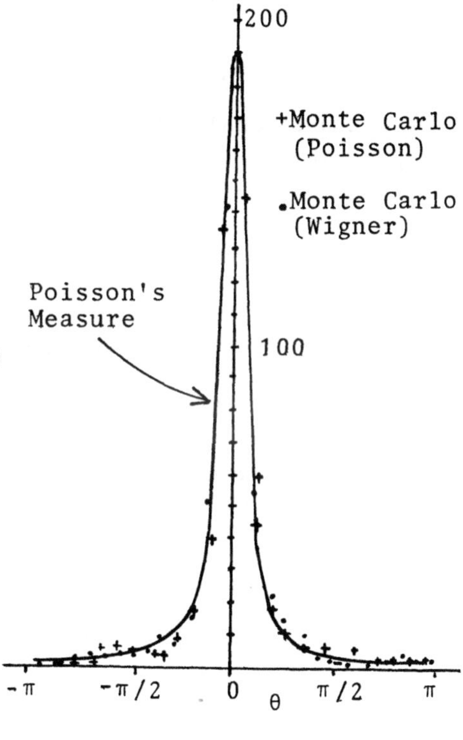

Fig. 2.4

We have thus shown that, at least in this problem, a *macroscopic* approach contains *the same information* as a *microscopic* one and, from "Occam's razor", the preference is obvious!

The multichannel case[8b,c]

Now the Argand diagram of Fig. 2 has to be generalized to include the axes $\mathrm{Re}\, S_{11}$, $\mathrm{Im}\, S_{11}$, $\mathrm{Re}\, S_{12}$, $\mathrm{Im}\, S_{12}$, ..., $\mathrm{Re}\, S_{NN}$, $\mathrm{Im}\, S_{NN}$ and S is restricted to move on the surface determined by unitarity, $SS^{\dagger} = I$, and symmetry, $S = S^T$.

In Fig.2.2, $d\vartheta$ was the natural differential to use; $d\vartheta$ remains invariant under an arbitrary shift ϑ_0 of the phase. In the multichannel case, too, one can find an invariant measure for unitary and symmetric matrices[5,6];we designate it by $d\mu(S)$ and it fulfills the requirement

$$d\mu(S) = d\mu(U_0 S U_0^T) \quad , \tag{2.14}$$

for a fixed unitary matrix U_0. The transformation $S \to U_0 S U_0^T$ transforms, for a fixed U_0, the space of unitary and symmetric matrices unto itself.

Similarly to Eq. (2.7), the differential probability $dP_{\bar{S}}(S)$ we are looking for will be written as

$$dP_{\bar{S}}(S) = p_{\bar{S}}(S) d\mu(S), \tag{2.15}$$

where $p_{\bar{S}}(S)$ will be referred to as the "probability density".

The analyticity-ergodicity (AE) requirement, or, equivalent, the reproducing property (2.12), has to be enforced here too. Mathematicians know a solution to this problem,which, for arbitrary dimensionality of the S-matrix, is again known as Poisson's kernel and is given by[5]

$$p_{\bar{S}}(S) = \frac{1}{V} \frac{[\det(I - \bar{S}\bar{S}^{\dagger})]^{\frac{N+1}{2}}}{|\det(I - \bar{S}^{\dagger}S)|^{N+1}} \quad , \tag{2.16}$$

where V is a normalization factor. We notice that for N=1 Eq. (2.16) reduces to (2.10).

In the case $N > 1$ there is, however, a very important difference with respect to the N=1 case, in that AE and reality of the solution do not fix $p_{\bar{S}}(S)$ uniquely. A physical example in which we know that (2.16) is *not* appropriate, is the case in which, besides a direct and a compound-nucleus (equilibrated) process, one has a contribution to the cross section coming from precompound processes, which correspond to early stages on the way to the formation of a compound nucleus. One knows, for instance, that whereas the optical matrix $\bar{S}$ is sufficient to parametrize the full distribution when there are no precompound contributions, it is not enough when the latter contributions are appreciable.

Out of all possibilities, though, Poisson's kernel (2.16) occupies a very special place: its information-entropy[7]

$$\mathcal{S}[p] \equiv -\int p_{\bar{S}}(S) \ln p_{\bar{S}}(S) d\mu(S) \tag{2.17}$$

is larger or equal to that of any other kernel satisfying AE and for the same $\bar{S}$. In a more qualitative way, we can say that Poisson's kernel represents an ensemble of S-matrices that is "as random as is allowed" by the conditions of unitarity, symmetry, AE and a fixed value of $\bar{S}$. In this sense we shall take (2.16) as a model for the problem in which one has a direct and a compound-nucleus contribution only.

The results[8c,9] that have been obtained thus far with Poisson's kernel (2.16) are indistinguishable from those obtained by numerical Monte Carlo calculations[10] of the type briefly discussed in connection with the one-channel problem.

By way of illustration, we show in Fig. 2.5 the ratio W of the elastic to the inelastic fluctuation cross section for N=2 equivalent channels, i.e. $\bar{S}_{11} = \bar{S}_{22}$ and with $\bar{S}_{12} = 0$, as a function of $1 - |\bar{S}_{11}|^2$. We observe that the macroscopic approach based on Poisson's kernel and the microscopic calculations give entirely consistent results.

We would also like to mention that in the limit of a large number of channels, $N \gg 1$, (2.16) gives the Hauser-Feshbach formula with the correct elastic enhancement factor, both in the absence and in the presence of direct reactions[11]. For instance, with no direct reactions one reproduces, for the compound-nucleus or fluctuation cross section, the well-known result[3b, 28]

$$\sigma_{ab}^{fl} = (1 + \delta_{ab}) \frac{P_a P_b}{\sum_c P_c} , \tag{2.18}$$

where $P_c = 1 - |\bar{S}_{cc}|^2$ is the transmission factor for channel c.

In summary, we can say that we have found the following minimal set of conditions that are necessary in constraining our ensemble of S-matrices:

1. Unitarity (flux conservation)
2. Symmetry (time - reversal invariance)
3. Analyticity-ergodicity (causality)
4. A value for $\bar{S}$ (dynamics).

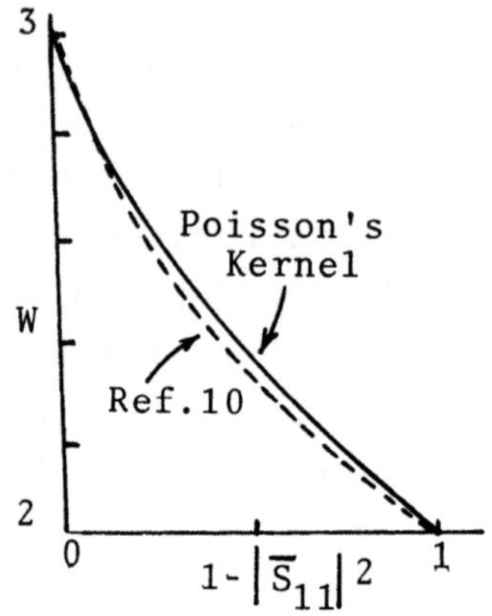

Fig. 2.5

All the rest is uncommitted in this approach. The agreement between this and other methods seems to indicate that the four above conditions contain the necessary physical content. We could thus say that there is the same physical information in the present macroscopic approach as in the more microscopic ones.

III. THE THEORY OF ONE-DIMENSIONAL DISORDERED CONDUCTORS

The familiar physical phenomena that occur in crystalline conductors give way to a host of new exciting facts when the system is amorphous[12,13].

For instance, quantum-mechanical interference by the randomly located scatterers in a one-dimensional conductor produces localization of the bound wave functions[14]. As a consequence, the elastic scattering contribution to the average resistance grows exponentially with length[15] : the familiar Ohm's law that describes a linear increase of resistance with length is recovered only for lengths that are very small compared with the localization length[15,16]. Elastic scattering is the only contribution to the resistance if the temperature is so low that the inelastic modes cannot be excited.

Another surprising fact is that as the length grows, the width of the resistance distribution grows faster than the mean[17]; as a consequence, there is really no "typical" resistance, and one needs the full distribution to describe the problem [16,18,19].

When measuring the resistance, one places the sample between two perfect conductors, and a potential difference between them is established.

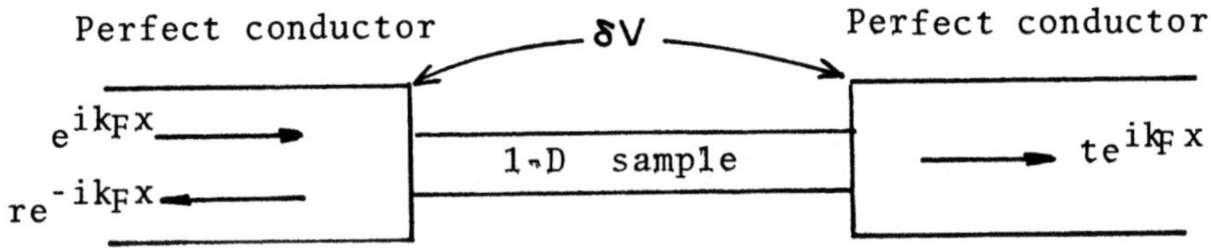

Fig. 3.1

Electrons with the Fermi momentum entering the sample from the left are partly reflected and partly transmitted. Landauer's theorem[15] shows that the resistance, measured in units of h/e^2, is given by

$$\rho = \frac{|r|^2}{|t|^2} . \tag{3.1}$$

The study of the resistance is therefore a study of the scattering properties of the system.

A convenient way to study scattering by a one-dimensional system is by means of the transfer-matrix formalism[13]. The general solution of the Schrödinger equation contains, on both sides of the system, a linear combination of waves travelling in either direction. The pair of coefficients (c,d) is related to the pair (a,b) by the so called <u>transfer matrix</u> i.e.

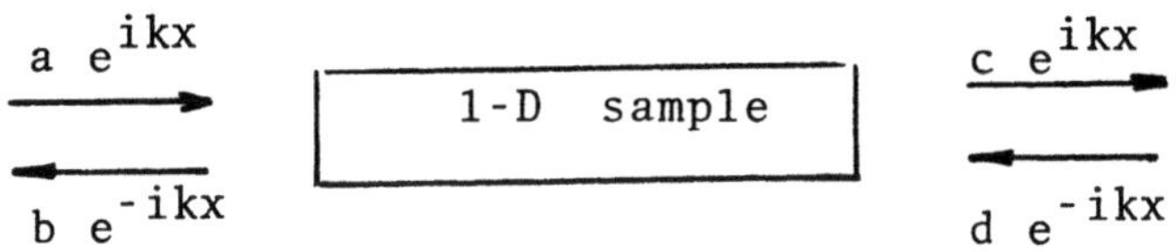

Fig. 3.2

$$\begin{pmatrix} c \\ d \end{pmatrix} = R \begin{pmatrix} a \\ b \end{pmatrix}, \qquad R = \begin{pmatrix} \alpha & \beta \\ \gamma & \delta \end{pmatrix}. \tag{3.2a,b}$$

An important property[13] of R is that from *flux conservation* and *time-reversal invariance*, R must be *pseudounitary*, i.e. $R\begin{pmatrix} 1 & 0 \\ 0 & -1 \end{pmatrix} R^{\dagger} = \begin{pmatrix} 1 & 0 \\ 0 & -1 \end{pmatrix}$ and *unimodular*, i.e. $\det R = 1$, and can thus be parametrized as[21]

$$R = \begin{pmatrix} \alpha & \beta \\ \beta^* & \alpha^* \end{pmatrix} = \begin{pmatrix} e^{-i\mu} & 0 \\ 0 & e^{i\mu} \end{pmatrix} \begin{pmatrix} \sqrt{1+\rho} & \sqrt{\rho} \\ \sqrt{\rho} & \sqrt{1+\rho} \end{pmatrix} \begin{pmatrix} e^{-i\nu} & 0 \\ 0 & e^{i\nu} \end{pmatrix}, \tag{3.3a}$$

where

$$|\alpha|^2 - |\beta|^2 = 1, \qquad -\pi \leq \mu, \nu \leq \pi, \qquad 0 \leq \rho < \infty. \tag{3.3b}$$

The parameter ρ occurring in (3.3) is the resistance of (3.1).

The product of two transfer matrices is again a transfer matrix; there is also an inverse and the unit element. The set of transfer matrices thus forms a group, the non compact group SU(1,1) [21-22], which is homomorphic to the Lorentz group SO(2,1).

A collection or ensemble of random one-dimensional conductors is then described by an ensemble of R-matrices, whose differential probability we write as

$$dP(R) = p(R)\, d\mu(R). \tag{3.4}$$

Here $d\mu(R)$ is the invariant or Haar's measure of the group SU(1,1), which remains invariant when all the matrices of the group are multiplied by the same fixed matrix R_0, i.e. $d\mu(R) = d\mu(R_0 R)$. In terms of the parameters μ, ρ, ν of (3.3), the invariant measure can be written as[21]

$$d\mu(R) = (2\pi)^{-2}\, d\mu\, d\rho\, d\nu. \tag{3.5}$$

The $p(R)$ of (3.4) will be referred to as the "probability density", and will have to satisfy the two following requirements[20]:

1) We specify the type of wires that we want to consider and consequently ask that the above probability density be only a function of R and the length of the wires, i.e. $p_\ell(R)$.

2) *The combination requirement.* Imagine each wire of length ℓ divided in two subsystems of lengths ℓ_1, ℓ_2 ($\ell_1+\ell_2=\ell$) and transfer matrices R_1, R_2 ($R=R_2R_1$). If R_1, R_2 are con-

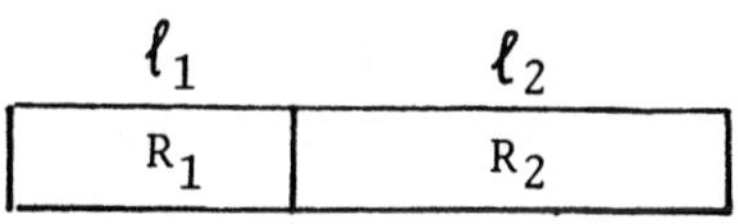

Fig. 3.3

sidered as statistically independent, the probability density $p_\ell(R)$ of the whole system is related to the $p_{\ell_1}(R_1)$ and $p_{\ell_2}(R_2)$ of the two sybsystems according to the convolution[23,26]

$$p_\ell(R)=\int p_{\ell_2}(RR_1^{-1})\,p_{\ell_1}(R_1)\,d\mu(R_1). \tag{3.6}$$

In particular, we can write

$$p_{\ell+\delta\ell}(R)=\int p_\ell(RR_1^{-1})\,p_{\delta\ell}(R_1)\,d\mu(R_1)\ , \tag{3.7}$$

where $\delta\ell$ will be taken in what follows as a small but still macroscopic length. Eq. (3.7) has the form of a Smoluchowski equation for Markovian processes[24,25], where $p_{\delta\ell}(R_1)$, which plays the role of a "transition probability", is now our building block.

We request[26] that $p_{\delta\ell}(R)$ *guarantees the recovery of Ohm's law as* $\delta\ell\to 0$. In other words, calling ρ_1 the resistance, we request

$$\langle\rho_1\rangle_{\delta\ell}=\delta\ell\ , \tag{3.8}$$

where $\delta\ell$ is measured in suitable units, so that the proportionality factor is unity.

We now propose an ansatz for $p_{\delta\ell}(R_1)$. We select the probability density that has *maximum entropy*[7], among all those that satisfy Ohm's law (3.8). The answer is

$$p_{\delta\ell}(R_1) = \frac{e^{-\rho_1/\delta\ell}}{\delta\ell} . \tag{3.9}$$

In the absence of any other information, our ansatz (3.9) is thus *isotropic* in the angles μ_1, ν_1 of Eq. (3.3). Additional requirements are needed in order to further restrict the choice of $p_{\delta\ell}(R_1)$ and thus provide a better treatment of the phases.

Expanding both sides of our combination requirement (3.7), using (3.9) and taking the limit $\delta\ell \to 0$, one finds the "Fokker-Planck or diffusion-type" equation[26]

$$\frac{\partial p_\ell(\rho)}{\partial \ell} = \frac{\partial}{\partial \rho}\left[\rho(\rho+1)\frac{\partial p_\ell(\rho)}{\partial \rho}\right] , \tag{3.10a}$$

to be solved with the initial condition

$$p_0(\rho) = \delta(\rho). \tag{3.10b}$$

When $\ell \ll 1$, one can easily check that (3.9) is indeed a solution of (3.10), so that we get consistency with our ansatz.

The diffusion equation (3.10) is the same equation obtained by N.Kumar[19] using an electronic potential related to a Gaussian random process.

From (3.10) one finds, for the first moment and the variance of the resistance[19]

$$\langle\rho\rangle_\ell = \frac{1}{2}\left(e^{2\ell} - 1\right) \tag{3.11}$$

$$(\mathrm{var}\,\rho)_\ell = \frac{1}{12}\left(2e^{6\ell} - 3e^{4\ell} + 1\right) , \tag{3.12}$$

showing that the mean resistance grows exponentially with length, while the width of the distribution grows faster than the mean.

From (3.10) one can also show that[26]

$$\langle -\ln T \rangle_{\ell} = \langle \ln(1+\rho) \rangle_{\ell} = \ell \quad , \tag{3.13}$$

where T is the transmission coefficient of the system. The variable $-\ln T$ thus scales *exactly* with the length of the conductor. For $\ell \gg 1$ one can also evaluate higher moments of $-\ln T$ and show that $-\ln T$ becomes a Gaussian variable with centroid ℓ (from (3.13)) and width $(2\ell)^{1/2}$.

The above results, which were obtained from *macroscopic* considerations, agree with those arising from sophisticated *microscopic* analysis[1,5-19]. A comparison with those treatments then shows that our quantity ℓ is the length of the conductor measured in units of the localization length.

The solution of the diffusion equation (3.10) can actually be reduced to quadratures. One can show that[26]

$$p_{\ell}(\rho) = \int_0^{\infty} f_s(\rho)\, e^{-(s^2+1/4)\ell}\, w(s)\, ds \quad , \tag{3.14}$$

where

$$f_s(\rho) = {}_2F_1\left(\tfrac{1}{2}+is,\ \tfrac{1}{2}-is\,;\,1\,;\,-\rho\right) \tag{3.15}$$

$$w(s) = 2s\, \mathrm{th}\, \pi s \quad , \tag{3.16}$$

${}_2F_1$ being the usual hypergeometric function. We remark that the function $f_s(\rho)$ of (3.15) is the $m=m'=0$ matrix element of a unitary irreducible representation of the group SU(1,1) belonging to the continuous class[21,27]. The solution (3.14) is thus the harmonic expansion[22,27] of $p_{\ell}(\rho)$ on SU(1,1).

Finally, we mention that the diffusion equation (3.10) is identical to the one found in Ref. 27 as a result of a central-limit theorem on the group SU(1,1). In that reference, the problem was analyzed from a microscopic point of view, starting from a distribution for the individual scattering centers (which was taken as isotropic in the angles). The limit was studied in which the number of the scattering centers grows and their individual effect becomes smaller, so that the average $\langle -\ln T \rangle = \langle \ln(1+\rho) \rangle$ for the whole system stays fixed and equal to ℓ ; it was then shown that the probability density for the resistance ρ of the macroscopic system tends to the result (3.14), *independently* of the individual distribution. It is thus this central-limit theorem which is ultimately responsible for the stability of the results.

In summary, we have proposed the following minimal set of conditions that have to be fulfilled by an ensemble of transfer matrices associated with disordered one-dimensional conductors of length ℓ :

1. Flux conservation
2. Time-reserval invariance
3. The combination requirement
4. Recovery of Ohm's law as $\ell \to 0$.

All the rest is uncommitted in the present approach.

The results so obtained for the resistance are consistent with those of microscopic calculations; but, in addition, we are able to provide the full resistance distribution. The above agreement suggests that, as far as the resistance goes, the microscopic calculations contain the same physical information as the present macroscopic approach.

IV. CONCLUSIONS

We summarize with the following conclusions:

1) We have presented two physical problems in which a *macroscopic* approach contains the same physical information as a *microscopic* one.

2) In the two problems that we have discussed, it is an *underlying central-limit theorem* that is responsible for the behaviour summarized in 1).

3) More physical ingredients that may occur in the microscopic calculation could be dealt with by appropriate constraints in the maximization of the entropy.

4) Perspectives for the future:

a) Concerning the problem of statistical nuclear

reactions, in Sect. II we restricted the analysis to "one-point functions", which are described by the ensemble distribution $dP_S(S)$, understood at a fixed energy. We may inquire as to the calculation, within the present philosophy, of the joint distribution $dP(S_1, S_2)$ of S_1 (at energy E_1) and S_2 (at energy E_2). That distribution would allow, for instance, the calculation of the cross section correlation function at two different energies. Previous analysis[28] has shown that the result should again depend on $\bar{S}$ only. However, more information has to be introduced to describe the energy dependence of S(E). This important problem certainly requires a closer analysis.

b) Concerning the problem of disordered conductors, Sect. III deals with strictly one-dimensional systems. The analysis presented here is so simple that it would be worthwhile trying to generalize it to treat samples with many channels[29]. Since a detailed knowledge of the group SU(1,1) was not needed in Sect. III, we expect a multichannel generalization to be feasible.

REFERENCES

1. B. Russell, "History of Western Philosophy", Vol. II, Part III, Ch.VIII.
2. P.v. Brentano, J.Ernst, O.Häusser, T. Mayer-Kuckuk, A.Richter and W.v. Witsch, Phys. Lett. 9 (1964) 48.
3. H. Feshbach, C.E. Porter and V.F. Weisskopf, Phys. Rev. 96 (1954) 448; M. Kawai, A.K. Kerman and K.W. McVoy, Ann. Phys. 75 (1973) 156.
4. G.López, P.A. Mello and T.H. Seligman, Z. Phys. A295 (1980) 247
5. L.K. Hua, "Harmonic Analysis of Functions of Several Complex Variables in the Classical Domains" (L.Ebner and A. A. Korányi, Tr.) Amer. Math. Soc., Providence, R.I., 1963.
6. F.J. Dyson, J. Math. Phys. 3 (1962) 140
7. R.D. Levine and R.B. Bernstein, in "Modern Theoretical Chemistry" (W.H. Miller, Ed.), Vol. III, Plenum, New York, 1976; Y. Alhassid and R.D. Levine, J. Chem. Phys. 67 (1978) 4321.
8. P.A. Mello, Phys. Lett. 81B (1979) 103; P.A. Mello and T.H. Seligman, Nucl. Phys. A 344 (1980) 489; P.A. Mello, P. Pereyra and T.H. Seligman, Ann. Phys. 161 (1985) 254.
9. P. Pereyra, Ph.D. thesis, Universidad Nacional Autónoma de México, unpublished.
10. H.M. Hofman, J. Richert, J.W. Tepel and H.A. Weidenmüller, Ann. Phys. 90 (1975) 403
11. W.A. Friedman and P.A. Mello, Ann. Phys. 161 (1985) 276.
12. "Physical Properties of Amorphous Materials" (D.Adler,

B.B. Schwartz and M.C. Steele, Ed.), Plenum, New York, 1985.
13. P. Erdös and R.C. Herndon, Adv. Phys. 31 (1982) 63
14. P.W. Anderson, Phys. Rev. 109 (1958) 1492.
15. R. Landauer, Phil. Mag.21 (1970) 863.
16. P.W. Anderson, D.J. Thouless, E. Abrahams and D.S. Fisher, Phys. Rev. B22, 3519 (1980).
17. E. Abrahams and M.J. Stephen, Journ. Phys. C13 (1980) L377
18. A.A. Abrikosov, Solid State Commun. 37 (1981) 997
19. N. Kumar, Phys. Rev. B31 (1985) 5513
20. N. Kumar and P.A. Mello, Phys. Rev. B31 (1985) 3109
21. V. Bargmann, Ann. Math. 48 (1947) 568
22. K.B. Wolf, in "Group Theory and its Applications in Physics" (1980 Latin American School of Physics, Mexico City), T.H. Seligman, ed., AIP Conf. Proc. No. 71 (AIP, New York, 1981).
23. K.B. Wolf, Rev. Mex. Fís. 21 (1972) 87.
24. S. Chandrasekhar, Rev. Mod. Phys. 15 (1943) 1, Sect. 4
25. R.L. Stratonovich, "Topics in the Theory of Random Noise", Vol. I., R.A. Silverman, tr., Gordon and Breach, New York, 1963, Ch. 4.
26. P.A. Mello, Phys. Rev. B, to be published
27. P.A. Mello, Jour. Math. Phys., to be published
28. D. Agassi, H.A. Weidenmüller and G. Mantzouranis, Phys. Rep. 22 (1975) 145
29. M. Büttiker, Y. Imry, R. Landauer and S. Pinhas, Phys. Rev. B31 (1985) 6207.

LBL-21380

STATISTICAL ORIGIN OF COMPLEX PARTICLES IN INTERMEDIATE ENERGY HEAVY ION REACTIONS

L.G. MORETTO
Nuclear Science Division, Lawrence Berkeley Laboratory,
University of California, Berkeley, California 94720

Abstract The compound nucleus mechanism for complex fragment emission is discussed theoretically and demonstrated experimentally. The role of the potential energy as a function of mass asymmetry is shown in experimental charge distributions. This process is followed from near the threshold up to bombarding energies of several tens of MeV A.

Complex particles have appeared very early and obscurely in the game of nuclear physics, ever since radiochemists bombarded a variety of targets with high energy protons and managed to fish out of the resulting soup a few light radioactive elements.[1] The same particles made their debut in the society of instrumental nuclear physics when they were detected by means of particle telescopes in the reaction of U, Ag with GeV protons,[2] although nobody paid much attention to them either. Recently a new interest, probably stirred by the abundance of their production in heavy ion reactions, has brought these particles in the limelight. At least two interpretations have been advanced regarding their origin. The first claims these particles to be the result of nuclear shattering or cold fragmentation.[3] The second sees them produced as droplets condensing out of a vapor at or near the critical temperature.[4] We felt it would be worth trying to clarify the picture by going way down in energy and in particular by checking whether the compound nucleus itself could not be one of their possible sources through a mechanism somehow intermediate between standard evaporation and fission.

In fact the experimental distinction between the processes of evaporation and fission in relatively heavy compound nuclei can be understood in terms of a specific topological feature in the liquid drop model potential energy surface $V(Z)$ as a function of mass asymmetry Z. This feature is a deep minimum at symmetry (fission region) flanked at greater asymmetries by the Businaro-Gallone mountains, which in turn descend at even larger asymmetries ("evaporation" region). The corresponding mass distribution expected for compound nucleus decay is approximately proportional to $\exp[-V(Z)/T_Z]$, where T_Z is the temperature of the conditional saddle point, and indeed shows a peak at symmetry (fission peak) and two wings at the extreme asymmetries (evaporation wings).

In progressively lighter compound nuclei, the potential energy surface undergoes a topological change as the fissionability parameter x

---

This work was supported by the Director, Office of Energy Research, Division of Nuclear Physics of the Office of High Energy and Nuclear Physics of the U.S. Department of Energy under Contract DE-AC03-76SF00098.

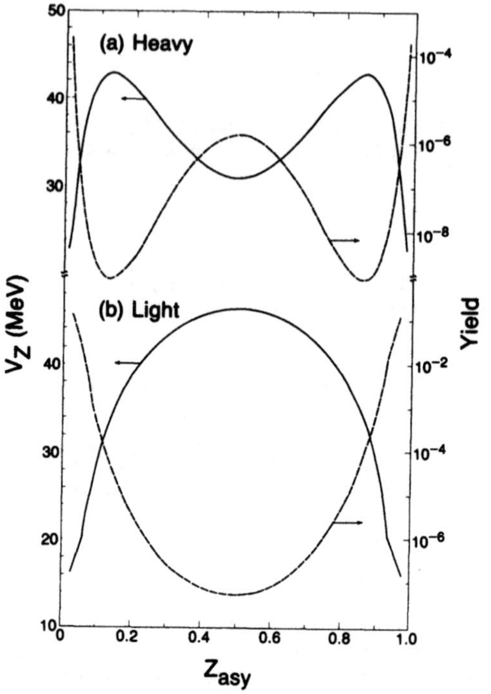

FIGURE 1 Comparison of the potential energy surfaces (solid curve) and expected yields (dashed curve) for a) a heavy CN (Au at $\ell = 0$ and $E^* = 97$ MeV) and b) a light CN (Ge at $\ell = 0$ and $E^* = 72$ MeV).

crosses the Businaro-Gallone point. At this point the second derivative of the potential energy with respect to the mass asymmetry coordinate evaluated at symmetry vanishes. Thus, below the Businaro-Gallone point, there is no longer a true fission saddle point, and the monotonically increasing potential energy towards symmetry determines the disappearance of fission as a process in any way distinct from evaporation. In other words the mass distribution should show the two evaporation wings extending as far as symmetry where a minimum should be observed. These features are illustrated in Figure 1.

Within this framework, we have developed a theory that describes, in a continuous way, the transition from light particle emission to fission.[5] This theory also predicts changes in the shapes of both the kinetic energy spectra and angular distributions of the emitted fragments as their masses increase from α-particles toward fission fragments. These predictions[5] are amenable to experimental test; unfortunately, there is an almost complete lack of experimental data to be compared with the above unified treatments or with more standard formalisms.

We obtained the first experimental evidence for the emission of complex nuclei from helium through fluorine by compound nuclei produced in the reaction 90 MeV ^{3}He + ^{nat}Ag.[6] The specific choice of ^{3}He as projectile was dictated by two reasons. On one hand one would like to have a relatively low velocity projectile in order to minimize preequilibrium losses, but massive enough to bring in sufficient energy. On the other, the mass of the projectile should be sufficiently smaller than those of the complex fragments of interest in order to rule out the ambiguity of projectile fragmentation or multi-nucleon transfer.

In order to determine the existence of an isotropically emitting source and its velocity, the laboratory energy spectra were transformed into invariant cross-section plots in velocity space which are presented

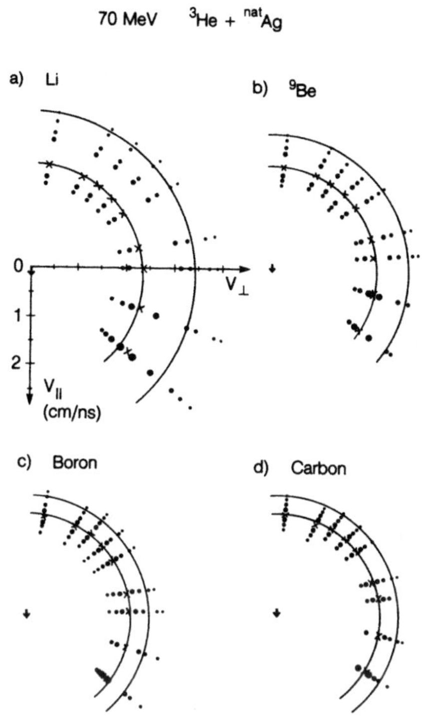

FIGURE 2 Invariant cross section plots $\left(\propto \frac{1}{V^2}\frac{d^2\sigma}{d\Omega dV}\right)$ for representative ejectiles (Li, ^{9}Be,B, and C). The diameter of the dots is proportional to the logarithm of the cross section and the X's indicate the peak of velocity distribution. The two large arcs are sections of circles centered on the c.m. velocity (center arrow) appropriate for complete fusion. The beam direction (0°) is indicated by the c.m. velocity vector.

in Figure 2. The peak cross section for a heavy complex fragment, such as carbon, has a constant value and occurs at the same c.m. velocity from 170° to 40° (as indicated by the position of the X's relative to the circular arc). At the most forward angle the peak cross section occurs at a slightly increased velocity. Similarly, the higher velocity region (the region near the arc with the larger radius) shows no significant change in the backward hemisphere, but does stretch out at forward angles. For a light complex fragment such as Li, the peak of the cross section occurs at a constant c.m. velocity for a smaller backward angle region (170° to 120°). Forward of 120° the peak increases both in cross section and in velocity. The slope of the high-energy tail does not change significantly for the three most backward angles, but the intensity of the tail increases as the scattering angle decreases. The ^{9}Be and B fragments show a behavior intermediate between that of Li and C. In general, the heavier ejectiles show patterns more consistent with the emission from a single source.

Two conclusions can be drawn from these invariant cross section plots. First, for all elements there is an angular region in the backward hemisphere where only a single component is observed, which can be characterized as compound nucleus emission. This angular region increases and extends to more forward angles as the ejectile mass

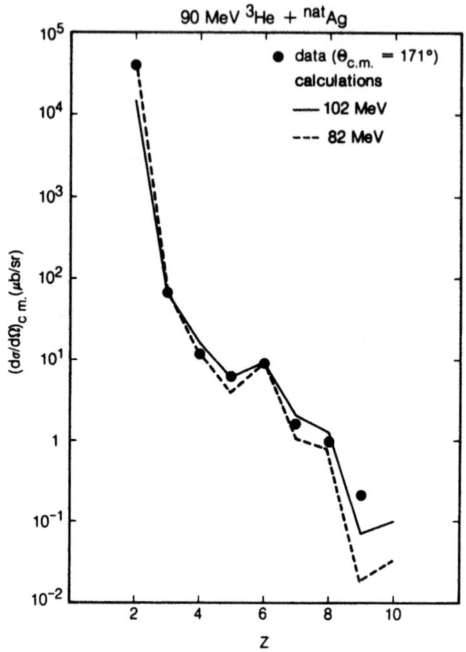

FIGURE 3 Experimental (circles) and theoretical yields versus ejectile atomic number (Z).

increases. Second, there is a component of non-compound nucleus emission that results in harder energy (or velocity) spectra at forward angles.

The mean energies of the spectra are Coulomb-like and increase as the charge of the fragment increases. The most interesting feature in the energy spectra of the equilibrium component is the evolution from a Maxwellian shape for α-particles or Li ions through a more symmetric shape for B or C to a symmetric shape for the heaviest ejectiles, as predicted in Ref. 5.

The experimental yields of the equilibrium component are shown in Figure 3. In order to minimize contributions from sources other than the compound nucleus, we have plotted the yields only for the most backward angle (171°). These yields drop precipitously in going from Z = 2 to Z = 3, after which they decrease more slowly. The one exception is the enhanced Z = 6 yield.

The yield from an equilibrium statistical emission process should be roughly proportional to a factor $\exp[-B_Z/T_Z]$, where B_Z is the emission barrier for fragment Z and T_Z is the temperature at the barrier. More quantitatively, the decay width is given by

$$\Gamma_Z \propto T_Z[E/(E - B_Z)]^2 \exp[2\sqrt{a(E - B_Z)} - 2\sqrt{aE}] \qquad (1)$$

$$\tilde{\propto} \exp[-B_Z/T_Z] \qquad . \qquad (2)$$

To calculate the theoretical yields, the following expression for the barrier was used

$$B_Z = U_1 + U_2 + \frac{Z_1 Z_2 e^2}{d} + U_{prox} - U_{CN} \qquad , \qquad (3)$$

where U_1 is the experimental mass of the light fragment, U_2, U_{CN} are the droplet model masses of the residual and compound nucleus, respectively, and U_{prox} is the proximity potential. The center-to-center distance d in the interfragment Coulomb term was taken to be $d = 1.225(A_1^{1/3} + A_2^{1/3}) + 2$ fm. The addition of 2 fm was done to obtain rough agreement with the energy spectra. The temperature (T_Z) was evaluated using $E - B_Z = aT_Z^2$. A compound nucleus excitation energy (E) of 102 MeV (the value for full momentum transfer) and a level density parameter (a) of A/8 were assumed. The calculated yields (Eq. 1) for each isotope were multiplied by 2I + 1 (where I is the ground state spin of the light fragment) and then summed. The theoretical ejectile yields were calculated as a ratio Γ_Z/Γ_6 and have been normalized to the data at Z = 6 in Figure 3.

The agreement between the data (circles) and this simple equilibrium statistical calculation (solid line) is exceptionally good for Z = 3-9. The calculation underpredicts the α-particle yield because it only takes into account first chance emission, whereas substantial amounts of higher chance α-emission occur. Precompound emission is expected to leave the compound nucleus with a broad excitation energy distribution with a most probable value of ~85 MeV. A calculation (not shown) with this lower excitation energy also reproduces the relative yields of the heavy fragments quite well but overpredicts the yield of first chance α-emission. More detailed comparisons between the data and theory require calculations that include precompound emission; however, the substantial agreement depicted in Figure 3 does indicate that an equilibrated process is responsible for the emission of these complex fragments.

Complete excitation functions obtained from the $^3He + {}^{nat}Ag$ measurements are shown in Figure 4 for a series of decay products. The measurements were restricted to the backward angles (120° - 160°) in order to insure measurement of only the equilibrium component.

With increasing bombarding energy, the cross sections (see Figure 4) rise rapidly and then flatten at higher energies. This is a characteristic signature of compound nucleus emission, and reinforces the assignment of compound nucleus decay that was made previously on the basis of data obtained at 90 MeV. The cross section for Z = 3 is a factor of 1000 lower than that for Z = 2, and for the heavier fragments it is even lower. In spite of these low cross sections, we were able to measure an excitation function over 2-3 orders of magnitude up to Z = 11, with a detection limit of about 50 nb.

The experimental excitation function data have been fitted using a transition state formalism, analogous to that used to fit fission excitation functions.[7] As shown in Ref. 5, the decay width for first-chance emission of a fragment of charge Z can be written as

$$\Gamma_Z = \frac{1}{2\pi\rho(E)} \int_0^{E-B_Z} \rho_Z^*(E - B_Z - \varepsilon)d\varepsilon \qquad (4)$$

where $\rho(E)$ is the compound nucleus level density, B_Z is the conditional barrier height, and $\rho_Z^*(E - B_Z - \varepsilon)$ is the level density at the conditional saddle with a kinetic energy ε in the decay mode. The neutron width Γ_n can be written as

$$\Gamma_n = \frac{2mR^2g}{2\pi\rho(E)} \int_0^{E-B_n} \varepsilon\rho(E - B_n - \varepsilon)d\varepsilon \qquad (5)$$

We make the assumption that the ratio of the decay widths, Γ_Z/Γ_n, is proportional to the ratio of the cross section for complex fragment emission, σ_Z, to that for complete fusion, σ_f, i.e.,

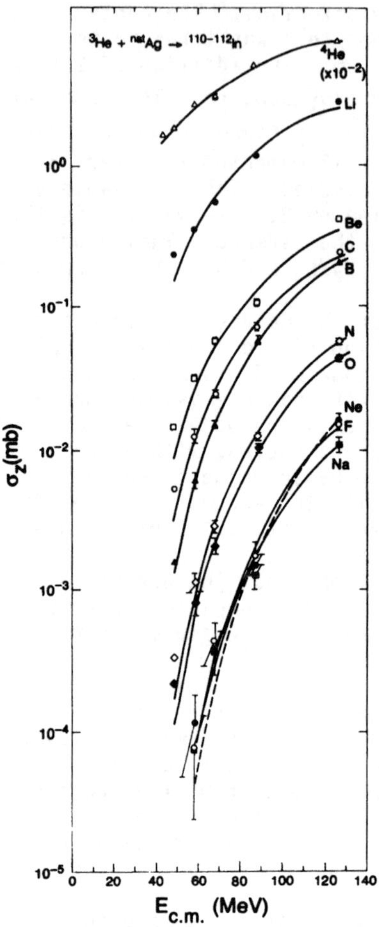

FIGURE 4 Dependence of the total integrated cross sections for emission of complex fragments on the center-of-mass energy, $E_{c.m.}$ in the reaction $^3He + {}^{nat}Ag$. The points and error bars correspond to the experimental cross sections. The curves are fits with the parameters of Figure 5.

$$\Gamma_Z/\Gamma_n \simeq \frac{\Gamma_Z}{\Gamma_t} \simeq \frac{\sigma_Z}{\sigma_t} \simeq \frac{\sigma_Z}{\sigma_R} \tag{6}$$

This is reasonable in this mass region because $\Gamma_n >> \sum_{Z\geq 1}^{\infty} \Gamma_Z$. One can then calculate $\Gamma_Z/\Gamma_n(E)$ using an appropriate choice for the level density expression. A Fermi gas level density was used because it gives an analytical expression for Γ_Z/Γ_n. A simple angular momentum dependence has been included by adding to the barriers the rotational energies appropriate to the ground and saddle point deformations.

Using the above expression for Γ_Z/Γ_n, the barriers B_Z, and the ratio a_Z/a_n, of the level density parameters were extracted from fits to the experimental data σ_Z/σ_R. These fits are shown by the solid lines in Figure 4. The agreement between the data and the fits is remarkably good for all Z-values and confirms that these products originate from compound nuclear decay.

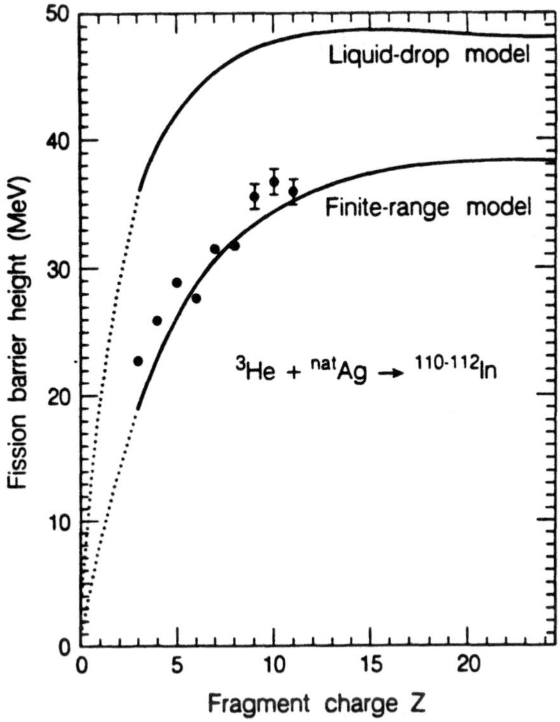

FIGURE 5 The emission barriers, B_Z, extracted in fitting the excitation mission of complex fragments functions as a function of fragment charge. The liquid drop model and finice range model calculations are from ref. 8.

The barriers extracted from the fits are shown by the circles in Figure 5 as a function of Z. The extracted barriers increase dramatically as the exit channel becomes more symmetric. Some evidence of shell effects in the exit channel is visible in the barrier for carbon emission, Z = 6, which is lower than those of the neighboring elements.

The barriers so obtained can be used to test modern corrections to the liquid drop model, like surface diffuseness and finite range, which become important for strongly indented saddle configurations like those presiding to the emission of complex fragments. A comparison of the standard liquid drop model prediction and of the model incorporating the corrections mentioned above[8] with our data is also shown in Figure 5. Clearly our data strongly support the introduction of surface diffuseness and finite range. It is also easy to understand how these and similar data may be very valuable in fixing the relevant parameters of the model.

As mentioned before, the sharp distinction between evaporation and fission in relatively heavy compound nuclei is a result of a specific topological feature of the liquid drop model potential energy surface V(Z) as a function of mass asymmetry Z. The potential energy shows a deep minimum at symmetry (fission region) surrounded by the Businaro-Gallone mountains which in turn descend at even larger asymmetries ("evaporation" region). The corresponding mass distribution from compound nucleus decay shows a peak at symmetry (fission peak) and two wings at the extreme asymmetries (evaporation wings). The qualitative dependence of the potential energy and of the mass yield vs. asymmetry is shown in Figure 1a for a heavy nucleus.

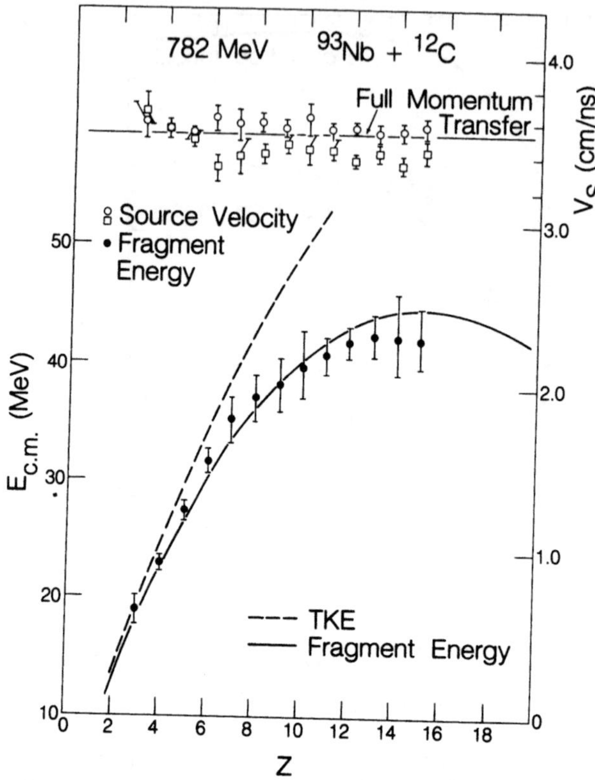

FIGURE 6 The deduced c.m. energies (filled circles) and source velocities (open symbols) for the $^{93}Nb + ^{12}C$ system. Source velocities were determined assuming that the product mass followed the line of β-stability (open circles) or the charge equilibration line (open squares). A Coulomb calculation for two spheres is shown both for the c.m. energy of the light fragment (solid line) and the total kinetic energy (dashed line). The value of the source velocity expected for full momentum transfer is indicated by the horizontal line.

With decreasing total mass the potential energy surface undergoes a qualitative change when the fissility parameter x crosses the so-called Businaro-Gallone point. At this point ($x_{BG} = 0.396$ for $\ell = 0$ and decreasing for larger ℓ values) the second derivative of the potential energy with respect to the mass asymmetry coordinate evaluated at symmetry vanishes. Thus below the Businaro-Gallone point there is no longer a traditional fission saddle point, and fission disappears as a process distinct from evaporation. Thus the mass distribution should show the two evaporation wings extending as far as symmetry where a minimum should be observed. This is illustrated in Figure 1b.

Such a transition has never been observed, as it requires the measurement of the entire mass distribution from symmetry to the extreme asymmetry of α,p evaporation for a series of systems straddling the Businaro-Gallone point. This measurement is made very difficult by the low yield for symmetric decay of the compound nucleus in this general mass region, and by the need to verify that the fragments were produced by a compound nucleus mechanism.[6]

We have measured[9] complete charge distributions from protons to symmetric splitting for a variety of nuclei and we have observed the Businaro-Gallone transition. Such a transition is inferred from the disappearance of the fission peak in the mass yield as the compound nucleus mass was decreased from ^{148}Eu, ^{102}Rh to ^{83}Kr.

The use of reverse kinematics (projectile heavier than the target) was crucial in performing these measurements. This technique virtually eliminates the problems associated with low cross section measurements due to the presence of light element target contaminants. Furthermore, reverse kinematics provides a large center-of-mass (c.m.) velocity which facilitates the verification of full momentum transfer and allows for easy identification of the fragment's atomic number at the higher lab energies. Finally the high energy solution at forward angles corresponds to very backward angles in ordinary kinematics. This enhances the observation of compound nucleus decay and virtually eliminates any possible deep-inelastic contamination.

The velocities of the emitting source and the energies in the source frame of reference for the $^{93}Nb + ^{12}C$ system are shown in Figure 6. The upper part of this figure demonstrates that, with two

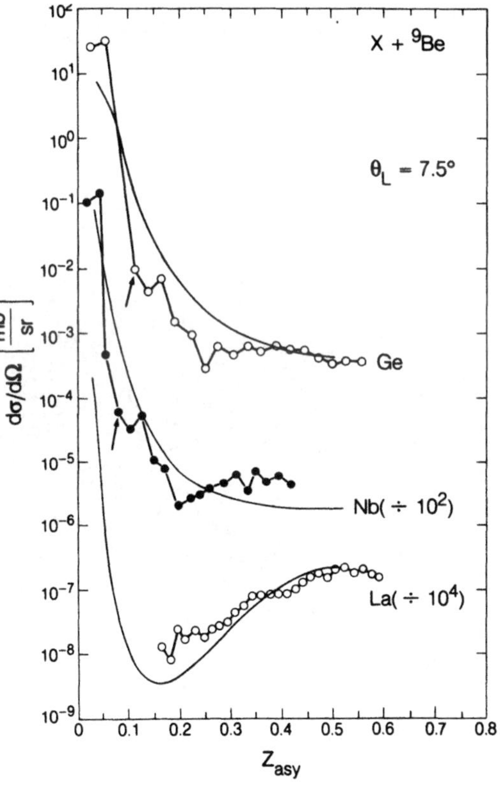

FIGURE 7 Center-of-mass cross sections for products from the ^{74}Ge, ^{93}Nb and $^{139}La + ^{9}Be$ systems detected at $\Theta_{Lab} = 7.5°$. The solid line is a liquid drop model calculation of the fragment yield at $\Theta_{c.m.} = 30°$. The arrows indicate the entrance channel asymmetry. See text. Data below $z_{asy} = 0.15$ were not obtained for the La + Be system, due to a limited dynamic range of the telescope.

extreme mass assumptions, all of the measured products result from the decay of a system with full momentum transfer. For the other systems studied, the extracted source velocities are also independent of Z within a few percent of the velocity expected for full momentum transfer. The deduced c.m. energies are shown in the lower portion of Figure 6. These energies are reproduced by a Coulomb calculation for two spheres with a surface separation of 2 fm. This same separation also reproduces the c.m. energies from the ^{74}Ge induced reactions; however a larger separation is required for the ^{139}La data. Both the full momentum transfer and the invariance with angle of the c.m. energies seen above are consistent with compound nucleus decay.

The experimental cross sections for 530-MeV ^{74}Ge, 782-MeV ^{93}Nb and 1157-MeV ^{139}La + ^{9}Be systems are shown in Figure 7. The cross sections are plotted as a function of charge asymmetry ($Z_{asy} = Z_{detected}/Z_{total}$). The lack of enhancement in yield near the target Z supports the compound nucleus origin of the products rather than a deep-inelastic origin. The yield from the ^{74}Ge + ^{9}Be system, with a fissility parameter of x = 0.31, decreases steadily as one moves towards symmetry. The yields from the ^{93}Nb + ^{9}Be system (x = 0.40) are

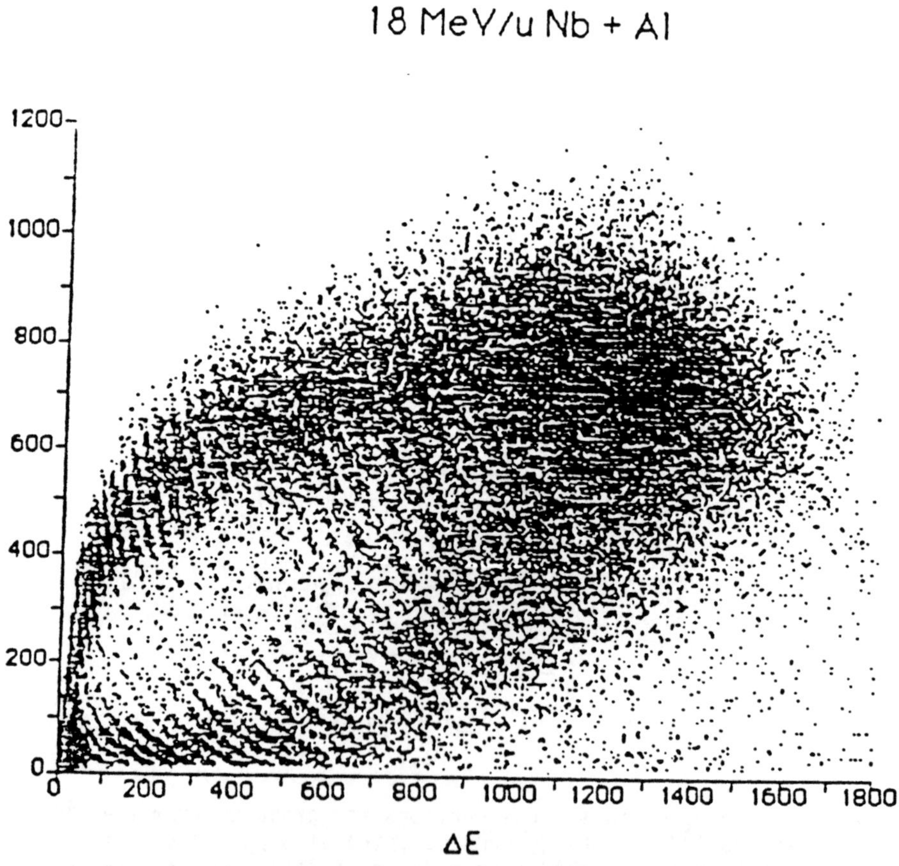

FIGURE 8 ΔE vs. E scatter plot for singles events produced in the reaction 18 MeV/u ^{93}Nb + ^{9}Be. The detector subtended angles from 4° to 12°. The two dark bands correspond to Coulomb emission from a compound nucleus forward and backward in the center-of-mass.

essentially constant from Z_{asy} = 0.2 to 0.4 while the yields from the $^{139}La + {}^{9}Be$ system (x = 0.50) show the characteristic fission peak at symmetry. These three systems clearly exhibit the qualitative trends expected from the topological changes in the potential energy surface predicted by the liquid drop model (see Figure 1).

A quantitative comparison between these data and a compound nucleus calculation based upon the liquid drop model is also shown in Figure 7. The agreement in absolute magnitude and in trend between this calculation and the data confirms the compound nuclear origin of these fragments.

In summary, we have shown that fragments with atomic numbers covering the entire range of the mass asymmetry coordinate are produced from the decay of an excited compound nucleus. The observed Z distributions indicate that the topological transition expected at the Businaro-Gallone point does indeed take place in the region of A ~ 100. The exact position of the Businaro-Gallone point and its angular momentum dependence can in principle be established by a systematic study of the Z

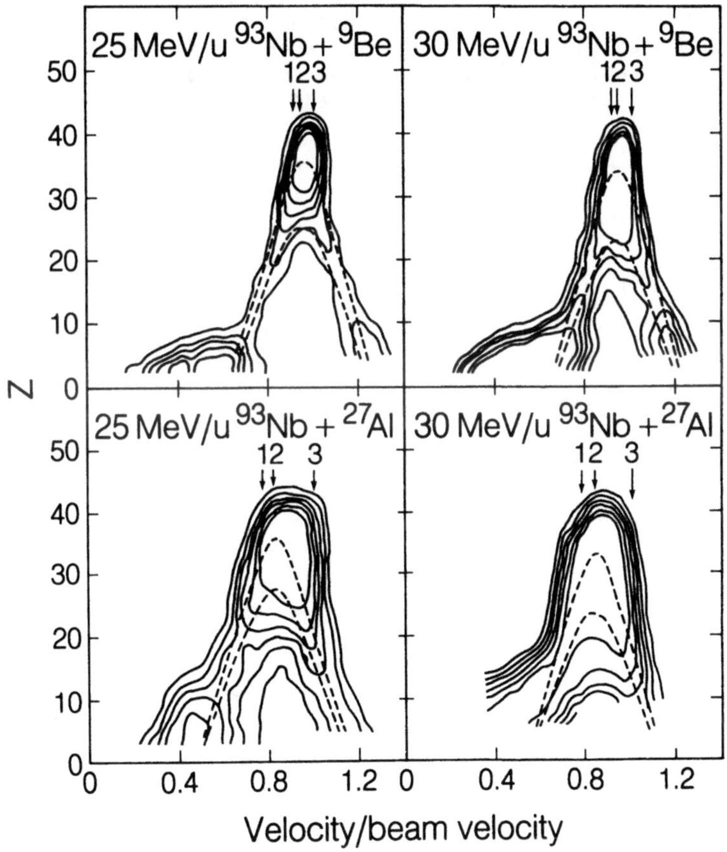

FIGURE 9 Singles distribution of reaction products plotted as logarithmic contours of invariant cross section $[(1/V^2)(\partial^2\sigma/\partial\Omega\partial V)]$ in the Z-velocity plane. The arrows indicate the velocities for 1) full momentum transfer 2) the experimentally determined momentum transfer and 3) the beam. Calculated (dashed lines) average velocities of complex fragments for the maximum and minimum lab angles of the telescope (3° and 8°) are indicated.

or A distributions as the fissility parameter x and the rotational parameter y are varied.

Having established that complex fragments can be emitted by compound nuclei, and that at low energies they are only emitted by compound nuclei, the decision naturally came to see what is in fact going on at higher energies. Should any kind of compound nucleus be formed, it would decay abundantly by complex fragment emission due to its high excitation energy. This is the inescapable conclusion provided by statistical mechanics. More quantitatively, one could expect cross sections as large as several tens of millibarns per Z. The corollary of this is that any additional "fancy" mechanism should ride on top of this already substantial compound cross section.

The reactions we chose are Nb + Be, C, Al in reverse kinematics from 8.5 to 30-40 MeV/u. As we shall see,the choice of a relatively light target simplifies the picture crucially because of the limitations in impact parameters and in the number of sources. While we explored first the upper part of the energy range at the Bevalac,[10] we shall begin with the lower energy data which we collected at GSI.[11]

Figure 8 demonstrates the advantages of reverse kinematics. In this picture we see the complex particles events displayed in the E-ΔE plane. The remarkable double ridge is due to a simple kinematic effect. A single source is emitting fragments in the center of mass with energies

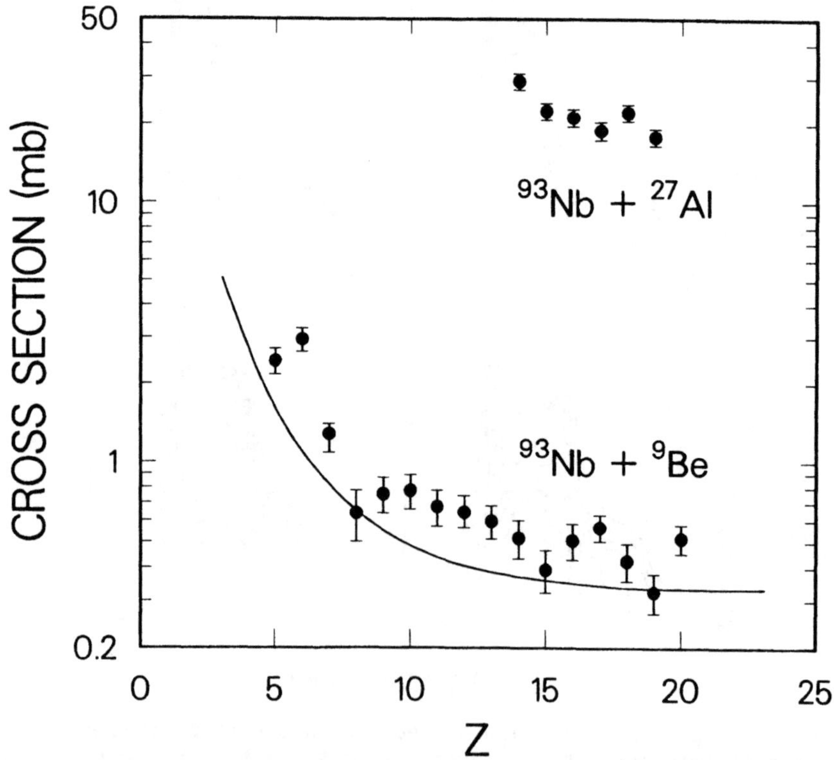

FIGURE 10 Angle-integrated cross sections (symbols) for complex fragments emitted from the reaction 30 MeV/u ^{93}Nb + ^{27}Al and ^{9}Be. Liquid-drop model calculation (solid line) of the fragment yield for the latter system.

independent of direction. Because of the large center-of-mass velocity, a given lab angle intersects the kinematic circle twice giving rise to a double solution. This simple observation allows us to conclude that the process is binary, especially when from the measured velocities one obtains a Coulomb-like Z dependence for the center-of-mass velocities. We also took coincidence data with another detector placed symmetrically on the other side of the beam. These data confirm that the process is indeed binary, and that an upper solution fragment in one detector is in coincidence with a lower solution fragment in the other.

The pattern seen in Figure 8 evolves regularly and smoothly with bombarding energy and target. This indicates that indeed we are observing the same kinematic circle, boosted by different velocities of the center-of-mass which are in approximate agreement with those expected from complete fusion. Furthermore, reverse kinematics allows us to verify that what you see is all there is. When we go to wider angles we lose the intersection with the kinematic circle and we see nothing. So there are no other processes than the one we have described, and we can conclude that, up to 18.5 MeV/u "nihil sub sole novum," nothing new under the sun but compound nucleus decay.

Proceeding to the better analyzed Bevalac data, we see more of the same. In Figure 9 the invariant cross sections plotted in the Z-V plane show that the double solutions are retained up to 30 MeV/u. Notice also that at very low Z's there is a trail of low velocity events which we call "big foot." This process is clearly target related, and may have to do with the onset of incomplete fusion. In this case the events are due to the target picking up a few nucleons for the projectile and a corresponding fraction of the momentum.

The velocities of the source clearly indicate a single source for all Z's with a velocity intermediate between the projectile velocity and

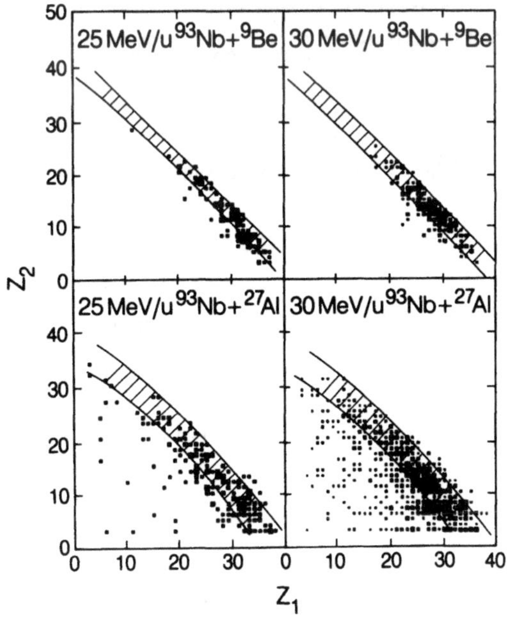

FIGURE 11 Scatter plots of coincidence events between the 5.5° telescope (Z_1) and the -11° telescope (Z_2). The shaded areas represent an estimation of regions where binary events should lie following sequential evaporation from the primary fragments.

the compound nucleus velocity but closer to the latter. The inferred incomplete momentum transfer in the direct kinematic solution is in good agreement with the standard momentum transfer systematics. Similarly the velocities in the center of mass are Coulomb-like.

An example of charge distribution is shown in Figure 10 together with an absolute calculation. The ability to fit the absolute cross sections vs. Z with a compound nucleus model is in our eyes very significant, because it implies a statistical branching ratio between complex fragment emission and the dominant n, p, ^{4}He decay.

The coincidence data are shown in Figure 11. The hatched bands are predicted on the basis of the incomplete momentum transfer, of the resulting excitation energy, and of the sequential evaporation from the binary fragments calculated from the code PACE.[12] The overall picture is consistent with binary decay. However notice that in the case of the Al target at 30 MeV/u a number of events falls outside the expected band, indicating perhaps three or more body decay. A better appreciation of the coincidence data and of the calculations is given in Figure 12 where the average sum of charges is plotted versus one of the charges. The dashed line is the primary sum inferred from the source velocity and the solid line is the calculation from PACE of the sequential charge evaporation. The excellent agreement indicates a solid understanding of the incomplete fusion process, of the energy deposition, and of the binary decay followed by sequential evaporation.

What we can state with a good degree of confidence is that up to the highest explored energy, compound nucleus decay is by far the main source of complex fragments. An additional source at low A's is the "big foot" which is target related and seems to be consistent with the target picking up a few nucleons for the projectile and decaying in its turn.

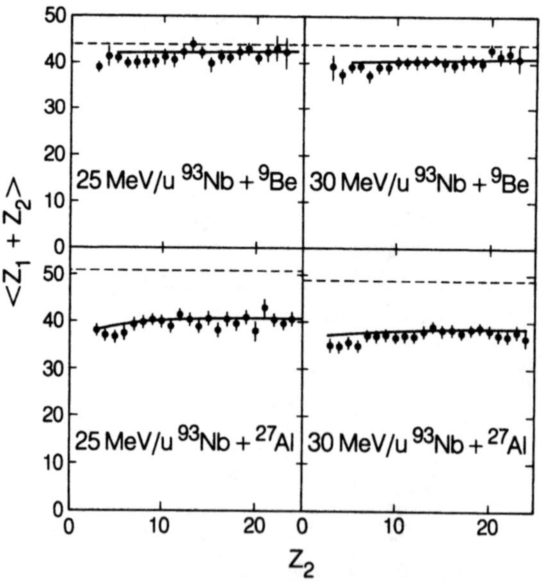

FIGURE 12 The mean sum, $\langle Z_1 + Z_2 \rangle$ of coincidence events (solid symbols) plotted as a function of Z_2. The dashed lines indicate the average charge of the compound system as estimated from the mass transfer. The charge loss of binary events, due to sequential evaporation, was estimated using the PACE[12] code and the residual $\langle Z_1 + Z_2 \rangle$ values are indicated by the solid curves.

The very thorough exploration of angle and energy "phase space" allowed by reverse kinematics does not leave much room for any other processes. Can we conclude then that they are not there? I think we must wait for ternary and quaternary events which will undoubtedly appear at higher energies. However even with these events one needs to be cautious. Ternary, quaternary and higher multiplicity events can originate from sequential binary decays. In fact, once one has a good excitation function for the binaries, it is a simple exercise to predict the rate of sequential ternaries and quaternaries. This will be the background, and it will not be small, on top of which we shall have to look in search of fancier mechanisms.

This work was supported by the Director, Office of Energy Research, Division of Nuclear Physics of the Office of High Energy and Nuclear Physics of the U.S. Department of Energy under Contract DE-AC03-76SF00098.

REFERENCES

1. E.K. Hyde, The Nuclear Properties of the Heavy Elements, Vol III: Fission Phenomena, Prentice-Hall, Inc., 1964, Ch. 11 and references therein
2. A.M. Poskanzer, G.W. Butler and E.K. Hyde, Phys. Rev. C3 (1971) 882; ibid C5 (1971) 1795.
3. J. Aichelin and J. Hüfner, Phys. Lett. 136B (1984) 15.
4. See for instance J.E. Finn, et al., Phys. Rev. Lett. 49 (1982) 1321.
5. L.G. Moretto, Nucl. Phys. A247 (1975) 211.
6. L.G. Sobotka, M.A. Padgett, G.J. Wozniak, G. Guarino, A.J. Pacheco, L.G. Moretto, Y. Chan, R.G. Stokstad, I. Tserruya, and S. Wald, Phys. Rev. Lett. 51 (1983) 2187.
7. M.A. McMahan et al., Phys. Rev. Lett. 54, 1995 (1985)
8. A.J. Sierk, Phys. Rev. Lett. 55, 582 (1985)
9. L.G. Sobotka, M.A. McMahan, R.J. McDonald, C. Signarbieux, G.J. Wozniak, M.L. Padgett, J.H. Gu, Z.H. Liu, Z.O. Yao and L.G. Moretto, Phys. Rev. Lett. 53, 2004 (1984)
10. R.J. Charity et al., Phys. Rev. Lett. 56, 1354 (1986)
11. R.J. Charity et al., to be published
12. A. Gavron, Phys. Rev. C21, 230 (1980)

QUANTUM STATISTICAL ASPECTS OF NON-EQUILIBRIUM PARTICLE EMISSION IN INTERMEDIATE-ENERGY NUCLEUS-NUCLEUS COLLISIONS

C.K. GELBKE

Department of Physics and National Superconducting Cyclotron Laboratory
Michigan State University, East Lansing, MI 48824, USA

Perhaps one of the most interesting open problems in nuclear physics concerns the question whether one can extract reliable information about the properties of nuclear matter at high densities and/or temperatures from nuclear collision experiments. This question cannot be answered trivially because of complications which arise from the fact that nuclear collision experiments correspond to the formation and decay of short lived systems which contain less than 500 nucleons: (i) Because of the finite number of particles, thermodynamic quantities may not be sharply defined and may be associated with sizeable fluctuations. (ii) Because of the small size of the reaction volume, surface effects may introduce considerable deviations from the thermodynamic behavior of bulk nuclear matter. (iii) Because of the small containment times, statistical equilibrium may not be achieved and the reaction dynamics may have to be understood in considerable detail before properties of nuclear matter can be inferred from experimental observations.

Heavy ion induced reactions at intermediate energies ($E/A \approx 20-500$ MeV) may be ideally suited to address fundamental questions concerning the determination of the nuclear equation of state from nucleus-nucleus collision experiments. Theoretically,[1-3] there is little doubt that a liquid gas phase transition should exist for nuclear matter at moderate temperatures of $T \lesssim 20$ MeV. Experimentally, the appropriate range of temperatures should be accessible - provided, at least, that local thermal equilibrium is reached during these collisions and that statistical concepts are still useful. Conceptually similar questions will have to be addressed at much higher energies where the quark deconfinement transition is expected to take place.

For intermediate energy nucleus-nucleus collisions, it has been clearly established that particle emission occurs prior to the attainment of full statistical equilibrium of the composite system.[4-10] In the absence of a complete dynamical treatment, recourse is often taken to models based on the assumption of statistical particle emission from highly excited subsets of nucleons[4-13] which are characterised by their average collective velocity, space-time extent, and excitation energy density or temperature. Clearly, the experimental characterisation of the collective and statistical properties of these subsets is important. In the following, I will discuss two-particle correlation measurements at small relative momenta which provide information about the space-time extent and temperature of the emitting system.

Most attempts to extract temperatures are based on analyses of the kinetic energy spectra of the emitted particles.[7] The interpretation of such spectra can, however, be complicated by sensitivities to the collective motion[14] and the temporal evolution of the emitting system.[9,15] An alternative determination of the "emission temperature", i.e. the temperature at the point at which the particles leave the equilibrated subsystem, is based on the relative population of states.[16-20] These "temperature measurements" can have large uncertainties whenever the primary population ratio is altered by secondary processes;[17,18,21] they should become insensitive to secondary processes if the level separation is much larger than the emission temperature.[18] These considerations suggest that accurate temperature determinations should be possible from the measurements of the relative populations of widely separated states. For the case of light nuclei, suitable states are generally particle unstable. The relative populations of particle unstable states can be expressed as:

$$\frac{d\,N(E)}{d\,E} = N_0 \cdot e^{-E/T} \cdot \frac{1}{\pi} \sum_{J,\ell} (2J+1) \frac{\partial\, \delta_{J,\ell}}{\partial\, E} \quad ,$$

where T denotes the temperature of the emitting source and $\delta_{J,\ell}$ is the scattering phase shift in the exit channel. If the energy dependence of the phase shifts is dominated by a series of resonances, one obtains:[18]

$$\frac{d\,N(E)}{d\,E} = N_0 \cdot e^{-E/T} \cdot \sum_i (2J_i+1) \frac{\Gamma_i/2\pi}{(E-E_i)^2 + \Gamma_i^2/4} \; .$$

Information about the space-time extent of the emitting system can be obtained from the measurement of two-particle correlation functions[22-26] at small relative momenta for which interactions between the emitted particles are important. The two-particle correlation function, R(q), is defined in terms of the singles yields, $Y_1(\vec{p}_1)$ and $Y_2(\vec{p}_2)$, and the coincidence yield, $Y_{12}(\vec{p}_1,\vec{p}_2)$, of particles 1 and 2:

$$Y_{12}(\vec{p}_1,\vec{p}_2) = C \cdot Y_1(\vec{p}_1) Y_2(\vec{p}_2)[1+R(q)] .$$

Here, $\vec{p}_1$ and $\vec{p}_2$ are the laboratory momenta of the two particles, q is the momentum of relative motion, and C is a normalization constant. The experimental correlation functions are obtained by inserting the measured yields into the equation and summing both sides over all energies and angles which satisfy a given gating condition and which correspond to a given relative momentum. To first order within the framework of equilibrium thermodynamics, the two-particle correlation function and the relative decay yields are independent functions of one variable each,[26] the source volume and temperature, respectively. For two non-identical particles of spins s_1 and s_2, the correlation function, R(q), can be approximated as[26]

$$R(q) = \frac{3}{(2s_1+1)(2s_2+1)\cdot 2r^3 q^2} \sum_{J,\ell} (2J+1) \frac{\partial \delta_{J,\ell}}{\partial q} ,$$

where r denotes the source radius. This formula has recently been shown[26] to be consistent with the original final-state-interaction model proposed by Koonin.[22]

In the thermal approximation, measurements of the two-particle correlation function and of the relative populations of states provide independent information on the temperature <u>and</u> the space-time evolution of the emitting system.[20,26]

As a first example, Figure 1 shows two-deuteron correlation functions measured for ^{40}Ar induced reactions on ^{197}Au by the MSU-GANIL collaboration. The dotted curve shows the correlation function predicted by calculations[24] based on deuteron-deuteron phase shifts extracted by the resonating group method.[27] These phase shifts exhibit a resonant behavior in the $\ell=1$ partial wave at energies below 2 MeV. The predicted maximum of the correlation function at low relative momenta is not observed experimentally. In fact, the experimental data can be reproduced rather nicely by calculations[24] based on a more recent set of phase shifts[28] extracted by the coupled channel R-matrix method,[28] see solid, dashed and dot-dashed curves. The calculations were performed by assuming an emitting source of Gaussian spatial density, $\rho(r)=\rho_0\cdot\exp(-r^2/r_0^2)$, and negligible lifetime. The sensitivity of the correlation function to the radius paramenter r_0 is illustratd by the dashed, solid, and dot-dashed curves which represent the correlations predicted for r_0=9, 7, and 5 fm, respectively. The experimental correlation functions were evaluated for several ranges of the sum kinetic energy: E_1+E_2= 25-75 MeV (solid circular points); E_1+E_2= 75-125 MeV (open points); E_1+E_2= 125-175 MeV (solid squares). The correlations become more pronounced with increasing energy of the emitted particles, indicating that more energetic particles may be emitted from subsets of nucleons which are more localised in space-time.

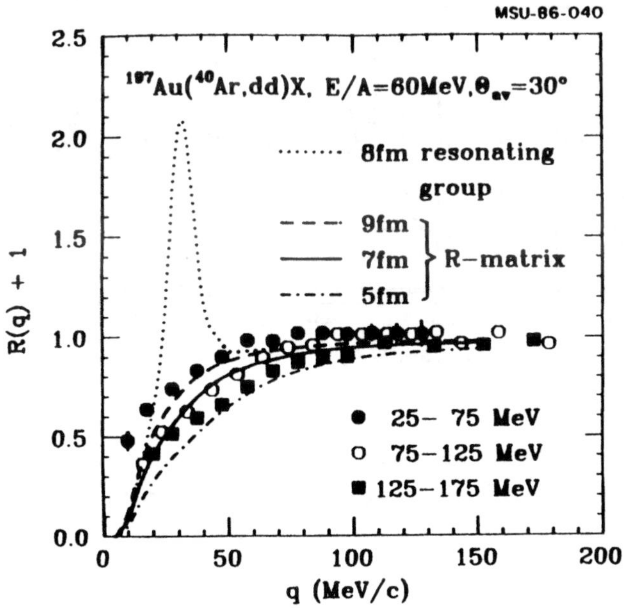

Fig.1: Two-deuteron correlation functions measured for ^{40}Ar induced reactions on ^{197}Au at E/A=30 MeV. A detailed discussion of the figure is given in text. (Data from MSU-GANIL collaboration.)

In order to give an example of the strong correlations which arise from the emission of particle unstable states, Figure 2 shows deuteron - alpha particle correlation functions measured[20] for ^{14}N induced reactions on ^{197}Au at E/A=35 MeV. The upper and lower parts of the figure show correlations measured at average laboratory angles of Θ=35° and 50°, respectively. The following gating conditions were applied: $E_\alpha \geq$ 40 MeV, $E_d \geq$ 15 MeV <u>and</u>: $E_\alpha+E_d$= 55-100 MeV (left hand part), $E_\alpha+E_d$= 100-150 MeV (center part), $E_\alpha + E_d$ = 150-220 MeV (right hand part). The α-d correlation functions exhibit two maxima corresponding to the T=0 state in ^{6}Li at 2.186 MeV ($J^\pi = 3^+$, Γ=24 keV, $\Gamma_\alpha/\Gamma_{tot}$=1.00) and the overlapping T=0 states at 4.31 MeV ($J^\pi = 2^+$, Γ=1.3 MeV, $\Gamma_\alpha/\Gamma_{tot}$=0.97) and at 5.65 MeV ($J^\pi = 1^+$, Γ=1.9 MeV, $\Gamma_\alpha/\Gamma_{tot}$=0.74). Qualitatively similar, but slightly less pronounced correlations were measured[18] for ^{40}Ar induced reactions on ^{197}Au at E/A=60 MeV.

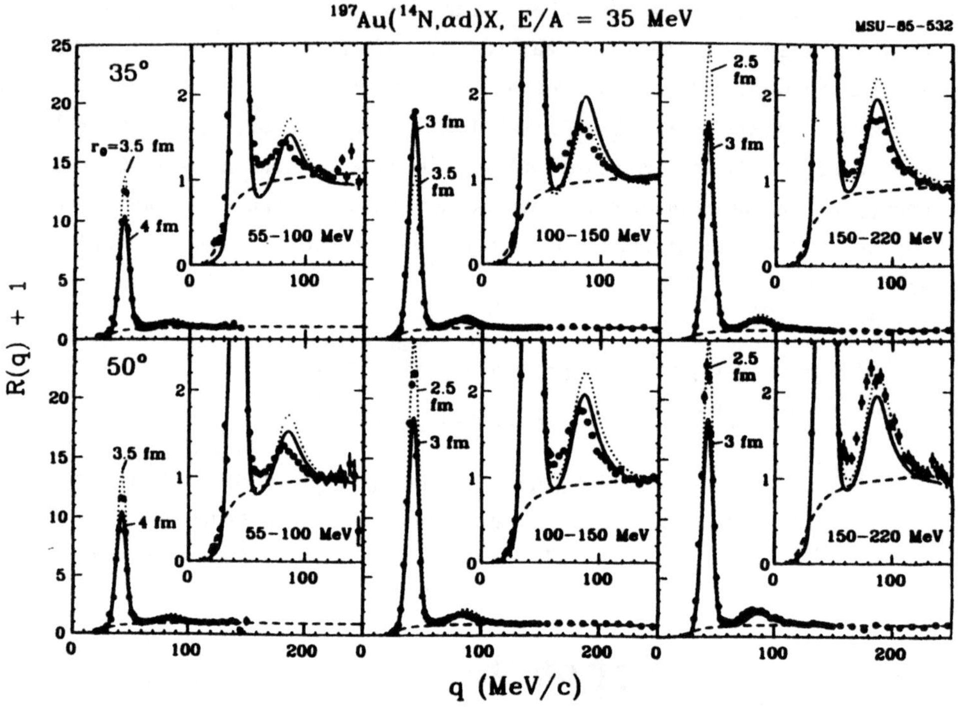

Fig.2: Correlation functions measured[20] for coincident deuterons and alpha particles for ^{14}N induced reactions on ^{197}Au at E/A=35 MeV. A detailed discussion of the figure is given in the text.

Calculations of the α-d correlation function based on a generalization[25] of the final-state interaction model of ref. 22 are shown by the solid and dotted lines in Figure 1. As before, a source of Gaussian spatial density and negligible lifetime was assumed. Similar to the trend observed for two-deuteron correlations, the α-d correlations become more pronounced with increasing energy of the outgoing particles indicating increasing space-time localisation. No significant angular dependence is observed. The extracted source radii are listed in Table 1.

Table 1:

Emission temperatures and source radii extracted from the decay $^6Li^* \to \alpha + d$.

constraint on	$\Theta = 35°$		$\Theta = 50°$	
$E_1 + E_2$	T(MeV)	r_0(fm)	T(MeV)	r_0(fm)
55 - 220 MeV	4	3.4	4	3.6
55 - 100 MeV	4	3.8	3	3.9
100 - 150 MeV	4	3.0	5	2.8
150 - 220 MeV	7	3.0	9	2.7

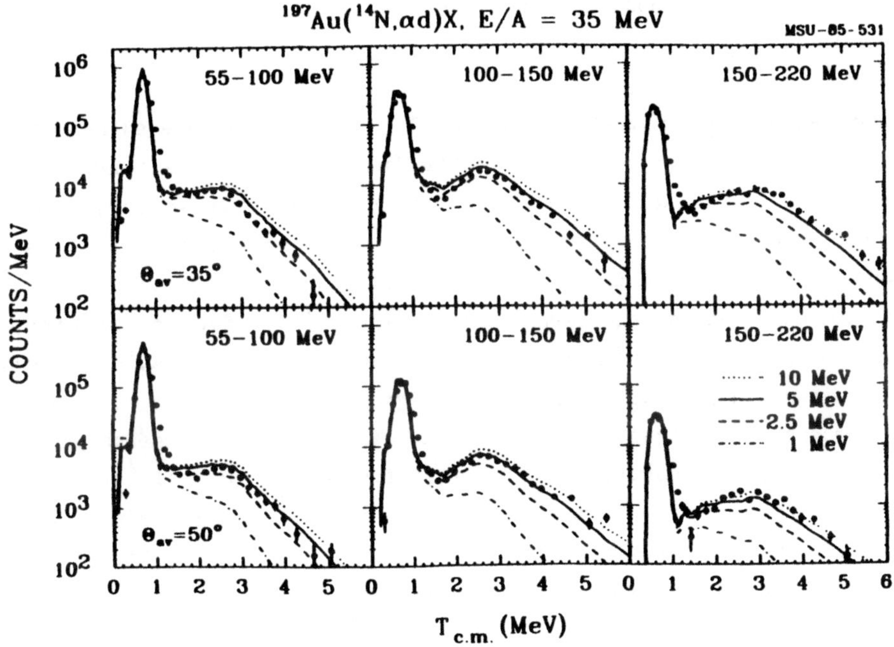

Fig.3: Energy spectra resulting from the decay $^6Li^* \to \alpha + d$. For a detailed discussion of the figure see text. (From ref. 20)

Emission temperatures can be obtained by comparing the experimental yield of particle unstable ^6Li nuclei with thermal calculations. The experimental yield of particle unstable decays $^6Li^* \to \alpha + d$, Y_c, was assumed to be given by $Y_c = Y_{\alpha d} - C \cdot Y_\alpha Y_d [1 + R_b(q)]$, where $R_b(q)$ denotes the background correlation function[18] shown by the dashed lines in Figure 2.

The resulting yields are shown in Figure 3 as a function of the kinetic energy, $T_{c.m.}$, in the ^6Li rest frame. The curves shown in the figure correspond to the theoretical coincidence yields resulting from the decays of thermally emitted particle unstable ^6Li nuclei. The calculations of the decay yields incorporate the appropriate decay branching ratios as well as the efficiency and resolution of the hodoscope; for a detailed discussion see refs. 18-20. The calculated yields are normalised to reproduce the experimental yield integrated over the energy range of $T_{c.m.}$ = 0.3 - 1.2 MeV. In order to extract emission temperatures, we have integrated the decay yields over the energy ranges of $T_{c.m.}$ = 0.25-1.45 and 1.5-6.25 MeV and compared the ratio of these yields to the corresponding theoretical ratio. The results are summarized in Table 1.

Higher emission temperatures and smaller source radii are extracted for higher kinetic energies of the emitted particles. These findings are consistent with particle emission from a subsystem which is in the process of cooling and expanding. Cooling and expanding subsystems of high excitation could arise from the equilibration of participant matter with the surrounding cold target nuclear matter[9] or from an isentropic expansion as expected from intranuclear cascade calculations.[29] Although these experimental results are highly suggestive, one should realize that the temperatures extracted from the α-d coincidence measurements shown in Figs. 2 and 3 have considerable uncertainties (≈25%, see ref. 20). These uncertainties are due to uncertainties in the α-d background correlation function, uncertainties concerning the importance of feeding from the sequential decay of heavier mass primary fragments[18-20] and due to the saturation of the coincidence yields at higher temperatures.

More accurate temperature determinations can be made by measuring the relative populations of states separated by significantly larger energy intervals;[19] e.g.: $^5Li_{0.0}\rightarrow\alpha+p$, $^5Li_{16.7}\rightarrow d+{}^3He$; $^8Be_{3.0}\rightarrow\alpha+\alpha$ and $^8Be_{17.6}\rightarrow{}^7Li+p$. As an example, Figures 4 and 5 show correlations[19] which can be attributed to states in ^5Li; these correlations were measured by the MSU-GANIL collaboration for ^{40}Ar induced reactions on ^{197}Au at E/A=60 MeV. Figure 4 shows the α-p correlation function. The broad peak near q≈50 MeV/c is due to the decay of $^5Li_{gs}\rightarrow\alpha+p$ in the Coulomb field of the heavy target residue.[30] (The sharp peak near q≈15 MeV/c is not related to a resonance in the mass five system:[30] it is caused by the three

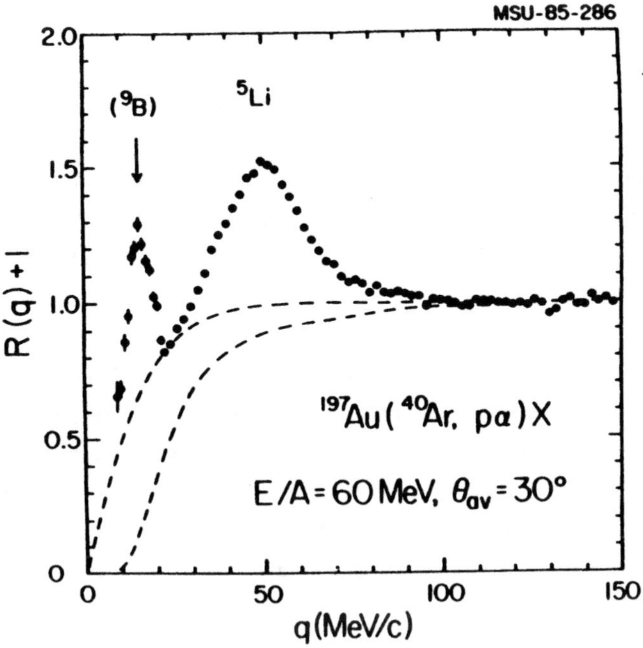

Fig. 4: Correlation function for coincident protons and alpha particles measured[19] for ^{40}Ar induced reactions on ^{197}Au at E/A=60 MeV. The dashed lines indicate the extremes within which the the background correlation function was assumed to lie.

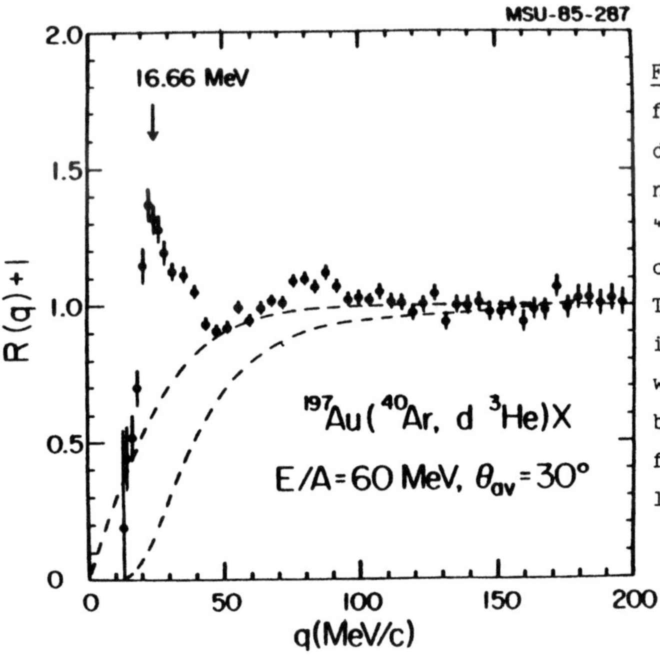

Fig. 5: Correlation function for coincident deuterons and ^{3}He nuclei measured[19] for ^{40}Ar induced reactions on ^{197}Au at E/A=60 MeV. The dashed lines indicate the extremes within which the background correlation function was assumed to lie.

particle decay of $^{9}B \rightarrow \alpha+\alpha+p$.) Figure 5 shows the d-$^{3}He$ correlation function. The most pronounced structure in this correlation function is due to the decay $^{5}Li^{*}_{16.7} \rightarrow d+^{3}He$. From the relative populations of these two states an average emission temperature of T≈4-5 MeV was extracted.[19]

Very similar emission temperatures were measured for ^{40}Ar induced reactions[18,19] at E/A=60 MeV and for ^{14}N induced reactions[20] at E/A=35 MeV. At present, this similarity between emission temperatures obtained for two very different reactions is not understood. One possibility is that the two measurements contain very different contributions from peripheral reactions. To eliminate such contributions, measurements must be performed in coincidence with central collision triggers. Clearly, more measurements are needed.

Detailed investigations of two-particle correlations at small relative momenta have only been performed very recently. The technique bears significant promise as a diagnostic tool for studies of nucleus-nucleus collisions at intermediate energies. It may also have useful applications in low energy nuclear physics whenever the nuclear interaction at energies close to the particle threshold is uncertain.

I would like to acknowledge many useful discussions with D.H. Boal, W.A. Friedman, W.G. Lynch, and J. Pochodzalla. This work was supported by the National Science Foundation under grants PHY 83-12245 and PHY 84-01845.

References:

1) M.W. Curtin, et al., Phys. Lett. 123B, 289 (1983)
2) P.J. Siemens, Nature, 305, 410 (1983)
3) H. Jaqaman, et al., Phys. Rev. C27, 2782 (1983)
4) T.C. Awes et al., Phys. Rev. C24, 89 (1981)
5) T.C. Awes, et al., Phys. Lett. 103B, 417 (1981)
6) T.C. Awes, et al. Phys. Rev. C25, 2361 (1982)
7) G.D. Westfall, et al., Phys. Lett. 116B, 118 (1982)
8) B.V. Jacak, et al., Phys. Rev. Lett. 51, 1846 (1983)
9) D.J. Fields, et al., Phys. Rev. C30, 1912 (1984)
10) G.D. Westfall, et al., Phys. Rev. Lett. 37, 1202 (1976)
11) J. Gosset, et al., Phys. Rev. C18, 844 (1978)
12) J. Knoll, Phys. Rev. C20, 773 (1979)
13) W.A. Friedman and W.G. Lynch, Phys. Rev. C28, 16 (1983)
14) P.J. Siemens and J.O. Rasmussen, Phys. Rev. Lett. 42, 880 (1979)
15) H. Stöcker, et al., Z. Phys. A303, 259 (1981)
16) D.J. Morrissey, et al., Phys. Lett. B148, 423 (1984)
17) D.J. Morrissey, et al., Phys. Rev. C32, 877 (1985)

18) J. Pochodzalla, et al., Phys. Rev. Lett. 55, 177 (1985)

19) J. Pochodzalla, et al., Phys. Lett. 161B, 275 (1985)

20) C.B. Chitwood, et al., Michigan State University preprint MSUCL-543, 1985

21) D.H. Boal, Phys. Rev. C30, 749 (1984)

22) B.E. Koonin, Phys. Lett. 70B, 43 (1977)

23) W.G. Lynch, et al., Phys. Rev. Lett. 51, 1850 (1983)

24) C.B. Chitwood, et al., Phys. Rev. Lett. 54, 302 (1985)

25) D.H. Boal and J.C. Shillcock, to be published in Phys. Rev. C

26) D.H. Boal, B.K. Jennings, and J.C. Shillcock, private communication

27) F.S. Chwieroth et al., Nucl. Phys. A189, 1 (1972)

28) G.M. Hale and B.C. Dodder, Few-Body Problems in Physics, edited by B. Zeidnitz, (Elsevier, Amsterdam, 1984), Vol. 2, p. 433

29) G. Bertsch and J. Cugnon, Phys. Rev. C24, 2514 (1981)

30) J. Pochodzalla, et al., Phys. Lett. 161B, 256 (1985)

STATISTICAL THEORY OF PRECOMPOUND NUCLEAR REACTIONS

H. Nishioka
Max-Planck-Institut für Kernphysik
Postfach 10 39 80, 6900 Heidelberg, Federal Republic of Germany

1. Introduction and Summary.

The purpose of my talk is to show the application of the Grassmann-integration method (or the graded-symmetry method) to a pre-equilibrium process in nuclear reactions. The Grassmann-integration method for random systems was first introduced by Efetov[1] and later largely extended and applied to nuclear physics by Verbaarschot, Weidenmüller and Zirnbauer[2] (referred to as VWZ). They have applied it to the equilibrium nuclear reactions; namely, the compound-nucleus reactions. It will be shown in this talk that this method is also applicable to non-equilibrium nuclear reactions.

Applying this method to precompound nuclear reactions (the work done together with J.J.M. Verbaarschot, H.A. Weidenmüller and S. Yoshida[3]), we have obtained the same expression of the cross-section as Agassi, Weidenmüller and Mantzouranis[4] (referred to as AWM) in the weak-coupling limit. In the general case our results show an important modification to AWM.

2. The precompound process; the experimental and theoretical situations

VWZ looked at the nuclear reaction in which the relaxation time in the bound-state space is much shorter than the particle-decay time. In this situation the bound states first reach an equilibrium and then decay. This is an equilibrium decay process; namely, the compound-nucleus reactions. The compound-nucleus reactions occur typically when a neutron bombards a heavy nuclear target with an incident energy up to a few MeV. When the energy of the incident particle goes up, the number of open channels exponentially increases and the decay time exponentially decreases. Finally the decay time becomes comparable to the relaxation time in the bound-state space, and we can no more treat the bound states as a single ensemble as in VWZ.

Experimental examples are typically provided by neutron emissions following neutron bombardments on a heavy target with the incident energies of ten to several tens of MeV. When the energy of the emitted neutron is near the incident energy, the excitation function shows sharp peaks. These peaks correspond to direct reactions, which I neglect in this talk. When the energy of the emitted neutron is very low and near the threshold, the excitation function is still described by an equilibrium-decay process. Although the excitation function falls off rapidly as the energy of the emitted neutron increases, there is still a continuous spectrum in the intermediate-energy region. This continuous spectrum cannot be explained by the equilibrium-decay process, and we expect that it comes from a preequilibrium-decay process, which is called as precompound reactions. I will discuss on the angle-integrated cross-sections and not on the angular distributions of the emitted neutrons.

A phenomenologically successful model for the precompound reactions was proposed by Griffin[5] and is called the exciton model. The model is based on several physically reasonable assumptions : The reaction proceeds via the two-body interaction through a chain of the particle-hole states. The number of particles and holes is called the exciton number. In each class of the exciton number, every partition of energy into the particles and holes occurs with equal probability. From every exciton-number class a particle can decay. With these assumptions the probability of particle emission in the energy range $\varepsilon \sim \varepsilon + d\varepsilon$ is given by

$$P(\varepsilon)\, d\varepsilon \propto \sum_{n=n_0,n_0+2,\ldots,\bar{n}} \varepsilon^{1/2}\, \frac{\rho_{n-1}(U)}{\rho_n(E)}\, d\varepsilon\ , \tag{1}$$

where $\rho_n(E)$ is the level density of the whole system with the exciton number n at the excitation energy E, and $\rho_{n-1}(U)$ is the level density of the residual nucleus with the exciton number n-1 at the excitation energy U. The summation over n is from the initial exciton number n_0 to the average exciton number of the equilibrated system. The level densities used in this model are unperturbed ones and do not include the coupling effects from other exciton-number states. They are calculated, for example, by the Fermi gas model.

AWM have layed down a quantum-mechanical foundation of the exciton model. They assumed several classes in the bound-state space. Each class is characterised by a certain exciton number. In each class the states are equilibrated and the eigenenergies are all equally distanced (the picket-fence model). The interaction strengths between states in different classes are random with Gaussian distribution around zero. Each bound state is coupled to decaying-channel states. Using a resummation technique of the perturbation series they obtained the fluctuating part of the cross-section; namely the ensemble average of the product of the S matrix element from the channel b to a with its complex conjugate from the channel c to d.

$$\overline{S^{fl}_{ab}\, S^{fl*}_{cd}} = (\delta_{ac}\,\delta_{bd} + \delta_{ad}\,\delta_{bc}) \sum_{m,n} T^a_m \; \Pi_{mn}\, T^b_n \quad , \tag{2}$$

$$\sum_k \; \Pi_{mk}\, \{\delta_{kn}\, 2\pi\, \rho_n\, (\Gamma^{ext}_n + \Gamma^{\downarrow}_n) - T^{ext}_{kn} - T^{int}_{kn}\} = \delta_{mn} \quad . \tag{3}$$

In eq. (2) T^a_m is the transmission coefficient from the class m to the channel a. In eq. (3) the width Γ^{ext} and the transition coefficient T^{ext} are due to the coupling between the bound states and the channels and depend on the transmission coefficients T^a_m. The $\Gamma^{\downarrow}$ and T^{int} are due to the coupling between the bound states in different classes and are given by

$$\Gamma^{int}_{kn} = 2\pi\, \rho_k\, \overline{V}^2_{kn}\, 2\pi\, \rho_n \; , \tag{4}$$

and

$$\Gamma^{\downarrow}_n = \sum_m \overline{V}^2_{mn} \; 2\pi\, \rho_m \quad , \tag{5}$$

where $\overline{V}^2_{mn}$ is the average of the square of the coupling strength between a state in the class m and a state in the class n. Here, the level densities ρ_n are unperturbed ones.

3. The Grassmann-integration method by Verbaarschot, Weidenmüller and Zirnbauer and the compound-nucleus reactions.

Before explaining our approach to the precompound reactions, I need to explain the work of VWZ on the Grassmann-integration method applied to the compound-nucleus reactions[2].

Their model is as follows. The bound states form a Gaussian Orthogonal Ensemble (GOE) and are coupled weakly to the channels. The Hamiltonian is given by

$$H = \sum_a \int_{\varepsilon_a}^{\infty} dE \, |\chi_a(E)> E <\chi_a(E)| + \sum_{\mu,\nu} |\mu> H_{\mu\nu} <\nu|$$

$$+ \sum_{\mu,a} \left\{ |\mu> \int_{\varepsilon_a}^{\infty} dE \, W_{\mu a}(E) <\chi_a(E)| + \text{h.c.} \right\} \quad . \tag{6}$$

In the first term $\chi_a(E)$ is the channel-state wavefunction in the channel a with the energy E, in the second term $|\mu>$ and $<\nu|$ are the bound states, and the last term expresses the coupling between the bound states and the channel states. Neglecting direct reactions the S-matrix elements are given by

$$S_{ab} = \delta_{ab} - 2i\pi \sum_{\mu\nu} W_{\mu a}(E) \, (D^{-1})_{\mu\nu} W_{\nu b}(E) \quad , \tag{7}$$

where D is the energy denominator and is given by

$$D_{\mu\nu} = E\delta_{\mu\nu} - H_{\mu\nu} + i \, \pi \sum_c W_{\mu c} W_{\nu c} \tag{8}$$

The hamiltonian matrix elements $H_{\mu\nu}$ form a GOE, which has the zero average value and the second moment of

$$\overline{H_{\mu\nu} H_{\mu'\nu'}} = \frac{\lambda^2}{N} (\delta_{\mu\mu'}\delta_{\nu\nu'} + \delta_{\mu\nu'}\delta_{\nu\mu'}) \quad , \tag{9}$$

where N is the total number of bound states. The coupling strengths $W_{\mu a}$ should satisfy the condition

$$\sum_\mu W_{\mu a} W_{\mu b} = \delta_{ab} v_a^2 \quad , \tag{10}$$

in the absence of the direct reactions.

In eq. (7) the $H_{\mu\nu}$ are in the denominator, and it is not possible to take the ensemble average exactly. Therefore using the generating function Z, eq. (7) is rewritten as

$$S_{ab} = \frac{\partial}{\partial J} \ln Z(J) \, |_{J=0} \quad , \tag{11}$$

$$Z(J) = \int \cdots \int_{-\infty}^{\infty} \prod_\mu dS_\mu \exp\left\{ \frac{i}{2} \sum_{\mu\nu} S_\mu (E\delta_{\mu\nu} - H_{\mu\nu} + i\pi \sum_c W_{\mu c} W_{\nu c} + M_{\mu\nu}(J)) \, S_\nu \right\} \tag{12}$$

The $M_{\mu\nu}(J)$ in eq. (12) is the generating term and is given by

$$M_{\mu\nu}(J) = 2\, i\pi\, (W_{\mu a}\, W_{\nu b} + W_{\mu b}\, W_{\nu a})\, J \quad . \tag{13}$$

The operation of eq. (11) means first to take the derivative of ln Z(J) with respect to J and then to set J = 0. The cross-section $(S_{ab}S^*_{cd})$ is also obtained from a generating function in the same way.

In eq. (12) the $H_{\mu\nu}$ is raised into the exponential. But the ensemble average of S_{ab} requires the ensemble average of ln Z, which is still difficult. If the condition

$$Z(J = 0) = 1 \ , \tag{14}$$

is satisfied, the ln Z in eq. (11) is replaced by Z, and the ensemble average is carried out easily. The Grassmann variables and the Grassmann integration were introduced in order to realize the condition (14).

The principle is simply illustrated by considering integration of a Gaussian function. With an ordinary commuting variable S,

$$Z_c = \int_{-\infty}^{\infty} dS\, \exp(-\tfrac{a}{2}\, S^2) = (\tfrac{2\pi}{a})^{1/2} \quad . \tag{15}$$

The Grassmann variables χ, χ^* are anticommuting variables and therefore,

$$\chi^2 = \chi^{*2} = 0 \tag{16}$$

The integration with the Grassmann variables are defined as

$$\int d\chi^* d\chi = 0, \tag{17a}$$

$$\int d\chi^* d\chi \cdot \chi\, \chi^* = (2\pi)^{-1} \tag{17b}$$

The integration of a Gaussian function with the Grassmann variables is performed by expanding the Gaussian function;

$$Z_A = \int d\chi^* d\chi\, \exp(-a\chi^*\chi) = \int d\chi^* d\chi\, [1 - a\chi^*\chi + (1/2!)(a\chi^*\chi)^2 + \ldots] \tag{18}$$

The third and higher terms in the square-brackets of eq. (18) are zero because of eq. (16), and the integration of the first term is zero because of the definition (17a). Therefore, only the second term survives in the integration and from the definition (17b),

$$Z_A = \frac{a}{2\pi} \tag{19}$$

Therefore, by multiplying two Z_c's and one Z_A,

$$Z = Z_{c1}\, Z_{c2}\, Z_A = 1, \tag{20}$$

is realised. The above argument applies also when the scalar a is a symmetrical matrix a_{ij}. Then we replace $(a/2\pi)$ by the determinant of the matrix $(\frac{a_{ij}}{2\pi})$.

Eq. (20) is the prescription for the generation function which satisfies eq. (14). This means that an additional index α is required. For the S-matrix element the α runs from 1 to 4 , where 1 and 2 correspond to commuting variables ($S_{1\mu}$, $S_{2\mu}$) and 3 and 4 to Grassmann variables (χ_μ^*, χ_μ). Eqs. (11) and (12) are replaced by

$$S_{ab} = \frac{\partial}{\partial J}\, Z(J)\, \Big|_{J=0} , \tag{21}$$

$$Z(J) = \int \ldots \int \prod_{\mu} \prod_{\alpha} d\phi_{\alpha\mu} \exp\left\{ \frac{i}{2} \sum_{\mu\nu} \sum_{\alpha} \phi^*_{\alpha\mu}(E\, \delta_{\mu\nu} - H_{\mu\nu} + i\pi \sum_{c} W_{\mu c} W_{\nu c} \right.$$
$$\left. + M_{\mu\nu}(J))\, \phi_{\alpha\nu} \right\} . \tag{22}$$

A similar equation is obtained for $S_{ab}S^*_{cd}$ with the index α running from 1 to 8; α = 1,2,5 and 6 correspond to commuting variables and α = 3,4,7 and 8 to Grassmann variables.

The average of eq. (22) over $H_{\mu\nu}$ with the Gaussian probability distribution is trivial, and as the result we obtain a ϕ^4-term. This ϕ^4-term is transformed into a ϕ^2-term by introducing new varables $\sigma_{\alpha\beta}$ (equal number of commuting and Grassmann variables) in the form of a 4×4 matrix for S_{ab} (8×8 matrix for $S_{ab}\; S^*_{cd}$), where the bar denotes the ensemble average. Then the integrations over $\phi_{\alpha\mu}$ is easily performed, and the ensemble-averaged generating function for S_{ab} becomes

$$\overline{Z}(J) = \int [d\sigma]\, \exp\left\{ -\frac{N}{4} \operatorname*{trg}_{\alpha} \sigma^2 - \frac{1}{2} \operatorname*{trg}_{\mu\alpha} \ln \mathcal{N}(J) \right\} , \tag{23}$$

$$\mathcal{N}^{\alpha\beta}_{\mu\nu}(J) = E\, \delta_{\mu\nu}\delta_{\alpha\beta} - \lambda\, \sigma_{\alpha\beta}\delta_{\mu\nu} + i\,\pi \sum_{c} W_{\mu c}\, W_{\nu c}\, \delta_{\alpha\beta} + M_{\mu\nu}(J)\, \delta_{\alpha\beta} , \tag{24}$$

where [$d\sigma$] means the integration over all the independent commuting and anti-commuting variables in the matrix σ , and the trg_{α} means to take the trace of the matrix with the sign + (−) when the α corresponds to a commuting-variable (Grassmann-variable) component.

Finally the $N \to \infty$ limit, where N is the number of bound states, is taken and the integration (23) is performed by the saddle-point approximation. The saddle point is obtained by taking the first derivative of the function inside the exponential in eq. (23) with respect to σ. In the compound-nucleus case, the coupling to the channels is small and is neglected in the saddle-point equation. The $M_{\mu\nu}(J)$ is also neglected. Then the saddle-point equation becomes

$$\sigma - \lambda(E - \lambda\sigma)^{-1} = 0 \quad . \tag{25}$$

For $\overline{S_{ab}}$ the solution of eq. (25) is uniquely determined and the value of $\overline{S_{ab}}$ is simply given by the saddle-point value of σ. For $\overline{S_{ab}\, S^*_{cd}}$, because of the special symmetry property of the matrix σ, the σ matrix which satisfies eq. (25) is not unique and the saddle-point is a continuous manifold. An exact integration was required on this saddle-point manifold, which was performed with a great effort in VWZ.

4. Precompound reactions with the Grassmann-integration method

Once one has understood VWZ in the compound-nucleus case, it is easy to understand our approach[3] to the precompound reactions. As in AWM the bound states are devided into several classes, each bound state $|m\mu\rangle$ carrying two indexes; one for the class m and the other for the running index μ in the class m. In each class the bound states form a GOE; therefore, the model is a chain of GOE's coupled with each other and also with the channels.

The Hamiltonian and the S-matrix element in the absence of direct reactions are given by

$$H = \sum_a \int_{\varepsilon_a}^{\infty} dE\, |\chi_a(E)\rangle\, E\, \langle\chi_a(E)| + \sum_{m\mu n\nu} |m\mu\rangle\, H_{m\mu,n\nu} \langle n\nu|$$

$$+ \sum_{m\mu,a} \left\{ |m\mu\rangle \int_{\varepsilon_a}^{\infty} dE\, W_{m\mu a}(E)\, \langle\chi_a(E)| + \text{h.c.} \right\} \quad , \tag{26}$$

and

$$S_{ab}(E) = \delta_{ab} - 2i\pi \sum_{m\mu n\nu} W_{m\mu a}(D^{-1})_{m\mu n\nu} W_{n\nu b} \quad , \tag{27}$$

$$D_{m\mu n\nu} = E\, \delta_{mn}\delta_{\mu\nu} - H_{m\mu n\nu} + i\,\pi \sum_{c} W_{m\mu c} W_{n\nu c} \quad . \tag{28}$$

The only difference from eqs. (6) ~ (8) in the compound-nucleus case is that there is an additional index m in eqs. (26) ~ (28).

The difference between average enérgies of different classes should be taken into account and, unlike the compound-nucleus case, the average of the hamiltonian matrix element cannot be set to zero, and

$$\overline{H_{m\mu n\nu}} = h_m\, \delta_{mn}\, \delta_{\mu\nu} \quad . \tag{29}$$

The distribution around the average value is characterised by the second moment as before and is given by

$$\overline{\delta H_{m\mu n\nu}\; \delta H_{m'\mu' n'\nu'}} = M_{mn}(\delta_{mm'}\, \delta_{\mu\mu'}\, \delta_{nn'}\, \delta_{\nu\nu'} + \delta_{mn'}\, \delta_{\mu\nu'}\, \delta_{nm'}\, \delta_{\nu\mu'}) \ , \tag{30}$$

where

$$\delta H_{m\mu n\nu} = H_{m\mu n\nu} - \overline{H_{m\mu n\nu}} \quad . \tag{31}$$

The average coupling strengths to the channels are diagonal in each class m and satisfy

$$\sum_{\mu} W_{m\mu a} W_{m\mu b} = \delta_{ab}\, N_m\, V_{ma}^2 \quad , \tag{32}$$

where N_m is the number of states in the class m.

The same procedure as in the compound-nucleus case is taken :

1) the S_{ab} , $S_{ab}\, S_{cd}^*$ are expressed in terms of the generating functions including the Grassmann variables so that Z(J=0) = 1 .
2) The ensemble average $\overline{Z}$ with respect to $H_{m\mu,n\nu}$ is performed : The $\overline{Z}$ includes a ϕ^4 term.
3) The new integration variables $\sigma_{\alpha\beta}^m$ are introduced for each class m. The ϕ^4 terms is tranformed into a ϕ^2 term.
4) The integration over ϕ is performed.

Then the average cross-section is given by

$$\overline{S_{ab}\, S_{cd}^*} = \frac{\partial^2}{\partial J_1 \partial J_2} \overline{Z(J_1,J_2)} \ , \tag{33}$$

$$\overline{Z(J_1,J_2)} = \int \cdots \int \prod_m [d\sigma^m] \exp \overline{\mathcal{L}} \ , \tag{34}$$

$$\overline{\mathcal{L}} = -\frac{1}{4} \sum_{m,n} g_{mn} \operatorname{trg}_{\alpha} (\sigma^m \sigma^n) - \frac{1}{2} \operatorname{trg}_{m\mu\alpha} \mathcal{N} (J_1,J_2) \quad , \tag{35}$$

where

$$\sum_k g_{mk} M_{kn} = \delta_{mn} \ , \tag{36}$$

and

$$\mathcal{N}^{m\mu n\nu}_{\alpha\beta} = (E - h_m) \delta_{mn} \delta_{\mu\nu} \delta_{\alpha\beta} + i \pi \sum_c W_{m\mu c} W_{n\nu c} L_{\alpha\beta}$$

$$+ M^{m\mu n\nu}_{\alpha\beta} (J_1,J_2) - \sigma^m_{\alpha\beta} \delta_{mn} \delta_{\mu\nu} \quad , \tag{37}$$

where L is a diagonal 8×8 matrix, with diagonal elements 1 for a = 1 ~ 4, and -1 for 5 ~ 8, which corresponds to complex-conjugate operation of the latter part.

In order to proceed further a consideration on the coupling strengths is required. The relaxation time within each class (denoted as t_1) should be naturally much shorter than the relaxation time between classes (denoted as t_2). In order to realize a pre-equilibrium process, the decay time (t_{decay}) should be comparable to t_2. We later take the limit $N_m \to \infty$ for all the classes m. Then there are two different cases. In one case,

$$\frac{t_1}{t_2} \sim \frac{t_1}{t_{decay}} \sim 0 \ (N_m^{-1}) \quad . \tag{38}$$

In the other case,

$$\frac{t_1}{t_2} \sim \frac{t_1}{t_{decay}} \sim 0 \ (1) \ll 1 \quad . \tag{39}$$

Eq. (38) means that the ratio is vanishing in the $N_m \to \infty$ limit, and eq. (39) means that the ratio is still non-zero in the limit, although it is much smaller than 1. In terms of the coupling strengths, eq. (38) means

$$\frac{M_{mn} \quad (n \neq m)}{M_{mm}} \sim \frac{\sum_c V^2_{mc}}{N_m \ M_{mm}} \sim O(N_m^{-1}) \ , \tag{40}$$

and this case is called the weak-coupling case.

Eq. (39) means

$$\frac{M_{mn}\ (n\neq m)}{M_{mm}} \sim \frac{\sum_c V_{mc}^2}{N_m\ M_{mm}} \sim O\ (1), \tag{41}$$

and is called the strong-coupling case.

First the weak-coupling case is considered. The interclass coupling g_{mn} and the coupling to the channels are negligible in the saddle-point equations, which become

$$\sigma_m - \lambda_m(E - h_m - \lambda_m\ \sigma^m)^{-1} = 0 \qquad m = 1,2,\dots\ , \tag{42}$$

where the M_{mm} is parametrised as λ_m^2/N_m and σ^m is rescaled to $\lambda_m\sigma^m$. This is exactly the same saddle-point equation as in the compound-nucleus case. For each m we have an independent saddle-point manifold and the integration should be carried over the manifolds. Instead of carrying out the exact integration for $S_{ab}\ S_{cd}^*$, we have taken the asymptotic expansion with respect to $(\sum_{m,c} T_m^c)^{-1}$. Then the first-order term is exactly the same as AWM in eqs. (2) and (3).

In the strong-coupling case, both the interclass coupling and the coupling to the channels should be taken into account in the saddle-point equations, which then becomes a set of coupled equations;

$$\sum_n g_{mn}\ \sigma^n = N_m(E - h_m - \sigma^m)^{-1} - \sum_c (E - h_m - \sigma^m)^{-1}$$

$$\times\ (1 + i\ L\ y_c)^{-1}\ i\ \lambda_m\ X_{mc}\ L(E - h_m - \sigma^m)^{-1}\ , \qquad m = 1,2,\dots \tag{43}$$

where
$$y_c = \sum_m \lambda_m\ X_{mc}\ (E - h_m - \sigma^m)^{-1}\ , \tag{44}$$

and
$$X_{mc} = \pi\ N_m\ V_{mc}^2/\ \lambda_m \tag{45}$$

The saddle points σ^m (m=1,...) are therefore coupled with each other and should be solved simultaneously. The resulting saddle points are diagonal matrices, with diagonal elements $(\tau^m,\tau^m,\tau^m,\tau^m,\tau^{m*},\tau^{m*},\tau^{m*},\tau^{m*})$, and there is no manifold.

The $\overline{S^{fl}_{ab} S^{fl*}_{cd}}$ is therefore given by the values τ^m, and is expressed a

$$\overline{S^{fl}_{ab} \; S^{fl*}_{cd}} = (\delta_{ac}\, \delta_{bd} + \delta_{ad}\, \delta_{bc}) \sum_{mn} T^a_m\, \Pi_{mn}\, T^b_n \;, \qquad (46)$$

$$\sum_k \Pi_{mk} \left\{ \delta_{kn}\; 2\pi\; \tilde{\rho}_n\; (\Gamma^{ext}_n + \tilde{\Gamma}^{\downarrow}_n) - T^{ext}_{kn} - \tilde{T}^{int}_{kn} \right\} = \delta_{mn} \quad , \qquad (47)$$

where

$$\tilde{T}^{int}_{kn} = 2\;\pi\;\tilde{\rho}_k\; \overline{V^2_{kn}}\;\; 2\;\pi\;\tilde{\rho}_n \quad , \qquad (48)$$

and

$$\tilde{\Gamma}^{\downarrow}_n = \sum_m \overline{V^2_{mn}}\; 2\;\pi\;\tilde{\rho}_m \quad . \qquad (49)$$

These equations are very similar to eqs. (2) ~ (5) of AWM. The only difference is that the unperturbed density ρ_m in eqs. (2) ~ (5) is replaced by an effective density given by

$$\tilde{\rho}_m = -\frac{N_m}{\pi}\; \text{Im}\; \tau^m \quad . \qquad (50)$$

Since all the saddle-point matrices σ^m (m = 1,2,...) are coupled with each other, the effective density (50) includes the effects of the coupling between classes.

5. Discussions

In the strong-coupling case, the effective level density (50) approaches to the unperturbed density when the couplings between classes and also to the channels are very weak. This means that the strong-coupling case includes the results of AWM in the limiting case and is a more general expression.

The weak-coupling case is not really a realistic one because of the following reasons : Firstly, the precompound processes occur only when the decay time becomes sufficiently short. This means that the couplings between classes and to the channels should be sufficiently strong. Secondly, studies on the level statistics and large shell-model calculations show that the couplings between different exciton-number states are not negligible at all. We have done a simple numerical calculation of effectice densities ($\tilde{\rho}_m$) with two classes. In the energy region of physical interest, the $\tilde{\rho}_m$ of the lower class (the number of states is smaller) is much larger than the corresponding unperturbed level density (ρ_m); whereas $\tilde{\rho}_m$ for the higher class (the number of states is larger) is nearly equal to the corresponding ρ_m. Numerical investigations using the exciton model cannot usually fit the high-energy tail of the neutron-emission excitation function. Increasing the level density of the lowest exciton-number states, one can increase the high-energy yield and fit the data. Our numerical investigation suggests that the replacement of ρ_m by $\tilde{\rho}_m$ in the exciton model is the right direction to achieve a better fit to the data.

Acknowledgement

My talk was based on the work done in collaboration with J.J.M. Verbaarschot, H.A. Weidenmüller and S. Yoshida. I express my sincere thankfulness to them.

References

1. K.B. Efetov, Adv. in Phys. 32 (1983) 53
2. J.J.M. Verbaarschot, H.A. Weidenmüller and M.R. Zirnbauer, Phys. Rep. 129 (1985) 367
3. H. Nishioka, J.J.M. Verbaarschot, H.A. Weidenmüller and S. Yoshida, submitted to Ann. Phys. (1986)
4. D. Agassi, H.A. Weidenmüller and G. Mantzouranis, Phys. Rep. 22 (1975) 145
5. J.J. Griffin, Phys. Rev. Lett. 17 (1966) 478.

LARGE SCALE NUCLEAR COLLECTIVE MOTION: FROM LOCAL RPA TO TRANSPORT THEORY.

Helmut Hofmann
Physik-Department, Technische Universität München
James-Franck-Straße, 8046 Garching
W.-Germany

and

Shuhei Yamaji
The Institute of Physical and Chemical Research,
Wako-shi, Saitama
Japan

1) INTRODUCTION

A motion we call collective when several particles participate coherently. Examples can be found in almost all many particle systems. As for a nucleus one knows that its density may rotate or vibrate around a stable configuration. These two examples usually are treated by linearized versions of the equations of motion. The vibrations, for instance, are commonly described by the random phase approximation (RPA). There are two other highly interesting examples of collective motion which differ from the previous ones in a twofold sense: nuclear fission and the collision of two heavy ions. Here, all nucleons participate in the process (so there can be no doubt about its collectivity), and it is higly non-linear. The second property does imply, for instance, that we have to be aware of dissipative effects. There may be transfer of energy from coherent motion into incoherent excitations, and this is seen experimentally, indeed. Among our main problems will then be to find the appropriate separation of variables, and to understand whether and how for a small system like a nucleus we may speak of dissipation with all its consequences, like for instance the creation of statistical fluctuations. Quite obviously it is at this place where a relation to the questions of chaos might be seen.

Clearly, when tackling these problems we must base our new development on the vast experience one has gathered in nuclear theories of timereversible collective dynamics. Among others we should mention: the definition of collective variables, the question of self-consistency, and, not least, the non-linearity problem.

2) COLLECTIVE VERSUS INTRINSIC DEGREES OF FREEDOM

For pragmatical reasons we do want to introduce collective variables explicitly. The benefit we gain is a tremendous reduction in numerical effort when finally we are to solve the dynamical equations. As the main disadvantage, this introduction has to rely on our intuition rather than on first principles. In nuclear physics it has proven very successful to relate collective motion to (volume conserving) shapes of the density distribution of nuclear matter [1]. For fission and heavy ion collisions this seems to be the obvious choice. In fig. 1 we have drawn schematically a sequence of possible shapes.

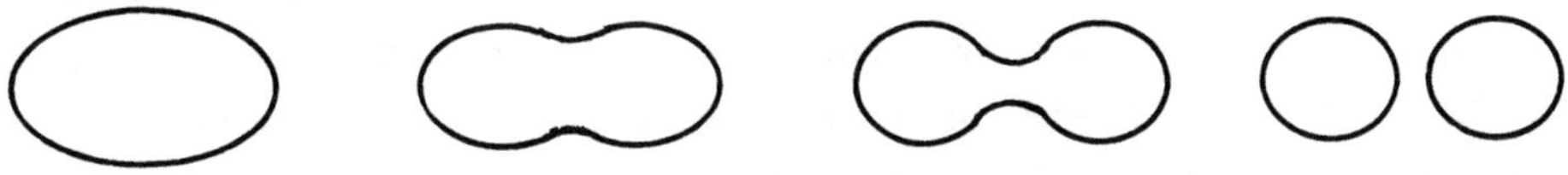

Fig. 1

For fission this figure should be read from left to right and in reverse direction for the entrance channel of a heavy ion collision. No doubt this is an oversimplification as the precise forms of these configurations are somewhat arbitrary. Furthermore, fission and a h.i.c. cannot really be characterized by exactly the same shapes. The figure shall only demonstrate the usefulness of our parametrization of collective variables in terms of shapes. (Clearly this will be possible for processes only where nuclear matter is not compressed beyond its saturation density).

Let us turn now to the motion of the nucleons which must be studied for a microscopic derivation of the macroscopic equations of motion. In first approximation the nucleons may be treated as independent particles, bound only to move in their own mean field. The latter will have to be deformed too, of course. This is due to the short range nature of the nuclear forces. For pragmatical reasons we shall assume that the mean field is related to a deformed shell model potential, which can be determined phenomenologically. Later on this picture shall only be modified in that we will consider residual interactions. They are introduced to mimic two-body collisons which become important for finite excitations.

3) THE AVERAGE LOCAL COLLECTIVE MOTION

We begin by studying the average collective motion which happens around a given shape configuration within some time interval $\delta t = t-t_0$. [For the sake of simplicity we will assume that the change of the shape can be parametrized by just one variable, $Q(t)$]. The δt shall be chosen much smaller than the typical time scale for $Q(t)$ at $t = t_0$: $|\delta t| \ll \tau_{coll}(Q_0)$. Let $\hat{h} = \hat{h}(\hat{x}_i,\hat{p}_i,Q)$ be a representative Hamiltonian for the nucleonic system $\{x_i,p_i\}$. The $\hat{h}$ does depend on Q as well but in a parametric way only: through the dependence of the mean field on this shape. For δt sufficiently small $Q(t)$ will be close to $Q_0 = Q(t_0)$ and we may thus expand like:

$$\hat{h} = \hat{h}(\hat{x}_i,\hat{p}_i,Q_0) + (Q-Q_0)\hat{F}(\hat{x}_i,Q_0) \tag{1}$$

with $F = \partial h/\partial Q_0$, and treat the second term as a low order perturbation. We may do this by exploiting linear response theory. This is appropriate, indeed, as finally we will be interested in quantities averaged over the intrinsic system. The induced local force on the collective variables, for instance, can be traced back to the average of $\hat{F}$:

$$\langle\hat{F}\rangle_t = \langle\hat{F}\rangle_0 - \int \tilde{\chi}(t-s)(Q(s)-Q_0)ds \quad , \tag{2}$$

with the causal response function being given by

$$\tilde{\chi}(t-s) = i\Theta(t-s)\langle[\hat{F}(\epsilon),\ \hat{F}(s)]\rangle_0 = 2i\Theta(t-s)\tilde{\chi}''(t-s) \quad . \tag{3}$$

The $\tilde{\chi}''(t)$ is the dissipative part of $\chi(t)$. The average in (2) and (3) are to be calculated from the hamiltonian $\hat{h}(Q_0)$. The same $\hat{h}(Q_0)$ does define also the time propagation for the F's of the commutator in (3).

Henceforth, we shall assume that the intrinsic system relaxes to a thermal equilibrium on a timescale τ shorter than τ_{coll}. In this sense we will then approximate the intrinsic density by the equilibrium operator $\hat{\rho}_0 = \hat{\rho}(\hat{h}(Q_0))$. For pragmatical reasons we shall further reduce this $\hat{\rho}_0$ to represent a canonical distribution rather than a microcanonical one. At first sight the latter might seem to be more adequate for a nuclear system. But we expect that this restriction does not cause too much harm for temperatures not too small, say above 1 MeV. Indeed, we have been able to verify this [2] in a simple model based on the random matrix ansatz for the essential matrix elements appearing in $\tilde{\chi}(t)$.

Clearly, through $\hat{h}$, our $\hat{\rho}_0$ will depend on Q_0 and so will the temperature T_0. Finally, the latter will be found by balancing the average total energy of intrinsic plus collective motion. Since the latter depends on the collective momentum P_0, we shall have our T to depend on the phase space variables Q,P.

4) SELFCONSISTENCY BETWEEN DENSITY AND MEAN FIELD

We have mentioned before that the matter density and the mean field should depend on each other in a selfconsistent way. Exploiting the simplification of the deformed shell model, it turns out that selfconsistency leads to a linear relation between the deformation of the potential, $Q-Q_0$, and the deformation of the density. If the latter is measured by the average of the single particle operator $\hat{F}$ we obtain:

$$k\langle\hat{F}\rangle_t = Q(t) - Q_0 \qquad (4)$$

This prescription is inherent to the vibrating potential model [1,3]. It can be used to describe vibrations both in the low as well as in the high frequency domain (giant resonances). One first considers (4) to hold in its static limit and determines k. Then it is used to obtain the vibrational frequency in the time dependent case. We will proceed similarly but generalize the scheme to large scale motion. Then Q_0 need not correspond to a static equilibrium. Furthermore, our frequency will correspond to that vibration which is to represent the true collective motion locally. It can be obtained [4] by inserting eq. (2) into (4) and shifting Q(t) by a constant amount Q_m. (This Q_m defines the extremum of the oscillator osculating the true static energy). Finally this frequency will be determined from the dispersion relation

$$\chi(\Omega_0) = \chi(-\Omega_0^*) = -\frac{1}{k} \quad . \qquad (5)$$

For the determination of k we use the static energy E(Q) of the intrinsic system. At zero temperature E(Q) can be computed in a Strutinsky type calculation. Its relation to k is given by [5], [4]:

$$k^{-1} = -\chi(0) - \partial^2 E / \partial Q_0^2 \quad , \qquad (6)$$

with $\chi(0)$ being the static response. (For finite temperature one must distinguish between free energy and internal energy). The first derivative of E appears to be given by:

$$\langle F\rangle_0 = \frac{\partial E}{\partial Q_0} \quad . \qquad (7)$$

As indicated in eq. (5) the frequencies come in pairs. Of course, there are many possible solutions of this equation, and we finally will have to select one pair. It shall be called Ω_0, $-\Omega_0^*$ henceforth.

Let us come back once more to the case of genuine vibrations. Experience shows that in this case eq. (5) (commonly used for undamped motion though, see [1], [3]) does represent well the experimental situation.

5) TRANSPORT COEFFICIENTS FOR THE AVERAGE MOTION

Let us assume first that eq. (5) has a solution Ω_0 which is small compared to any typical frequency of the response function itself. We may then find this Ω_0 by expanding (5) to second order. Using eq. (6) we obtain:

$$\Omega_0{}^2 \frac{1}{2} \frac{\partial^2 \chi'}{\partial \omega} \Big|_{\omega=o} + i\, \Omega_0 \frac{\partial \chi''}{\partial \omega} \Big|_{\omega=o} - \frac{\partial^2 E}{\partial Q_0{}^2} = 0 \qquad (8)$$

(with χ' being the reactive part of the response function). The interpretation of (8) is straightforward: it is the secular equation of a damped harmonic oscillator. Its inertia and friction are given by:

$$M = \frac{1}{2} \frac{\partial^2 \chi'}{\partial \omega^2} \Big|_{\omega=o} = -\frac{1}{2} \int \tilde{\chi}'(t)\, t^2 dt \qquad (9)$$

$$\gamma = \frac{\partial \chi''}{\partial \omega} \Big|_{\omega=o} = i \int \tilde{\chi}''(t)\, t\, dt \quad . \qquad (10)$$

The stiffness simply is given by the second derivative of the static energy (free energy, for finite temperature).

In general the zero frequency limit does not apply and the solutions Ω_0, $-\Omega_0{}^*$ have to be found directly from (5). In practice this can be done [6,7,4] by computing numerically the total response function

$$\chi_{coll}(\omega) = \frac{\chi(\omega)}{1+k\chi(\omega)} \qquad (11)$$

for real frequencies, where it shows more or less pronounced peaks. We may pick one. Its position ϵ and its width $\Gamma/2$ determine Ω_0 (and $-\Omega_0{}^*$). From these two quantities together with the height of the peaks we again are able to deduce the oscillator parameters M,γ and c. They can be written as:

$$M = \frac{1}{2\epsilon} \frac{\partial \chi}{\partial \omega} \Big|_{\Omega_0} \quad , \quad \gamma = M\Gamma \quad , \quad C = M\,(\epsilon^2 + \frac{\Gamma^2}{4}) \qquad (12)$$

The derivation of (11) is straightforward. One simply has to add to eq. (1) an external coupling $q_{ext}\, \hat{F}$. The χ_{coll} measures the response to this hypothetical external field. It is to be calculated within our mean field approach by carefully considering applied and induced fields. Different to the unperturbed response function introduced in section 3) this χ_{coll} does contain collective effects, hence the subsript "coll". Notice that for undamped motion it is just the RPA response function for the schematic interaction $k/2\ \hat{F}\cdot\hat{F}$ (see [1]).

6) THE RESPONSE OF THE INTRINSIC SYSTEM

The central object in our formulation is the intrinsic response function $\chi(\omega)$. It contains all information necessary to compute the coefficients for macroscopic transport. Clearly, it will tell us whether or not the macroscopic motion is damped. But this question is intimately related to the problem of relaxation of the intrinsic excitations: In order to have γ finite the microscopic response function $\tilde{\chi}(t)$ must go to zero for $t \to \infty$.

The behaviour of $\tilde{\chi}(t)$ at large times depends on the complexity of the system. Suppose first that within our model we do neglect completely all kind of residual interactions. In this way the $h(Q_0)$ entering the right hand side of (3) will just be a one body operator and $\tilde{\chi}(t)$ will be given by a sum over all 1p-1h excitations which for a given temperature our on-body operator F will allow for. Neglecting true scattering states, our system will show some kind of recurrence with an approximate Poincaré-time being given by $\hbar/d$. Here d will be the average spacing of these 1p-1h excitations. It is finite and, therefore, we need to introduce irreversibility by hand. But this is allright as long as we can motivate it physically. Indeed, take a heavy ion collision, for instance. There will always be a finite uncertainty in energy (typically of the order of one MeV) and be it only because of a finite reaction time. We may therefore compute an energy averaged response function $\overline{\chi(\omega)}$, with its Fourier transform showing a proper decaying behaviour. Clearly, the final result should not depend on the width of the energy-smearing. There are cases where this situation is given [8]. But this need not be and will not be so in general. Then it will become difficult to justify an irreversible behaviour within a pure single particle picture.

On the other hand, since we are dealing with systems at <u>finite</u> intrinsic excitation the reduction of the dynamics to this single particle picture is an oversimplification. We should expect the presence of residual interactions to act incoherently. Certainly, such interactions will make it extremely difficult to compute the response function explicitly. Eventually one might attack this problem by using the random matrix model [9] (see also [2]). We chose a different way [6,10]. We rewrite the response function in terms of a two body Greens function and then assume that the latter can be calculated within a factorization approach. In this way we trace back the computation of $\chi(\omega)$ to folding products of one body Greens functions. The latter represent the propagation of particles and holes in the medium. All we then need to do is to find an approximation for their selfenergies. Here, we argue that the essential excitations will be close to the Fermi surface, and, secondly, we use a low temperature expansion. In this way the self-energy is expressed by:

$$\mathrm{Im}\ \Sigma(\omega) = \frac{1}{\Gamma_0} \frac{[(\omega-\lambda)^2 + \pi^2T^2]}{1 + \frac{[(\omega-\lambda^2 + \pi^2T^2]}{c}} \tag{13}$$

Here λ is the Fermi energy and the cut off factor c is chosen as to reproduce the expected behaviour of the effective mass of the nucleons. The parameter Γ_0 is chosen in accord with microscopic computations in nuclear matter [11].

Using a self-energy like (13) the intrinsic system will behave irreversibly of course, and, hence, $\tilde{\chi}(t)$ will vanish for large times. Clearly, inherent to formula (13) there is an energy average too (only think of the definition of the optical potential in nuclear physics). But the essential difference is that now one deals with configurations of a complexity much higher than just 1p-1h. And for 2p-2h, 3p-3h excitations etc. the average level spacing is much smaller.

We have tried to exhibit the intimate connection between an energy average, the behaviour of the response function at large times and the question of macroscopic irreversibility. Is this the same averaging procedure which is discussed in connection to quantum chaos? (see [12]) Probably yes, but closer relations have to be studied further. At this stage we would like to draw the attention to a semiclassical version [13] of the linear response formulation of nuclear dissipation. It is obtained by describing the motion of the nucleons inside their container (with infinitely high and sharp walls) by means of classical trajectories rather than quantum states. In order to obtain finite friction (within the zero frequency limit) a certain limit to an infinite system had to be performed. In this way the result for γ became identical to the wall formular of [14]. However, in the spirit of our discussion above we would argue that residual interactions cannot be neglected, and those help to make the response function go to zero at large times. These interactions are effective on a scale somewhat smaller than τ_{coll}, but definitely larger than our microscopic time τ (see [7]). The latter can be associated with the decay of the first peak, and typically comes out to be around $2\cdot10^{-22}$ sec.

7) AN APPLICATION TO FISSION

In this section we want to show some results which were obtained recently for the case of nuclear fission. We will limit ourselves to discuss briefly inertia and friction as a function of one collective variable: the distance z = Q between the centers of mass of the two fragments. The study of the dependence on parameters like T and Γ_0 will be published elsewhere [5], together with more details about the numerical calculation.

The distance z is introduced via the two center shell model of [6]. The shape looks like shown in fig. 2.

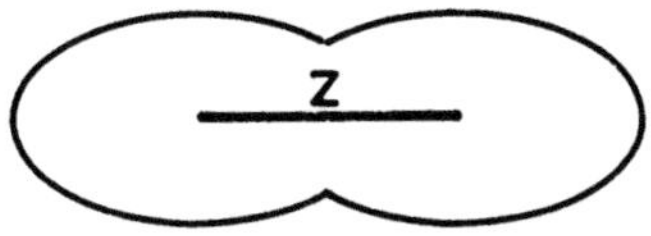

Fig. 2

It consists of two spheroids of fixed deformation ϵ(=0.35) penetrating each other. (This value corresponds to the saddle point deformation of the nucleus ^{212}Po). Within this particular model we compute the transport coefficients as outlined before. The results for M and γ can be seen in figs. 3 and 4 as a function of our one collective variable.

In each case we compare the results obtained both in the zero frequency limit as well as for finite Ω_0. The importance of the extension to finite Ω_0 is clearly visible, at least for deformations around and beyond the saddle point. In [6,7] such a difference was present also for more compact shapes. In fig. 3 the intertia is compared with the one of the liquid drop model with irrotational flow. It is seen that the latter can reproduce only roughly the average trend. In fig. 4 we compare our friction with the one of Nix' modified wall formula [7].

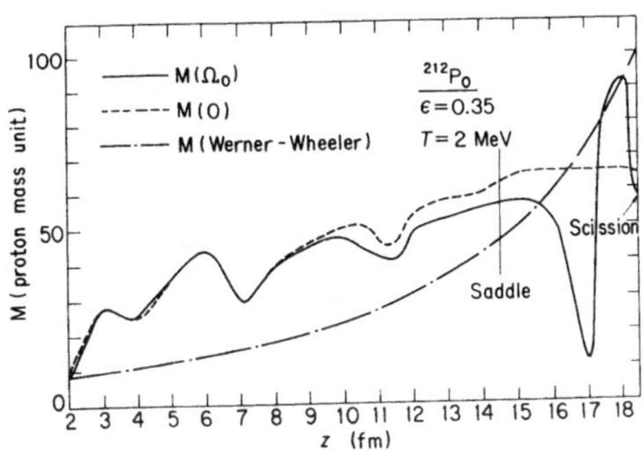

Fig. 3

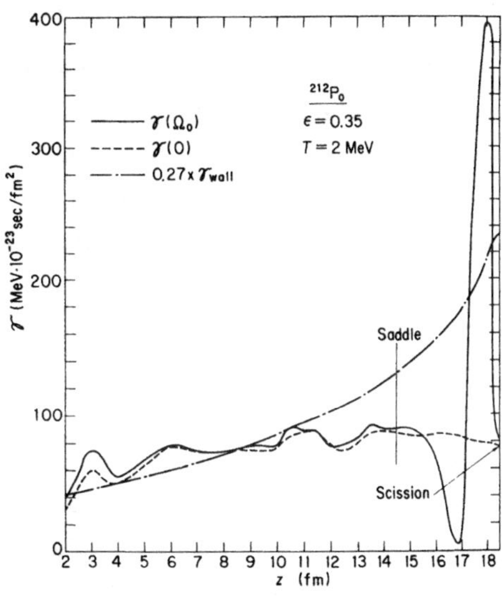

Fig. 4

From these values for the transport coefficients we may estimate the importance of damping. We may do this by calculating the ratio $\gamma/2\sqrt{M|C|}$. It can be interpreted as an inverse quality factor, characterizing locally the harmonic motion. We get values definitely smaller than one, thus proving that nuclear collective motion is underdamped [6,7,5]. (Close to the inflection point, of course, this number does loose its precise meaning).

8) THE FLUCTUATIONS

Up to now we have been treating the collective degree of freedom as a c-number variable. It represents some average quantity of our nuclear system (see e.g. eq. (4)). But there are several reasons to look for fluctuations. First of all, since we have frictional forces there must be fluctuating forces as well. This is a mere consequence of the fluctuationdissipation theorem. Secondly, for several examples of collective motion we must expect quantum features to hold and, therefore, we must be able to treat quantum fluctuations as well. Usually, in nuclear theory about reversible dynamics the latter are dealt with by first deriving classical equations of motion and then quantizing them.

In the presence of friction this is not a promising procedure. Simply because one does not know of how to quantize such a system and be able to treat the fluctuations at the same time. Therefore we choose a different way. We first establish a Hamiltonian for the total system: microscopic plus collective. Hereby we want to treat the collective degrees of freedom as dynamical variables (having an appropriate canonical momentum, as well). Of course, there will be too many degrees of freedom and we have to have a subsidiary condition. Clearly, this condition as well as the Hamiltonian has to be set up in such a way that we recover the average behaviour discussed before. We again will try for our locally harmonic approximation only. For the Hamiltonian this implies to look for a form which is quadratic in $(\hat{Q}-Q_0)$ and $\hat{P}$. (To a good approximation we may assume that the nonlinearities mainly come through Q rather than P). As for the subsidiary condition, we expect that on its time-dependent average it will turn over into eq. (4), with $Q(t) = \langle\hat{Q}\rangle_t$. Amazingly enougth such a procedure does exist in the literature. It is the method established in the early 50's by Bohm and Pines [8] to treat the Electrongas. Before 1982 [9] it was never exploited in nuclear physics in connection to dissipation.

Having the Hamiltonian for the full system we derive an equation for the reduced density operator in the collective phase space. The reduction is achieved by averaging the total density operator over the intrinsic degrees of freedom. An appropriate means to achieve this goal is to use the Nakajima-Zwanzig projection procedure. Different to common use we must not apply simple second order perturbation theory when approximating the integral kernel by an expression local in time. For the average trajectory, for instance, low order perturbation would not lead to our desired dispersion relation. Instead we perform a kind of partial summation when treating the impact of the coupling on the collective part. However, its impact on the intrinsic system still is handled in the framework of linear response theory.

Finally we need to mention the implication of the subsidiary condition. It appears as an operator equation at first, and in this sense might as well influence the fluctuations. So far we have not studied this effect, but treated the subsidiary condition on its time dependent average only.

In such procedure we finally end up with a diffusion equation for the reduced density. It is "Markoffian" in the sense that it is local in time, but to some extend it is "non-Markoffian" in that the transport coefficients do depend on the (local) collective frequency in a selfconsistent way. The latter is determined from the dispersion relation (5), and thus from the local collective motion itself. As transport coefficients we first recover the ones from section 5 for the average trajectory. But in addition we have the diffusion terms: $D_{QQ} \frac{\partial^2}{\partial Q^2} + D_{QP} \frac{\partial^2}{\partial Q \partial P} + D_{PP} \frac{\partial^2}{\partial P^2}$. The diffusion coefficients D_{ij}, too, depend on the frequency Ω_0. Also for them there does exist a zero frequency limit. But now there appears to be an additonal parameter to compare with, namely the temperature T. It is only for $|\Omega_0| \ll T$ that the sum of the diffusion terms turns over into the famous form $\gamma\, T \frac{\partial^2}{\partial P^2}$.

9) AN EXTENSION TO LARGE SCALE MOTION

So far we have limited our discussion to propagation over small regions both in time ($|\delta t| \ll \tau_{coll}$) as well as in phase space (around some Q_0, P_0, with phase space being introduced through a Wigner transform). To overcome this restriction we suggest the following scheme:

a) Solve the local diffusion equation for particular initial conditions, namely:

$$\lim_{\delta t \to 0} K_{\delta t}(Q,P,Q_0,P_0,\delta t) = \delta(Q-Q_0)\delta(P-P_0) \quad . \qquad (15)$$

b) Use this solution as a propagator for the physical distribution, like

$$d(Q,P,\delta t) = \int dQ_0 dP_0 \; K_{\delta t}(Q,P,Q_0,P_0,\delta t) \; d(Q_0,P_0,t_0) \qquad (16)$$

if the time laps may still be considered small.

c) For arbitrary $t-t_0$ repeat this folding procedure, like

$$K(Q_N,P_N,Q_0P_0,t_N-t_0) = \prod_{i=1}^{N} \int dQ_i dP_i \; K(Q_i,P_i,Q_{i-1}\; P_{i-1},\delta t_i) \qquad (17)$$

to obtain a finite time propagator. The d(Q,P,t) may then be obtained by inserting (17) into an equation like (16).

This scheme reminds of Feynman's path integrals. The difference is that our "short time" propagator $K_{\delta t}$ behaves as such on the macroscopic scale only. The limit $\delta t \to 0$ cannot be performed on the microscopic level. (Our equation of motion must be non-local in time for δt smaller than the microscopic relaxation time τ). This will restrict our scheme as far as all quantum features are concerned. But surely some important quantum effects will survive and be it only on the level of WKB. After all, with respect to the macroscopic diffusion we do not really perform a semiclassical limit - beyond the one which goes with our locally harmonic approximation.

It remains to be said that this scheme can be used to solve actual diffusion equations, even for such non-trivial problems as an escape over a barrier. This has been shown in [20] for the case of Kramers' equation to describe fission. Of course, the procedure (15) to (17) is more general in the sense that we do not necessarily need to write down the differential equation for large scale diffusion. All we need is the information about $K_{\delta t}$. Eq (16) and (17) can then be treated numerically for instance by Monte Carlo techniques.

The authors would like to express their gratitude for financial support to the Deutsche Forschungsgemeinschaft (H.H.) and the Bundesministerium für Forschung und Technologie, a careful reading of the manuscript by R. Blümel is well acknowledged.

REFERENCES

[1] A. Bohr and B. Mottelson, Nuclear Structure, Vol. II (Benjamin, London, 1975)

[2] H. Hofmann, FIZIKA 9(1977) Supplement 4, 441

[3] D.J. Rowe, Nuclear Collective Motion, Models and Theory (Methuen, London 1970)

[4] H. Hofmann, "Damped Motion in Collective Phase Space" p. 306 in: "Phase space approach to nuclear dynamics", Ed.: M. di Toro, W. Nörenberg, M. Rosina an S. Stringari, World Scientific, Singapore (1986)

[5] P.J. Siemens, A.S. Jensen and H. Hofmann, Nucl. Phys. A409, 135c, (1983)

[6] A.S. Jensen et al., Phys. Lett. 117B (1982) 157, Physica Scripta, Vol. T5 (1983) 186

[7] H. Hofmann and A.S. Jensen, Nucl. Phys. A 428 (1984) 1

[8] P.J. Johannsen et al., Nucl. Phys. A288 (1977) 152

[9] H.A. Weidenmüller, Progress of Particle and Nuclear Physics, 3, 49 (1980); and S. Ayik and W. Nörenberg, Z. Phys. A288, 401 (1978)

[10] P.J. Siemens et al., Nucl. Phys. A 441 (1985) 410

[11] J.P. Jeukenne, A. Lejeune and C. Mahaux, Phys. Rep. 25C (1976) 83

[12] N.G. van Kampen: "Quantum Chaos as Basis for Statistical Mechanics" and E.H. Heller and R.L. Sundberg: "Quantum ergodicity and intensity fluctuations", in "Chaotic Behaviour in quantum Systems", Ed.: G. Casati, Plenum, 1985 (Nato Adv. Study Inst. Series B 120)

[13] S.E. Koonin, J. Randrup .and R.L. Hatch, Nucl. Phys. A283 (1977) 87

[14] W. Swiatecki, Progr. in Particle and Nuclear Physics, ed. by D. Wilkinson, Vol. 4 (1980), and references therein.

[15] S. Yamaji and H. Hofmann, to be published

[16] S. Yamaji and A. Iwamoto, Z. Phys. A313 (1983) 161

[17] J.R. Nix, private communication

[18] D. Bohm and D. Pines, Phys. Rev. 32, 609 (1963)

[19] H. Hofmann, Phys. Lett. 119B (1982) 7
P.J. Siemens, Nucl. Phys. A387 (1982) 247

[20] F. Scheuter and H. Hofmann, Nucl. Phys. A394, 477 (1983)

STATISTICAL FEATURES OF THE EFFECTIVE HAMILTONIAN AND THE MANY BODY BOUND STATE PROBLEM

D. Boosé*
Max Planck Institut für Kernphysik, Postfach 103980, 6900 Heidelberg 1 (RFA)
J. Richert
Centre de Recherches Nucléaires et Université Louis Pasteur
Groupe de Physique Nucléaire Théorique BP 20, 67037 Strasbourg Cedex (France)

Abstract

We propose a new approach for the derivation of an effective Hamiltonian in the framework of the many body bound state problem by imposing statistical features on some of the states which span the physical space. We discuss the formal problems as well as the physical aspects which come up along the derivation.

1. Introduction

During the last years there has been much progress in the detailed study of the properties of many body systems, from the methodological as well as from the munerical point of view [1]. Given a Hamiltonian which fixes the dynamics these approaches allow in principle for a rigorous derivation of the properties of the many-body systems. However these methods often lead to intricate and cumbersome resolution algorithms and it is not always easy to gain a clear insight into the role played by the many different contributions of the many-body correlations which built up the physical behaviour of the system.

On the other hand, especially in nuclear physics,spectra show specific features which may be interpreted in a statistical framework [2,3]. They seem to possess properties of universal character such as local properties which are reproduced in the framework of the GOE (Gaussian Orthogonal Ensemble) as well as global properties which can be interpreted in the EGOE (Extended Gaussian Orthogonal Ensemble) [4].

It is however not yet firmly and definitely established that these statistical

*On leave of absence from the Centre de Recherches Nucléaires, Strasbourg, France.

properties apply to all the members of the spectra. The analysis of states of fixed spin and parity seem to confirm the universality of this behaviour over the whole sequence of levels [5]. Recent investigations [6] seem to indicate that this might not always be the case, different local behaviour is observed in even-even and odd-odd nuclei for different sets of low-lying levels. The confirmation of these derivations would then raise the question whether one can characterise the states which spoil the statistical description and describe them.

We propose here a model which is adapted to this situation. We construct an effective Hamiltonian in a finite subspace spanned by physical states which require a detailed treatment. The complementary subspace contains those states to which a statistical treatment can be assigned. We shall give some criteria in order to justify such a decomposition. These criteria are of course of no absolute significance since we lack a precise definition of the boundaries which separate the two types of states.

In the present work we first recall the formal framework which defines an effective Hamiltonian in a given model space and set the secular equation leading to the eigenvalue and projected eigenvector solutions. Further we introduce the statistical assumptions on the matrix elements defined in the complementary subspace and work out the expression of the ensemble averaged propagator (one-point function) which enters the effective Hamiltonian. We discuss the existence of physical solutions and present the algorithm which leads to them. We mention some specific properties of the one-point function and finally develop some arguments about physical aspects of the approach, open problems and possible extensions which lead beyond the framework in which the problem is actually formulated.

2. Effective Hamiltonian and secular equation

We recall here the well known projection formalism [7]. The Hamiltonian

$$H = H_o + V$$

where H_o is an unperturbed part and V a residual interaction is defined in the space $\mathcal{H}$ which is divided into a finite model space $P\mathcal{H}$ spanned by a basis of states $\{|\Phi_a\rangle\}$ and a complementary space $Q\mathcal{H}$ spanned by N basis states $\{|\Phi_\mu\rangle\}$. In the sequel N may tend to ∞ as it is the case in the physical Hilbert space. The basis of states is fixed by

$$H_o|\Phi_a\rangle = E_{oa}|\Phi_a\rangle$$

$$H_o|\Phi_\mu\rangle = \varepsilon_\mu|\Phi_\mu\rangle$$

If

$$|\Psi_P^{(i)}> = P|\Psi^{(i)}> = \sum_{a\varepsilon P} c_a^{(i)} |\Phi_a>$$

is the ith projected eigenvector defined in $P\mathcal{H}$,the eigenvalues and eigenvectors are given as the solutions of

$$\left[\mathcal{H}_{ab}(E_i) - E_i\delta_{ab}\right] c_b^{(i)}(E_i) = 0 \tag{1}$$

with

$$\mathcal{H}_{ab}(E) = E_{oa}\delta_{ab}+V_{ab} + \sum_{\mu\nu} V_{a\mu}(E-QHQ)^{-1}_{\mu\nu} V_{\nu b}$$

The difficulty which appears in the practical derivation of the matrix elements of the effective Hamiltonian $\mathcal{H}$ lies in the calculation of the matrix elements of the propagator $(E-QHQ)^{-1}$ (one-point function) in Q space.

3. Statistical assumptions and the derivation of the propagator.

a) Properties of the Q-space states

The matrix elements of H in Q

$$(QHQ)_{\mu\nu} = \varepsilon_\mu\delta_{\mu\nu} + V_{\mu\nu}$$

are supposed to be characterised by gaussian distributed matrix elements $V_{\mu\nu}$ with ensemble averaged first and second moments [9]

$$\overline{V_{\mu\nu}} = 0$$

$$\overline{V_{\mu\nu}V_{\mu'\nu'}} = \frac{y^2}{N} (\delta_{\mu\mu'}\delta_{\nu\nu'}+\delta_{\mu\nu'}\delta_{\nu\mu'})$$

when N is the number of basis states in Q .

The ensemble average of the matrix elements which couple $P\mathcal{H}$ to $Q\mathcal{H}$ are given by

$$\overline{V_{\mu a}V_{\mu b}} = \delta_{\mu\nu}.f(a,b)$$

when f(a,b) expresses the correlation between states a and b. The statistical assumptions introduced above lead to

$$\mathcal{H}_{ab}(E) = E_{oa}\delta_{ab}+V_{ab}+f(a,b)\bar{G}(y;E) \tag{2}$$

with

$$\bar{G}(y;E) \equiv \overline{\sum_{\mu=1}^{N} \left[E-\varepsilon_\mu-QVQ\right]^{-1}_{\mu\mu}} = \overline{\mathop{Tr}_{\mu}\, 0^{-1}(E)}$$

with $0_{\mu\nu}(E) \equiv (E-\varepsilon_\mu\delta_{\mu\nu}-V_{\mu\nu})$

b) Integral expression of $\bar{G}$.

Given the generating function

$$Z(E,J) = \int_{-\infty}^{+\infty} \prod_{\mu=1}^{N} d\phi_\mu \, e^{-\frac{i}{2}\sum_{\mu\nu} \phi_\mu O_{\mu\nu}(E)\phi_\nu - iJ\sum_\mu \phi_\mu^2}$$

where ϕ_μ are real variables and J a real parameter,

$$\left.\frac{i\partial \ell n Z(E,J)}{\partial J}\right|_{J=0} = \mathop{\mathrm{Tr}}_\mu O^{-1}(E) = i \lim_{n\to 0} \frac{\partial}{\partial J} \left.\frac{(Z^n-1)}{n}\right|_{J=0}$$

The last step first introduced by Edwards and Anderson [8,9] (replica trick) allows for a manageable expression of $\bar{G}$ in terms of n replicated variables $\{\phi_\mu^k \ ; \ k=1,\ldots,n \ ; \ \mu=1,\ldots,N\}$.

This leads to

$$\bar{G}(y;E) = i \lim_{n\to 0} \sum_{\mu=1}^{N} \int \prod_{k,\lambda} \left[d\phi_\lambda^k\right] \left(\phi_\mu^\ell\right)^2 e^{-\frac{i}{2}\overline{\sum_{k,\mu,\nu} \phi_\mu^k O_{\mu\nu} \phi_\nu^k}}$$

Using the statistical properties of $\{V_{\mu\nu}\}$ and introducing real symmetric auxiliary variables $\sigma_{kk'}$ $(k,k'=1,\ldots,n)$ it is easy to eliminate $\{\phi_\mu^k\}$ by integration [10] and to obtain

$$\bar{G}(y;E) = \frac{N}{y} \lim_{n\to 0} c(n) \int \Pi\left[d\sigma\right] \sigma_{\ell\ell} e^{\mathcal{L}(\sigma)}$$

with $\lim_{n\to 0} c(n) = 1$

$$\mathcal{L}(\sigma) = -\frac{N}{4} \mathop{\mathrm{Tr}}_k \sigma^2 - \frac{1}{2} \mathop{\mathrm{Tr}}_k \sum_\mu \ell n \left[(E-\varepsilon_\mu)\delta_{kk'} - y\sigma_{kk'}\right] \qquad (3)$$

c) Saddle point prescription and 1/N expansion of $\bar{G}$.

It is now possible to work out the expression of $\bar{G}$ in terms of an infinite sum of quadratures starting from a saddle point prescription

$$\sigma_{kk'} \equiv \sigma_o \delta_{kk'} + \sigma_{kk'}^{(1)}$$

$$\left.\frac{\delta \mathcal{L}}{\delta\sigma}\right|_{\sigma=\sigma_o} = 0$$

which leads to the dispersion relation

$$\sigma_o = \frac{y}{N} \sum_\mu \frac{1}{E-\varepsilon_\mu - y\sigma_o}$$

and

$$\mathcal{L}(\sigma) = \mathcal{L}(\sigma_o) + \sum_{p=2}^{\infty} \frac{1}{p!} \left.\frac{\delta^P}{\delta\sigma^P}\right|_{\sigma=\sigma_o} \mathop{\mathrm{Tr}}_k \left[\sigma^{(1)}\right]^P = \mathcal{L}(\sigma_o) + \tilde{\mathcal{L}}(\sigma^{(1)}) \qquad (5)$$

In the limit $n \to 0$, $\mathcal{L}(\sigma_o) \to 0$ and $\bar{G}$ can be expressed in terms of an asymptotic series in $1/N$

$$\bar{G}(y;E) = \sum_\mu \frac{1}{E-\varepsilon_\mu - y\sigma_o} + \frac{N}{y}\left[\frac{c_3}{4c_2^2} + \frac{3c_5}{4c_2^3} + \theta\left(\frac{1}{N^3}\right)\right] \tag{6}$$

with

$$c_2 \equiv \frac{N}{2}\left[1 - \frac{1}{N}\sum_\mu\left(\frac{y}{E-\varepsilon_\mu - y\sigma_o}\right)^2\right]$$

$$c_p \equiv \sum_\mu \left(\frac{y}{E-\varepsilon_\mu - y\sigma_o}\right)^p \qquad p \geqslant 3$$

In the limit $N \to \infty$ $\bar{G}$ reduces to the first term in (6).

d) Properties of the one-point function $\bar{G}$

One may verify that the saddle point prescription fixes $\bar{G}$ uniquely when $N \to \infty$ [10]. From (5)

$$\bar{G}(y;E) = \frac{N}{y}\left[\sigma_o + \int \Pi\left[d\sigma^{(1)}\right]\sigma^{(1)}_{\ell\ell}\; e^{\tilde{\mathcal{L}}(\sigma^{(1)})}\right] \tag{6a}$$

Any further application of the saddle point procedure to the integral term in (6a)

$$\sigma^{(1)}_{kk'} = \sigma_1\delta_{kk'} + \sigma^{(2)}_{kk'}$$

leads to an expression of $\bar{G}$ which is invariant in the transformation $\sigma_o \to \sigma_o + \sigma_1$, $\sigma^{(1)} \to \sigma^{(2)}$, $\tilde{\mathcal{L}}(\sigma^{(1)}) \to \tilde{\mathcal{L}}(\sigma^{(2)})$ and this is the case for any number of steps. Hence the saddle point solutions work as fixed points under these transformations.

For big but finite N the saddle point prescription might not be the most appropriate choice since it does not necessarily minimise the lowest order contribution to the asymptotic $1/N$ expansion. One may expand σ around an arbitrary point, $\sigma_{kk'} = \tau_o\delta_{kk'} + \sigma^{(1)}_{kk'}$. Again the expression of $\bar{G}$ is invariant under further iterations $\sigma^{(1)} = \tau_1 + \sigma^{(2)}, \ldots$ [10] and its expression can be worked out

$$\bar{G}(y;\tau_o) = \frac{N}{y}\left\{\tau_o + \frac{1}{2c_2}(c_1 - N\tau_o) + \frac{1}{4c_2^2}\left[c_3 + \frac{1}{2}(c_1 - N\tau_o)^3\right] + \theta\left(\frac{1}{N^2}\right)\right\}$$

where $c_1 \equiv \sum_\mu \frac{y}{E-\varepsilon_\mu - y\tau_o}$

A possible choice of τ_o is dictated by the minimisation of the third term $[\theta(1/N)]$ of this expression. This might be an ackward prescription since τ_o is furthermore constrained, like σ_o, by further conditions which are discussed in the next section.

4. Resolution of the eigenvalue problem

a) Constraints on the saddle point solutions

The figure shows the N-1, N or N+1 solutions of (4). All the asymptotes have to verify

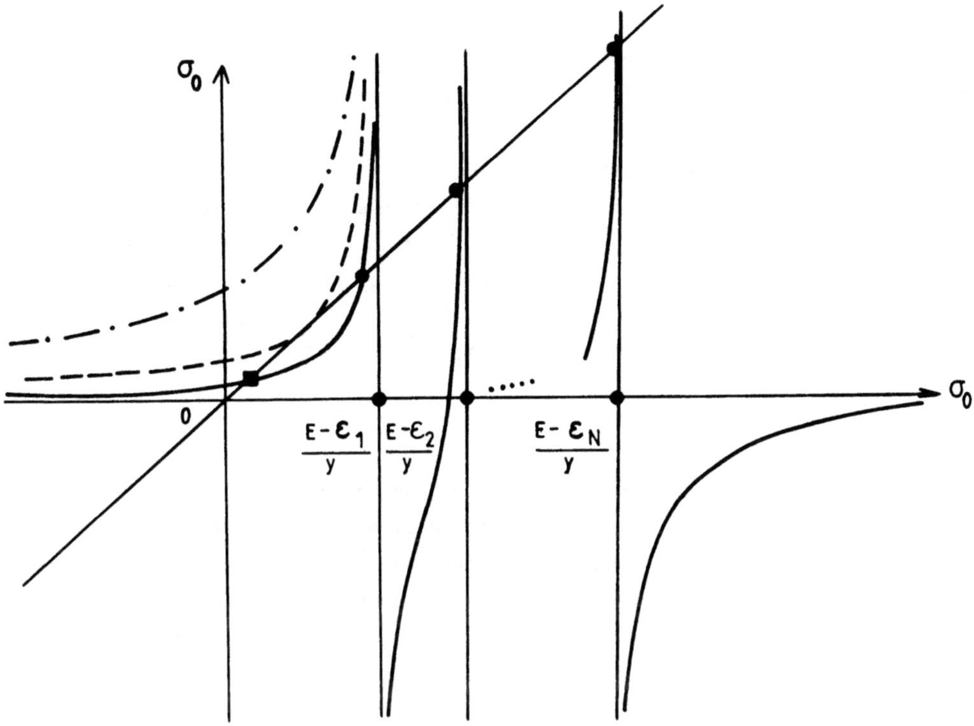

Figure : The solutions of eq. (4) correspond to the intersection of the lim σ_0 with the branches of the hyperbola. The physical solution corresponds to the intersection of the full left branch with σ_0 closest to the origin (N+1) solutions of (4). The dashed branch corresponds to the case c_2=0 (N solutions of (4)), the dashed-dotted branch to no physical solution (N-1 solutions of (4)).

$$\frac{E-\varepsilon_\mu}{y} > 0, \ \forall \varepsilon_\mu \tag{7a}$$

for the logarithm appearing in $\mathcal{L}(\sigma_0)$ (eqs. (3) and (5)) to make sense. Furthermore

$$c_2 > 0 \tag{7b}$$

This eliminates all the solutions, except, if it exists, for the one which lies closest to the origin on the branch at the extreme left of the graph. Notice that in principle one may encounter the situation where the two lowest solutions merge into a single degenerate solution (dashed curve in the figure). In this case the asymptotic expansion diverges term by term (see eq. (6)) because $c_2 = 0$.

Conditions (7a) and (7b) put stringent constraints on the choice of the model space. Only the eigenvalues for which $E>\{\varepsilon_\mu\}$ or $E<\{\varepsilon_\mu\}$ are acceptable. Hence unperturbed states of the complementary space $Q\mathcal{H}$ should not intrude into the spectrum of the model space $P\mathcal{H}$.

There might exist different ways to overcome this problem. One way which we did not yet investigate consists in an analytical continuation of the equations into the complex energy plane. The other uses a double partitioning of the complementary subspace $Q\mathcal{H} = Q_1\mathcal{H} + Q_2\mathcal{H}$ [11] where $Q_1\mathcal{H}$ is spanned by the intruder states and $Q_2\mathcal{H}$ by all the others.

b) Solutions of the secular equations.

The eigenvalue and eigenvector problem is solved in a selfconsistent way. Given $\{E_{oa}, V_{ab}, f(a,b)\ ;\ y\ ;\varepsilon_\mu\}$, choose $E^{(o)}$ which verifies (7a), solve eq. (4) with (7b) in order to get $\sigma_0^{(o)}$, work out the matrix elements given by eq. (2) and diagonalise the matrix (1). Start again from one of the eigenvalues $E_i^{(1)}$ which verifies (7a) and follow the algorithm up to the point where $E_i^{(p+1)} = E_i^{(p)}$.

5. Summary and conclusions.

We have developped a new approach of the many body bound state problem. The description introduces statistical features into the derivation of the effective Hamiltonian which defines the many-body system in the model space. Using the well known replica technique [8,9] one gets a compact expression of the effective interaction in the limit where the number of states in the complementary space goes to infinity. The resolution of the eigenvalue and eigenvector problem in the model space is performed self consistently by means of a secular equation which is coupled to a saddle point equation fixing an average field contribution. The procedure induces constraints on the choice of the eigenstates in the model space. We discussed possible ways to overcome these constraints.

The practical implementation requires the resolution of further points. In particular it is necessary to fix up the residual interaction which acts in the model space, the second moments of the matrix elements in the complementary space as well as the coupling matrix elements. It might also be worthwhile to estimate the fluctuations induced on the eigenvalues by the statistical behaviour of the complementary space states.

Except for specific constraints on the relative sequence of states in the model space vs. those in the complementary space which are imposed by the formalism and might be overcome in one way or another, the choice of the model space is open. How should one choose this space ? There exists no clear cut answer to this question. A reasonable recipe might be : keep together in the model space all the states which strongly couple to each other through the residual interaction. A more satisfactory answer would only be obtained if a satisfactory criterion for the distinction between states behaving statistically and not statistically could be found.

References

1) See : P.J. Siemens, R.A. Smith, J. Zabolitzky, Proceedings of the Fourth International Conference on Recent Progress in Many-Body Theories, San Francisco August 85, Springer Verlag, Heidelberg
2) O. Bohigas, M.J. Giannoni, C. Schmidt, Phys. Rev. Lett. 52 (1984) 1
3) H.A. Weidenmüller, Contribution to the Niels Bohr Centennial, Nuclear Structure 1985, R.A. Broglia, G.B. Hagemann, B. Herskind editors ; Elsevier Science Publishers B.V., 1985
4) K.K. Mon, J.B. French, Ann. Phys. (NY) 95 (1975) 90
5) T.A. Brody, J. Flores, J.B. French, P.A. Mello, A. Pandya, S.S.M. Wong, Rev. Mod. Phys. 53 (1981) 385
6) A. Abul-Magd, H.A. Weidenmüller, Phys. Lett. 162B (1985) 223
7) H. Feshbach, Ann. Phys. (NY) 5 (1958) 357
8) S.F. Edwards, P.W. Anderson, J. Phys. F5 (1975) 865
9) J.J.M. Verbaarschot, H.A. Weidenmüller, M.R. Zirnbauer, Phys. Rev. Lett. 52 (1984) 1597
10) D. Boosé, J. Richert, J. Phys. G11 (1985) 701
11) H.M. Hofmann, S.Y. Lee, J. Richert, H.A. Weidenmüller, Ann. Phys. (NY) 85 (1974) 410

Regularity of Motion and Spectra: Polyatomic Rotation and Vibration

William G. Harter
Department of Physics
J. William Fulbright College
University of Arkansas
Fayetteville, AR 72701

1. Molecular rotation and precession

The study of molecular dynamics and spectra has been closely related to the problem of distinguishing regular or quasi-periodic motion from that which is irregular or chaotic. Much of this study has been devoted to vibrational motion in nonlinear or anharmonic force fields such as those provided by Henon-Heiles or Barbannis potentials. Here I would like to focus on molecular rotational motion which also has interesting dynamics and is practically unavoidable in most real laboratory experiments. I would also like to show a theoretical framework based on upon spinors which provides a connection between anharmonic vibrational behavior such as local mode formation and the recently observed spectral phenomenon of clustering in high resolution rotational spectra of molecules.

It is only recently that the possible effects of chaos in classical and quantum dynamics have been discussed with regard to molecular rotation. The dynamics of kicked diatomic rotors have been discussed in this volume by Casati and Guarneri and experimental tests have been proposed by Smilansky. In the following I shall be concerned with just the beginnings of understanding of the theory and experiments connected with poly-atomic rotors. Polyatomic rotors are capable of a more complicated motion called precession or "rotation of rotation." This motion is associated with what is called rotational fine structure in high resolution molecular spectra.

Two examples of fine structure are seen in Fig.1. infrared spectra by Patterson,*et.al.*[1] arising from angular momentum (J=30) multiplets in Fig.1a for tetrafluorosilane (SiF_4) and in Fig. 1b for cubane (C_8H_8) by Pine,*et.al.*[2]. The fine structure pattern called P(30) in the cubane spectra is similar to the R(30) pattern for SiF_4. It turns out that the same pattern scaled up or down appears in the quantum levels of both the ground and excited vibrational states of the theoretical models of these and other molecules such as SF_6. Let us consider just the theory of the rotor ground levels for a given total angular momentum J=30.

The pure rotational levels for a tetrahedral, octahedral, or cubic rotor may generally be derived numerically using the Hecht Hamiltonian [3].

$$H = B(J_x^2+J_y^2+J_z^2) + 10\, t_{044}(J_x^4 + J_y^4 + J_z^4 - (3/5)J^4) \qquad (1.1)$$

The numerical eigenvalue problem is complicated for high quanta, and modern experiments have observed fine structure for J=140 and above. The numerical output is also complicated, and considerable effort has been made to simplify and understand the results.[4-7] A graphic semi- classical way to understand the spectra is described now.[8-9] We consider the anatomy of the J=30 fine structure and related dynamics of the SF_6 molecule which are graphically depicted in Fig.2.

To understand the dynamics it helps to imagine the SF_6 molecule as a spacecraft with its payload hung on elastic supports. Spacecraft occupants can track their rotational motion using the trajectories of stars traced on a globe fixed to the center of their spacecraft. Of particular interest would be the path of a "north star" located on the axis of their total angular momentum vector **J**. In the absence of external torque the **J** vector will be fixed with respect to the stars no matter how wobbly is the rotation of the spacecraft. However, the "north star" and **J** may move around on the internal globe, and this motion is called precession.

The precessional motion is completely determined if the shifting payload follows **J** adiabatically. This is an assumption one makes when accepting an effective centrifugal Hamiltonian such as (1.1). The **J**-vector is constrained classically to conserve energy (H) and magnitude J=|**J**|, and this confines **J** to one of the topography lines on the rotational energy (RE) surfaces drawn in the upper portion of Fig.2. The rate of precession is proportional to the RE surface gradient and normal to it. The RE surface is obtained by writing Hamiltonian (1.1) in a polar coordinate form.

$$H = BJ^2 + 10\, t_{044}\, J^4 (\cos^4\gamma \sin^4\beta + \sin^4\gamma \sin^4\beta + \cos^4\beta - (3/5)) \qquad (1.2a)$$

Here the molecular frame polar coordinates are (θ=-β,ϕ= -γ) for the angular momentum vector **J**, where (α, β,γ) are the rotor Euler angles.

$$J_x = -J \cos\gamma \sin\beta \qquad J_y = J \sin\gamma \sin\beta \qquad J_z = J \cos\beta \qquad (1.2b)$$

The rotational energy is plotted radially as a function of the direction in the molecular frame of **J** subject to the constraint that its magnitude |**J**|=J is fixed. Then topography lines represent **J** trajectories of constant energy and J. The trajectories in the three-fold symmetric valleys are lower in energy because the centrifugal force is more effective when the rotation axis is in between the x, y, or z bond axes of SF_6. It is much easier to bend an SF bond than to stretch it. Therefore, the molecule is more rigid when rotating around, or perpendicular to, the four-fold symmetry axes which contain the SF bonds. The latter axes correspond to fixed points centered on each of the six peaks of the surfaces in Fig. 2.

The top five trajectories surrounding the peaks correspond to the top five eigenlevels shown at the bottom of Fig.2, and to corresponding peaks in the spectral patterns of Fig.1. One can also see two peaks on the other side of the spectral patterns and these correspond to the two lowest valley trajectories indicated on the right side of Fig.2. The magnifying circles in Fig.2 draw attention to the fact that most levels are actually clusters of various combinations of symmetry singlets (A_1 or A_2), doublets (E), or triplets (T_1 or T_2) belonging to the cubic symmetry species. For example, the third highest (K_4=28) level contains a singlet (A_1), doublet (E), and a triplet (T_1) , or six levels in all. The clusters on the right hand of Fig. 2 all contain six rotational levels because there are six equivalent trajectories on the energy surface for each of the energy levels labeled K_4=25-30. Similarily, there are eight valley trajectories and eight levels in each of the two clusters labeled K_3=29-30 on the lefthand side of Fig 2. The only place where the energy levels are not clustered and evenly spaced in in the neighborhood of the dotted separatrix trajectory. The separatrix crosses the saddle points of the energy surface and encloses the respective regions of ordered spectral clusters.

The spacing between the clusters such as the 5.3 MHz interval between the top two clusters in the lower righthand portion of Fig.2 is called *fine structure* splitting. Fine structure splitting is approximately equal to the classical precessional frequency. (Note that the classical *rotational* frequency corresponds to the

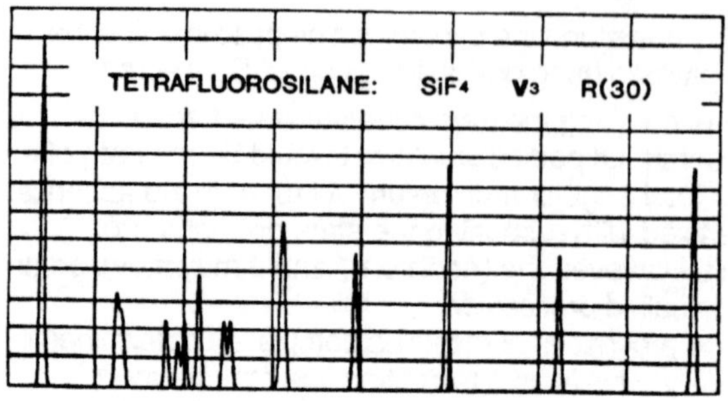

Fig.1 *Infrared spectra showing rotational fine structure.*

(a) *Tetrafluorosilane* SiF_4 *(Patterson, et.al. [1])*

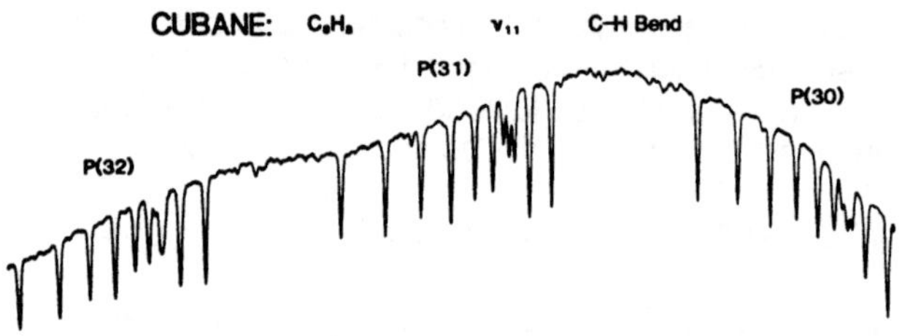

(b) *Cubane* C_8H_8 *(Pine, et.al. [2])*

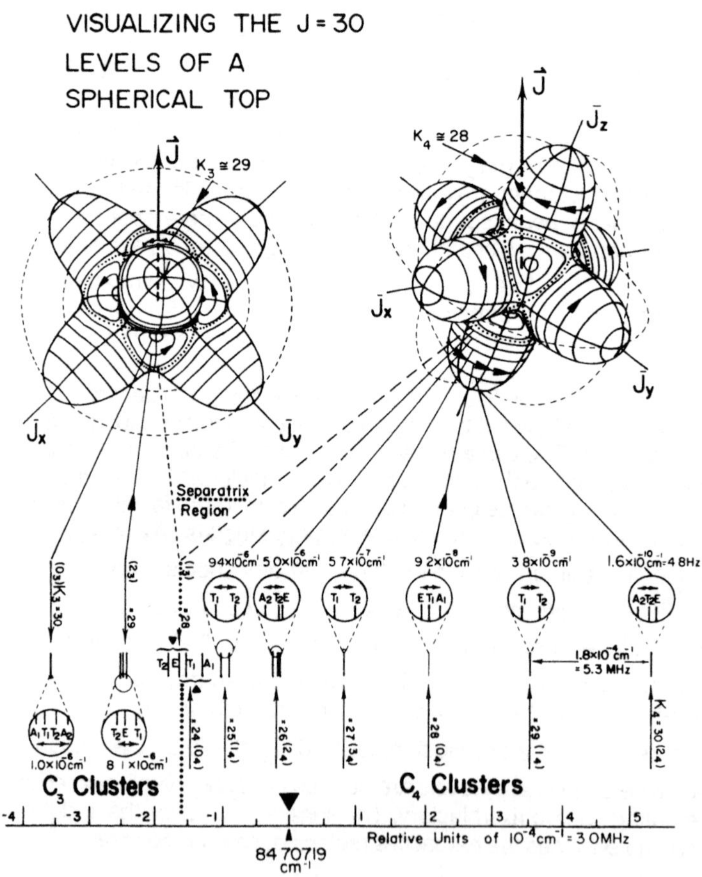

Fig.2

J=30 energy level clusters and associated RE surface trajectories.(B=0.091 083 cm^{-1}*,* t_{044} *= 5.44 Hz) Radial scale of RE surfaces is exaggerated . (Harter [9])* interval between different J-surfaces.) The intra-cluster splitting, such as the 4.8Hz splitting of the K_4=30 cluster, is called *superfine* structure. This is related to the rate for quantum tunneling between equivalent semi-classical trajectories. Approximate semi-classical formulas for fine and superfine splittings can be constucted using simple Bohr quantization schemes based on action integrals A_n as described in Refs. 8 and 9. The classical fine structure splitting (ν_c) is the action derivative $\nu_c = (dA_n/dE_n) = \delta E/h$, while the quantum superfine splitting ($S = \nu_c e^{-|\theta|}$) varies exponentially with an imaginary action integral $|\theta|$ evaluated between closest approach points of a pair of quantizing trajectories to the saddle point. (Methods developed by Wilkinson in this volume may be useful for extending the calculations to multiple energy surface geometries.)

As long as the equivalent trajectories are well separated (as in the case of the K_4=26-30 clusters in Fig.1.2) the tunneling rate S is extremely slow compared to precessional rate , and superfine structure is much finer than fine structure. However, in the neighborhood of the separatrix the precession rate slows down while tunneling increases dramatically. In this region the classical precession and quantum tunneling cannot be distinguished as separate processes, nor are the energy levels or spectra grouped into orderly clusters. The separatrix region is the theater for the most irregular classical or quantum dynamics. Trajectories and eigensolutions near to separatrices are most sensitive to perturb- ations and changes of parameters or initial conditions and are delocalized. (Note that the three patterns P(32), P(31), and P(30) in Fig. 1b differ mainly in their separatrix regions.) The separatrix divides localized dynamics in C_4-symmetric regions from a completely different dynamics localized in C_3-symmetric regions, and therefore extraordinary behavior may exist there.

Still the classical motion described by the adiabatic effective Hamiltonian can never be chaotic since it has only one effective degree of freedom.The situation may be different if one perturbs the rotor with external torques or internal dynamics of spins, vibrations, etc. which respond non-adiabatically to the precessional motion. This possibility will be discussed after we consider an analogy between spins, rotor dynamics, and vibrational motion.

2. Spinors, rotors, and coupled oscillators

The quantum theory of spin-1/2 and other 2-state systems is well known as the Rabi-Ramsey-Schwinger or Feynman-Vernon-Hellwarth formulation. There one solves the Schrodinger equation

$$\begin{aligned} i\, d\psi_1/dt &= A\ \psi_1 + (B-iC)\ \psi_2 \\ i\, d\psi_2/dt &= (B-iC)\ \psi_1 + D\ \psi_2 \end{aligned} \qquad (2.1)$$

in terms of a phased rotation.

$$\Psi(t) = \exp[-iH_R t]\ \Psi(0) = \exp[-i(\omega_0 \mathbf{1} + \omega\cdot\mathbf{J})t]\ \Psi(0) \qquad (2.2a)$$

The spinor expansion of the Hamiltonian is well known.

$$H_R = \omega_0\ \mathbf{1} + \omega_x\ \mathbf{J_x} + \omega_y\ \mathbf{J_y} + \omega_z\ \mathbf{J_z}$$

$$\begin{pmatrix} A & B-iC \\ B+iC & D \end{pmatrix} = (A+D)/2 \begin{pmatrix} 1 & 0 \\ 0 & 1 \end{pmatrix} + 2B \begin{pmatrix} 0 & 1/2 \\ 1/2 & 0 \end{pmatrix} + 2C \begin{pmatrix} 0 & -i/2 \\ -i/2 & 0 \end{pmatrix} + (A-D) \begin{pmatrix} 1/2 & 0 \\ 0 & -1/2 \end{pmatrix} \qquad (2.2b)$$

The spinor evolution corresponds to a 3-space rotation around an angular velocity vector ω=(2B,2C,A-D) and to an overall phase angular frequency ω_o=(A+D)/2. One may write ω in polar coordinates.

$$\omega_x = \omega \cos\phi \sin\theta = 2B \qquad \omega_y = \omega \sin\phi \sin\theta = 2C \qquad \omega_z = \omega \cos\theta = A-D \tag{2.3}$$

One may also express each spinor state as an Euler rotation of spin up, i.e., $|\Psi\rangle = \mathbf{R}(\alpha\beta\gamma)|1\rangle$. This is represented as follows, using standard SU(2) representations.

$$\begin{vmatrix}\psi_1\\ \psi_2\end{vmatrix} = \begin{pmatrix} e^{-i(\alpha+\gamma)/2}\cos\beta/2 & \cdot \\ e^{-i(\gamma-\alpha)/2}\sin\beta/2 & \cdot \end{pmatrix}\begin{vmatrix}1\\0\end{vmatrix} = \begin{vmatrix} e^{-i\alpha/2}\cos\beta/2 \\ e^{\,i\alpha/2}\sin\beta/2\end{vmatrix} e^{-i\gamma/2} \tag{2.4}$$

This suggests that a 3-dimensional rigid rotor with Euler coordinates ($\alpha\beta\gamma$) corresponds to each spinor state. The rotor axis will be the spin expectation vector **S**=<**J**>.

$$S_x = \langle\Psi|J_x|\Psi\rangle = 1/2 \cos\alpha \sin\beta \qquad S_y = \langle\Psi|J_y|\Psi\rangle = 1/2 \sin\alpha \sin\beta \qquad S_z = \langle\Psi|J_z|\Psi\rangle = 1/2 \cos\beta \tag{2.5}$$

The Euler angles (α,β) are the polar angles of the rotor axis while γ represents a twist around that axis. The 3-dimensional rotor precesses around the ω-vector according to (2.2) as the spinor state evolves.

Alternatively, one may relate the spinor state evolution to that of the two coupled oscillators with the following phasor variables:

$$\psi_1 = x_1 + ip_1 = \sqrt{I}\, e^{-i(\alpha+\gamma)/2}\cos\beta/2 \quad (2.6a) \qquad \psi_2 = x_2 + ip_2 = \sqrt{I}\, e^{-i(\gamma-\alpha)/2}\sin\beta/2 \quad (2.6b)$$

An intensity I or normalization factor $\sqrt{I}$ is included to be a fourth variable. Let the coupled Coriolis harmonic oscillator Hamiltonian be defined by

$$H_{HO} = (A/2)(p_1{}^2+x_1{}^2) + (D/2)(p_2{}^2+x_2{}^2) + B(p_1p_2+x_1x_2) + C(x_1p_2-x_2p_1). \tag{2.7}$$

Then the classical equations of oscillator motion are as follows:

$$dx_1/dt = Ap_1+Bp_2-Cx_2 \qquad -dp_1/dt = Ax_1+Bx_2+Cp_2$$
$$dx_2/dt = Bp_1+Dp_2+Cx_1 \qquad -dp_2/dt = Bx_1+Dx_2-Cp_1 \tag{2.8}$$

These are identical to the quantum equations of motion resulting from substituting phase variables (2.6) into the spinor Schrodinger equation (2.1). In general, an n-level quantum system is identical to an n-dimensional classical harmonic oscillator, and this idea has existed at léast as long as the Planck and DeBroglie quantum hypotheses.

Also, the spinor characterization of the coupled oscillator trajectories is related to the theory of polarization ellipsometry given by Stokes[10] over a century ago. Phasor amplitudes ψ_1 and ψ_2 correspond to complex electric field amplitudes E_x and E_y, respectively. The spin expectation vector <J> corresponds to a Stoke's vector **S** which defines a state of elliptical polarization or oscillator trajectory as shown in the figure below.

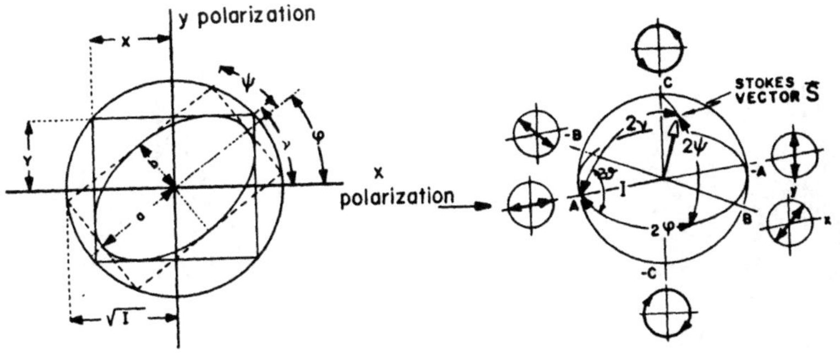

Fig. 3 *Elipsometric characterization of oscillator trajectories.* ***(a)*** *Spinor or SU(2) space.* ***(b)*** *Vector or R(3) space.*

The elliptical trajectory is plotted on the left in the real part of the spinor 2-space where the coordinates (x_1, x_2) are labeled (x,y). The corresponding **S** vector is plotted on the right in the vector 3-space in which the coordinates (S_z, S_x, S_y) are labeled (A,B,C). The ellipse may be characterized by axis ratio $b/a=\tan\psi$ and tipping angle φ. The latitude and longitude of **S** relative to the C-axis is 2ψ and 2φ, respectively. Relative to the A-axis, the polar angles of **S** are $(\alpha=2\vartheta, \beta=2\nu)$, and this is the appropriate choice for a description based upon linearly polarized states $|x\rangle=|x_1\rangle$ and $|y\rangle=|x_2\rangle$. Given the state

$$\begin{aligned} |\Psi\rangle &= (\; X \;\; e^{-i\vartheta}|x\rangle + \;\; Y \;\; e^{i\vartheta}|y\rangle) \\ &= (\cos\nu \;\; e^{-i\vartheta}|x\rangle + \sin\nu \;\; e^{i\vartheta}|y\rangle)\sqrt{I}\, e^{-i\Theta} \end{aligned} \tag{2.9}$$

the expectation value of the **J** operators (2.2b) gives the polar form of **S**, and (2.6) gives its components in terms of phase variables.

$$\begin{aligned} S_x &= \langle\Psi|\mathbf{J_x}|\Psi\rangle = x_1x_2 + p_1p_2 &&= (I/2)\sin 2\nu \;\cos 2\vartheta \\ S_y &= \langle\Psi|\mathbf{J_y}|\Psi\rangle = x_1p_2 - x_2p_1 &&= (I/2)\sin 2\nu \;\sin 2\vartheta \\ S_z &= \langle\Psi|\mathbf{J_z}|\Psi\rangle = (p_1^2+x_1^2-p_2^2-x_2^2)/2 &&= (I/2)\cos 2\nu \\ J_o &= \langle\Psi|\mathbf{1/2}|\Psi\rangle = (p_1^2+x_1^2+p_2^2+x_2^2)/2 &&= (I/2) \end{aligned} \tag{2.10}$$

Then the classical oscillator Hamiltonian (2.7) may be written in an action angle form.

$$\begin{aligned} H_{HO} &= [(A+D)/2]\, J_o + B \;\; S_x + \;\; C \;\; S_y + [(A-D)/2] S_z \\ &= \;\; \omega_o \;\; J_o \;\; + (\omega_x/2) S_x + (\omega_y/2) S_y + (\omega_z/2)\, S_z \\ &= \;\; \omega_o \;\; J_o \;\; + \;\; \Omega_x \;\; S_x + \;\; \Omega_y \;\; S_y + \;\; \Omega_z \;\; S_z \end{aligned} \tag{2.11}$$

Comparing this to (2.2b) we see that the angular rates ω_α of the 3-dimensional rotor are double the corresponding angular rates Ω_α of the 2-dimensional oscillator. This is a common feature of spinor-to-vector mappings as seen in the angle doubling ($\varphi \Rightarrow 2\varphi$ etc.) in Fig.3.

Examples of the mapping between oscillator and rotor motion are shown in Fig.4. Two archtypical cases are used. The first case corresponds to isotropic (A=D), uncoupled (B=O), Coriolis oscillators, or Foucault pendulum with ω along the C-axis. ($\omega = C\mathbf{e}_c$) The resulting motion is called Faraday rotation in polarization optics. Spin vectors parallel or anti-parallel to the C axis correspond to left or right circular polarization, and are not moved. However, elliptical states all precess in the 2-plane with angular frequency $\Omega_c=C/2$, while their spin vectors rotate around the

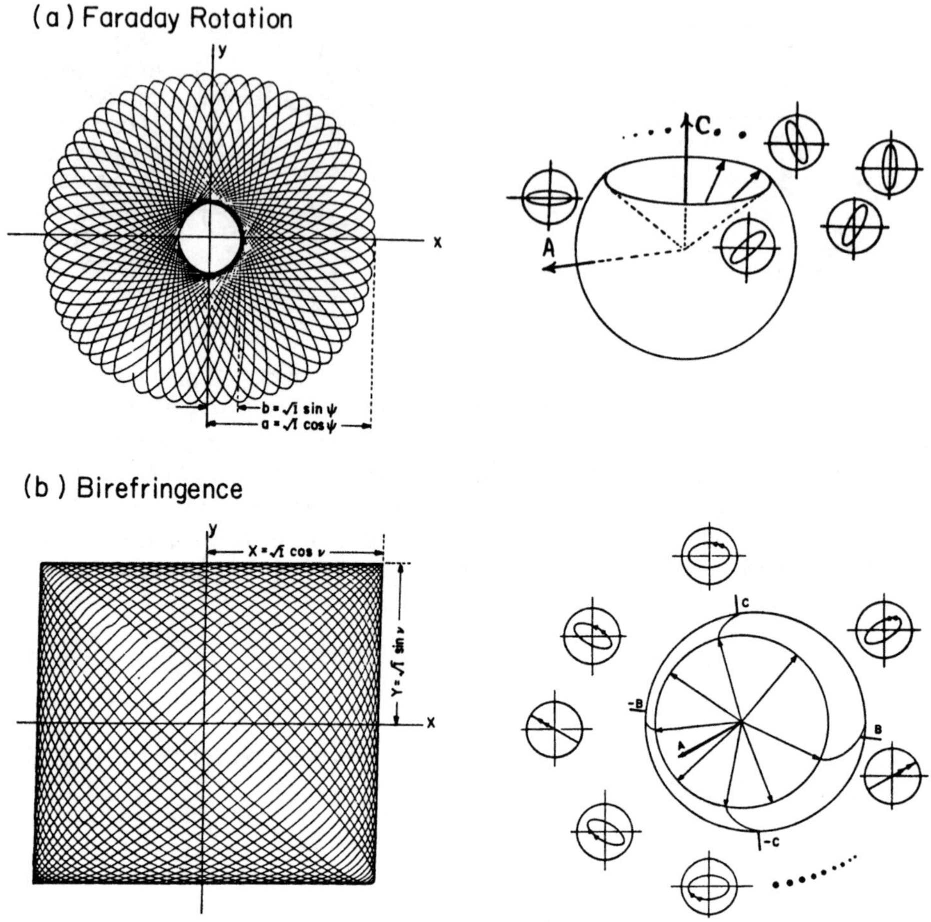

***Fig.4** Oscillator 2-space trajectories and corresponding 3-space rotation.*
***(a)** C-axial rotation.* ***(b)** A-axial rotation.*

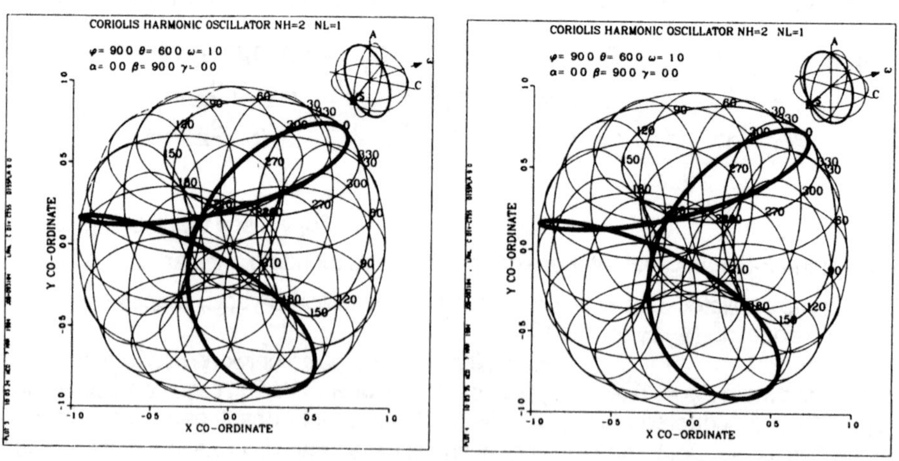

***Fig.5** Phase space torus and trajectory for mixed (ABC) rotation.*

C-axis with frequency ω_C=C, which is double the oscillator precession rate in agreement with (2.11).

The second case corresponds to an anisotropic (A≠D), diagonalized (B=0), Coriolis-free (C=0) oscillator normal modes. The motion is called beating or birefringence and corresponds to a 3-space rotation around the A axis (ω = (A-D)$\mathbf{e}_A$). Spin vectors which are not parallel to A or -A axes correspond to oscillator states which are not normal modes and they are rotated. Rotation of the **S**-vectors occurs with angular frequency (ω_A=A-D) which is commonly called the beat frequency. (It's the Rabi-frequency in the rotating wave model.) However, careful analysis shows that the true oscillator modulation or beat frequency is Ω_A = (A-D)/2, i.e. half that of the rotor as predicted by (2.11).

Rotation around the B-axis corresponds to birefringent oscillator motion (like Fig.4b) that is tipped by $\pi/4$ in the 2-plane. However, it is convenient to choose oscillator coordinates $[x_1,x_2]$ and the (±A) rotor axes to correspond to symmetric (A_1) and anti-symmetric (A_2) normal modes of a harmonic diatomic molecule. Then $\pi/4$-tipped coordinates $[(x_1+x_2)/\sqrt{2},(x_1-x_2)/\sqrt{2}]$ and the (±B) axes correspond to local modes in which one (or the other) atom has all the vibrational motion. The A-rotation shown in Fig.4b causes the atoms to take turns vibrating as the S-vector rotates from B to -B and back to B during each half-beat period $2\pi/(A-D)$. As the S-vector passes through C or -C the atoms are both vibrating, but the phase of one is $\pi/2$ behind the phase of the other and resonant energy transfer is maximum. We will see that anharmonic perturbations can spoil this resonance and trap the spin vector near the B or -B axes.

The 3-vector picture and spin rotor analogy helps one to visualize at once a variety of possible oscillator Hamiltonians and the range of possible motions allowed by each one. On the other hand the spinor 2-space or oscillator phase 4-space gives a more detailed physical picture of the dynamics. For comparison we show in Fig. 5 stereo pictures of the 4-space and 3-space together for typical harmonic motions in which both B and C components are present. This figure should be viewed with sufficient ocular decoupling that the left and right images go to the left and right eyes, respectively. It provides one of the few opportunities to see an invarient torus plotted in a 3-dimensional subspace $[x_1,x_2,p_1]$ of its phase 4-space!

3.Vibrational energy surfaces and anharmonic motion

The analogy between rotor dynamics and vibrational motion given in Sec.2 suggests that the RE surface picture described in Sec.1 might be a useful way to picture vibrational dynamics and spectra. Indeed, the harmonic oscillator Hamiltonian (2.7) or (2.11) yields the simplest type of vibrational energy VE surface since it is linear in the angular monemtum **S** or **J**. THe topography lines or S-trajectories are parallel circles on a quasi- spherical surface which is displaced in the direction of ω vector (2B,2C,A-D). Examples of **S** trajectories for harmonic normal modes and modes with Coriolis perturbation are pictured in stereo diagrams of Fig.6 a and b, respectively. The radial energy coordinate of the VE surface is obtained by writing (2.11) in polar form.

$$H_{HO} = [(A+D) +2B \cos\alpha \sin\beta + 2C \sin\alpha \sin\beta + (A-D)\cos\beta](I/4) \qquad (3.1)$$

Each trajectory traces the same curves in (ABC)-space which would be drawn in diagrams such as Fig.4. (Note that the A-axis is vertical and the C-axis points to the right in Fig.6) As long as I and <J>=<S>=I/2 are constants of the motion along with total energy E, the spin vector trajectories are congruent to the VE level curves. The angular frequencies for precession around each of the circular paths has the same value: $\omega = |\omega|$. Harmonic oscillator frequencies depend upon the harmonic constants and not upon the initial conditions.

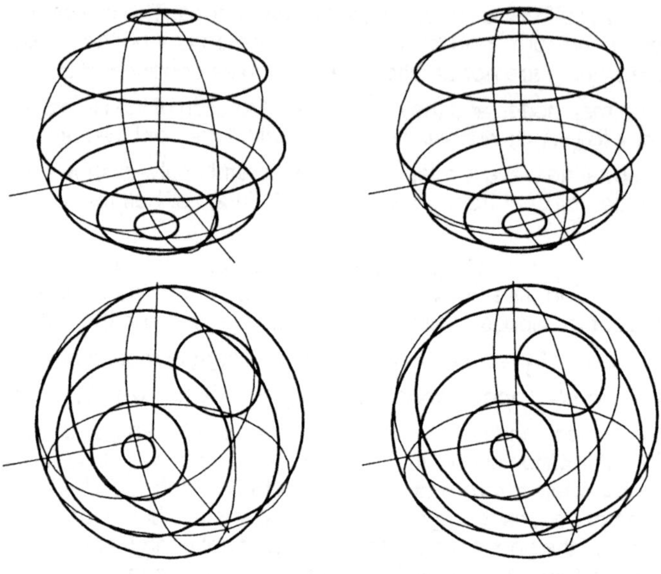

Fig.6
Vibrational energy (VE) surfaces and topographical paths (Harmonic cases)

(a) *A=4.0, D=2.0, B=0.0, C=0.0 (Normal modes)*

(b) *A=4.0, D=2.0 B=0.0, C=1.0 (Coriolis modes)*

A similar geometrical picture can be made for a class of anharmonic coupled oscillator Hamiltonians which conserve I. An example is the following:

$$\begin{aligned} H_{AO} &= H_{HO} + a_x S_x{}^2 = H_{HO} + a_x (I^2/4)\cos^2\alpha \sin^2\beta \\ &= H_{HO} + a_x\,(x_1{}^2 x_2{}^2 + p_1{}^2 p_2{}^2 + x_1 x_2 p_1 p_2 + p_1 p_2 x_1 x_2) \end{aligned} \tag{3.2}$$

Hamiltonians similar to this have been introduced in algebraic theories of Kellman [11] and interacting boson theory of VonRoosmalen, Benjamin, and Levine [12] for quantun models of local mode behavior in YXY polyatomic molecules. The classical variables in (2.6) and (2.10) are related to quantum operators by replacing phase variables $\psi_k = x_k + ip_k$ and $\psi_k{}^*$ by boson operators $\mathbf{a}_k$ and $\mathbf{a}_k{}^\dagger$, respectively.

Energy surfaces show how the harmonic oscillator dynamics and spectra are modified for anharmonic Hamiltonians such as (3.2). Plots shown in Fig.7 a,b, and c show the VE surfaces for anharmonicity values a_x=1.0,2.0,and 3.0, respectively, while harmonic values are fixed to a Coriolis-free normal mode case (A=4, B=0=C, D=2). The **S**-paths are no longer perfectly circular except possibly in the neighborhood of a fixed point. Also, one notes in Figs.7 b and c the appearance of two maxima along the AB meridian which approach the ±B axes as anharmonicity a_x increases. These correspond to local mode motions and the ellipsometry of the (ABC)-space precisely defines the shape of the oscillator (x_1,x_2)-space trajectory for each local mode extremum. According to (3.1-2) the extrema are located along the $(\alpha=2\vartheta=0)$ meridian for B=0.0=C. The elliptical aspect angle $(\beta=2\nu)$ must satisfy

$$(dH_{AO}/d\beta) = 0 = -(I\omega/4)\sin\beta + a_x(I^2/2)\sin\beta\cos\beta \ ,$$

or

$$\beta = 0,\ \pi,\ \pm\cos^{-1}\,[\omega/(2a_x I)]\ , \qquad (\omega = A-D)\ . \tag{3.3}$$

Two of the solutions correspond to normal modes: $\beta(A_1)=0$ and $\beta(A_2)=\pi$, and they are the only allowed values until a_X is great enough that $\omega/(2a_X I)<1$. Then one of the normal modes becomes unstable as its **S**-vector becomes the center of a saddle point for a figure-8 separatrix surrounding two new stable "local modes." These new modes are not really local until the new fixed points approach the B axis. If the product of the anharmonicity a_X and principle quantum number I are several times the true-beat frequency Ω= (A-D)/2, then the new modes are effectively localized classical motions.

Larger $a_X I/\Omega$ would imply the appearance of doublet quantum level clusters on one side of a separatrix region while the singly degenerate "normal" mode spectrum would occupy the other side. The doublet splitting due to tunneling between equivalent local mode trajectories in Fig.7 is precisely analogous to the superfine intra-cluster splitting discussed after Fig.2. This provides a transparant and elementary example of vibrational dynamical tunneling and a comparison of localized dynamics near classical fixed points with delocalized motion near separatrices.

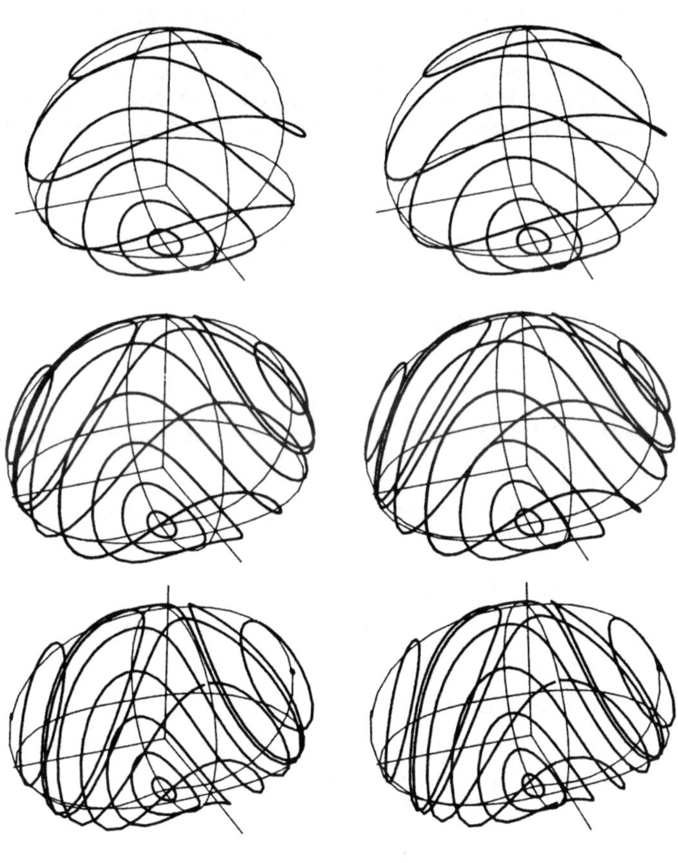

Fig.7
Vibrational energy (VE) surfaces and topographical paths. (Anharmonic cases) I=1.0, A=4.0, D=2.0

(a) *a_X=1.0*

(b) *a_X=2.0*

(c) *a_X=3.0*

4. Classical Rotor Chaos

The models described by rotor or vibrator Hamiltonians like (1.1) or (3.3) may exhibit classical chaos if extra degrees of freedom or external perturbations are introduced. However, as long as J(I) are conserved, the classical equations, however sensitive, are still integrable. Can one construct an easily visualizable rotor model in which classical chaos is present?

Some interesting possibilities involve a free asymmetric rotor carrying an asymmetrically coupled spin or a rotor in an asymmetric external field or asymmetrically coupled spins. Each of those are described by Hamiltonians of the following form:

$$H = \sum_j (A_j J_j + B_j S_j) + \sum_{j,k} (A_{jk} J_j J_k + B_{jk} J_j S_k + C_{jk} S_j S_k) \tag{4.1}$$

where $\qquad [J_i, S_j] = 0$

Obviously higher order terms such as in (1.1) could be considered. However, (4.1) contains more than enough to make chaos. Feingold and Peres [13] have shown irregular motion occurs with just $H = J_3 + S_3 + B_{11} J_1 S_1$.

Understanding Hamiltonians of the form (4.1) is extremely important since they show up in many atomic and molecular problems. Molecular electronic, vibrational,and nuclear spin Coriolis dynamics are modeled by (4.1) and so are caged-rotor and crystal field motions. There is also a model of atomic Rydberg diamagnetism which has this form. It appears that it may be possible to visualize the behavior of this type of model by appealing to multiple RE eigensurfaces [9]. It is relatively simple to obtain exact quantum solutions for some of these models for comaparison. The semiclassical picture allows one to draw a smooth connection between extreme cases or generalized Hund's cases.[7] The results of this approach will be reported elsewhere.

Acknowledgement

This research is supported in part by the National Science Foundation through Grant PHY 86-96052.

References

1. C.W.Patterson, R.S. McDowell, N.G. Nereson, B.J. Krohn, J.S. Wells ,and F.R. Peterson, J. Mol. Spectrosc. **91**, 416 (1982).
2. A.S. Pine, A.G. Maki, A.G. Robiette, B.J. Krohn, J.K.G. Watson, and Th. Urbanek, J. Am. Chem. Soc. **106**, 891 (1984).
3. K.T. Hecht, J. Mol. Spectrosc. **5**, 355 (1960).
4. K.R. Lea, M.J. M. Leask, and W.P. Wolf, J. Phys. Chem. Solids. **23**, 1381 (1982).
5. A.J. Dorney and J.K. G. Watson, J. Mol. Spectrosc. **42**, 1 (1972).
6. W.G. Harter and C.W. Patterson, J. Chem. Phys. **66**, 4872 (1977).
7. W.G. Harter, C.W. Patterson, and F.J. da Paixao, Revs. Mod. Phys. **50**, 37 (1978)
8. W.G. Harter and C. W. Patterson, J. Chem. Phys. **80**, 4241 (1984).
9. W.G. Harter, J. Stat. Phys. **36**, 749 (1984).
10. G. Stokes, Proc. R. Soc. (London) **11**, 545 (1862). For an SU(2) based description of polarization see W.G. Harter and N.dos Santos, Am. J. Phys. **46**,251 and 264 (1978).
11. M.E. Kellman, J. Chem Phys. **76**, 4528 (1982); **83**, 3843 (1985).
12. O.S. van Roosmalen, I. Benjamin, R.D. Levine, J. Chem Phys. **81**, 5986 (1984).
13. M. Feingold and A. Peres, Physica **9D**, 433 (1983).

RESONANCES AND RESONANT STATES IN NON-LOCAL POTENTIALS

A. Mondragón and J.M. Velázquez-Arcos
Instituto de Física, UNAM
Apdo. Postal 20-364, 01000 México, D.F.
MEXICO

I. INTRODUCTION

In this work we are concerned with the study of Gamow states in non-local potentials and with their practical use in the description of resonances in collisions and reactions between composite systems with an internal structure.

Gamow states are the eigenstates of the time independent Schrödinger equation with purely outgoing boundary conditions. This condition makes the eigenvalue problem non-self adjoint. Hence, Gamow states are a generalization of bound states in that they belong to complex wave numbers, $k_n=K_n-i\gamma_n$, with $K_n>\gamma_n>0$, instead of purely imaginary ones. This generalization leads out of the Hilbert space based quantum mechanics. Therefore, the usual quantum mechanical rules for normalization, orthogonality and completeness do not apply. Extending the formulation of quantum mechanics to a rigged Hilbert space, Gamow states are accomodated in the theory on a sound mathematical basis(1). Rules for the normalization of Gamow states, as well as orthogonality and completeness relations have been derived in configuration space representation(2) and in momentum space representation(3). However, in most of these studies Gamow functions are defined in local short ranged potentials(2) or in non-local separable potentials of finite rank(3). Since the effective potential that describes the interaction of composite systems with an internal structure usually is non-local and has no sharp cut off radius, in this work we extend the validity of the normalization rule and completeness relations to the more general case of resonances and resonant states in a non-local, non-separable potential with non-compact support. An example of the use of Gamow states in a realistic problem— the description of the resonant behaviour of the cross section for the direct radiative capture of ^{3}He by ^{3}H to the first excited state of ^{6}Li-is also briefly discussed.

2. RESONANT STATES IN A NON-LOCAL POTENTIAL

The non-relativistic scattering of a particle by a non-local potential is described by the solution of a Schrödinger equation which, in this case, is an integro-differential equation. When the non-local potential is rotationally invariant, the wave function is expanded in partial waves and one is left with the radial equation

$$\frac{d^2u_\ell(k,r)}{dr^2} + k^2 u_\ell(k,r) - \frac{\ell(\ell+1)}{r^2} u_\ell(k,r) - \int_0^\infty U_\ell(r,s)u_\ell(k,s)ds=0 \qquad (2.1)$$

The potential $U_\ell(r,r')$ is the sum of a direct and an exchange term

$$U_\ell(r,r')=\frac{2m}{\hbar^2}(V_d^{(\ell)}(r)\delta(r-r') + V_{exch}^{(\ell)}(r,r') \qquad (2.2)$$

The scattering solution $\phi_\ell(k,r)$ of (2.1) vanishes at the origin and asymptotically it behaves as the sum of a free incoming wave plus a free outgoing wave

$$\phi_\ell(k,0)=0 \qquad (2.3)$$

$$\lim_{r\to\infty}[\phi_\ell(k,r) - (\hat{h}_\ell^{(-)}(kr) + S_\ell(k)\hat{h}_\ell^{(+)}(kr))]=0 \qquad (2.4)$$

In this expression $\hat{h}_\ell^{(-)}(kr)$ and $\hat{h}_\ell^{(+)}(kr)$ are Riccati-Hankel functions that describe incoming and outgoing waves respectively, $S_\ell(k)$ is the collision matrix.

The resonant state or Gamow function represents a decaying state of the physical system in a situation in which there are no particles incident. Therefore, the Gamow function is a solution of (2.1) which vanishes at the origin and, asymptotically, it behaves as a purely outgoing wave

$$u_{n\ell}(k_n, 0) = 0 \tag{2.5}$$

$$\lim_{r\to\infty} \left[\frac{du_{n\ell}(k_n, r)}{dr} - ik_n u_{n\ell}(k_n, r) \right] = 0$$

These boundary conditions cannot be satisfied for all values of the wave number k, but only for some complex values k_n

$$k_n = K_n - i\gamma_n \tag{2.7}$$

When

$$K_n > \gamma_n > 0 \tag{2.8}$$

the solution $u_n(k_n, r)$ is a Gamow state.

The bound state solutions, $v_{m\ell}(r)$, of (2.1) also satisfy the boundary conditions (2.5) and (2.6), but in this case, the real part of the wave number k_m is equal to zero and the imaginary part is positive, which means that, asymptotically, the wave function $v_{m\ell}(r)$ decreases exponentially with r and the energy eigenvalue is real and negative.

It may be shown, by an elementary argument involving nothing more than the use of Green's theorem, that the Gamow function $u_{n\ell}(k_n, r)$ is also a solution of the homogeneous Lippmann-Schwinger equation

$$u_{n\ell}(k_n, r) = \int_0^\infty K_\ell^{(+)}(k_n; r, s) u_{n\ell}(k_n, s) ds \tag{2.9}$$

The kernel $K_\ell^+(k_n; r, s)$ that appears in this expression is

$$K^{(+)}_\ell(k; r, r') = \int_0^\infty G^{(+)}_{o\ell}(k; r, s)\, U_\ell(s, r') ds \tag{2.10}$$

$G^{(+)}_{o}(k;r,s)$ is the Green's function of the radial equation for a free particle

$$G^{(+)}_{o\ell}(k;r,r') = -\frac{(-1)^{\ell}}{k}\hat{j}(kr_<)\hat{h}^{(+)}_{\ell}(kr_>) \qquad (2.11)$$

$\hat{j}_{\ell}(kr)$ is the Riccati-Bessel function.

After substitution of the explicit expression (2.11) for $G^{(+)}_{o\ell}(k;r,r')$ in (2.10), the integral equation (2.9) may be written in a more explicit form

$$u_{n\ell}(k_n,r) = -\frac{(-1)^{\ell}}{k_n}\{h^{(+)}_{\ell}(k_n r)\int_0^r ds\hat{j}_{\ell}(k_n s)\int_0^{\infty}dtU_{\ell}(s,t)u_{n\ell}(k_n,t)$$

$$+\hat{j}_{\ell}(k_n r)\int_r^{\infty}ds\,\hat{h}^{(+)}_{\ell}(k_n s)\int_0^{\infty}dtU_{\ell}(s,t)u_{n\ell}(k_n,t)\} \qquad (2.12)$$

From this expression, and the properties of $\hat{j}_{\ell}(kr)$ and $h^{(+)}_{\ell}(kr)$, it is obvious, that when the integrals exist, $u_{n\ell}(k_n,r)$ vanishes at the origin and, asymptotically, it behaves as an outgoing wave.

When the imaginary part of the wave number is negative, $\hat{j}_{\ell}(k_n r)$ and $h_{\ell}^{(+)}(k_n r)$ behave asymptotically as outgoing waves of exponentially increasing amplitude

$$\hat{j}_{\ell}(k_n r) \xrightarrow[r\to\infty]{} \frac{1}{2}\, e^{\gamma_n r}\, e^{i(k_n r-\frac{1}{2}(\ell+1)\pi)} \qquad (2.13)$$

$$\hat{h}^{+}_{\ell}(k_n r) \xrightarrow[r\to\infty]{} \frac{i}{2}\, e^{\gamma_n r}\, e^{i(k_n r-\frac{1}{2}(\ell+1)\pi)} \qquad (2.14)$$

From these expressions and (2.12), it follows that the Gamow function $u_n(k_n,r)$, also behaves asymptotically as an outgoing wave which oscillates between envelopes that increase exponentially with r.

This property of $u_{n\ell}(k_n,r)$ produces no divergencies in the right hand side of the integral equation (2.12), since the same conditions on the non-local potential that guarantee the existence of scattering and bound state solutions of the radial equation (2.1), make the product of the non-local potential and the Gamow function an exponentially decreasing function of r.

To be more precise, it will be assumed that the potential $U_\ell(r,r')$ is invariant under time inversion and that its local and non-local parts satisfy the following conditions:

$$\int_0^\infty \int_0^\infty ds\,dt\, F_\alpha^{-2}(s)\, |V_{exch}(s,t)|^2 F_\alpha^{-2}(t) = C_1 < \infty \qquad (2.15)$$

where

$$F_\alpha(r) = e^{-\alpha r}\, r^{-1+\varepsilon_0}\, (1+r)^{-\varepsilon_0-\varepsilon_\infty} \qquad (2.16)$$

with $\alpha \geq 0, \varepsilon_0 > 0$ and $\varepsilon_\infty > 0$.

The direct term in the potential is square integrable in any closed interval in the real positive axis R_+.

$$\int_a^b |V_d(r)|^2\, dr < C_d \qquad (2.17)$$

with $0 \leqslant a \leqslant b < \infty$, and asymptotically it behaves as

$$V_d(r) \underset{r\to 0}{\longrightarrow} r^{-2-\eta_0} \qquad (2.18)$$

$$V_d(r) \underset{r\to \infty}{\longrightarrow} e^{-2\alpha r}\, r^{-2-\eta_\infty} \qquad (2.19)$$

with $\eta_0 > \varepsilon_0$ and $\eta_\infty > \varepsilon_\infty$.

When this conditions are satisfied, the integrand in the integral equation (2.12) is convergent, since, according to (2.15) and (2.19), when

$$\mathrm{Im}\, k_n = -\gamma_n > -\alpha \qquad (2.20)$$

the non-local potential $U_\ell(s,t)$ damps out the exponentially growing oscillations of $\hat{j}_\ell(k_n s)$, $\hat{h}_\ell^{(+)}(k_n s)$ and $u_{n\ell}(k_n,t)$, making the integrals finite. It will be shown below that the energy eigenvalues E_n and the corresponding wave numbers k_n of relevance to this work are precisaly those for which condition (2.20) is satisfied.

3. THE GAMOW FUNCTION AND THE FREDHOLM EIGENFUNCTION

N. von der Heydt[5] has shown that, when the potential satisfies the conditions stated above and Im $k \geq -\alpha$, the trace of the kernel $K_\ell^{(+)}(k;s,t)$, and the Fredholm determinant

$$\Delta_\ell^{(+)}(\eta,k)=\det(\mathbb{1}-\eta\, K_\ell^{(+)}(k)) \qquad (3.1)$$

exist.

Solutions of the homogeneous Lippmann Schwinger equation (2.12) exist when η is set equal to one and the Fredholm determinant vanishes.

This may be seen from the following argument:Fredholm's first minor $M_\ell(k;r,r_0)$ satisfies the integral relation sometimes called Fredholm's second fundamental relation

$$M_\ell(k;r,r_0)=\eta\Delta_\ell^{(+)}(\eta,k)+\eta\int_0^\infty K_\ell^{(+)}(k;r,s)M_\ell(k;s,r_0)ds \qquad (3.2)$$

when $\Delta^{(+)}_\ell(k)$ vanishes, this equation reduces to the homogeneous Fredholm equation

$$w_\ell(k,r)=\eta(k)\int_0^\infty K_\ell^{(+)}(k;r,s)w(k,s)ds \qquad (3.3)$$

The vanishing of $\Delta_\ell^{(+)}(k)$ defines η as a function of k

$$\Delta_\ell^{(+)}(k) = \prod_{i=1}^{N}(1-\eta(k)\alpha_i(k)) = 0 \qquad (3.4)$$

the $\alpha_i(k)$ that appear in this equation are the eigenvalues of the kernel $K_\ell^{(+)}(k)$.

First, let us keep k fixed with Im $k \geq -\alpha$, when $\Delta_\ell^{(+)}(\eta, k)$ has a simple zero, $\eta(k)$ is equal to the inverse of one of the eigenvalues, say $\alpha_i^{-1}(k)$. If $K_\ell^{(+)}(k)$ has N different eigenvalues, $\eta(k)$, as function of k has N branches.

Now, let us set η equal to one in (3.4), then this condition is satisfied only for some values k_n of the wave number k, with $k_n \geq -\alpha$.

$$\Delta_\ell^{(+)}(1, k_n) = 0 \tag{3.5}$$

and, at $k = k_n$, one of the eigenvalues is equal to one

$$\alpha_i(k_n) = 1 \tag{3.6}$$

In this case, equation (3.3) reduces to the homogeneous Lippmann-Schwinger equation (2.12), showing that the Gamow function $u_n(k_n, r)$ is equal to the first Fredholm minor evaluated at $k = k_n$. When the zero of $\Delta_\ell^{(+)}$ at $k = k_n$ is simple, there is only one eigenvalue $\alpha_i(k_n)$ that satisfies (3.6). The corresponding Fredholm eigenfunction $w_{\ell i}(k_n, r)$ is then equal to the Gamow function

$$u_{n\ell}(k_n, r) = w_\ell(k_n, r) \tag{3.7}$$

Hence, to each simple zero of $\Delta_\ell^{(+)}(k)$ corresponds one Gamow state. On the other hand, since equation (3.5) may have many different roots k_n, and the same eigenvalue $\alpha_i(k)$ may take the value one more than once, there may be many Gamow functions obtained from the same Fredholm eigenfunction when k is equal to any of the k_n's that make $\alpha_i(k)$ equal to one. We might say that all these Gamow functions belong to the same branch of $\eta(k)$.

4. ADJOINT OF THE GAMOW FUNCTION

The adjoint of the Gamow function is readily obtained

from the symmetries of the kernel $K_\ell^{(+)}$ $(k;r,s)$. Let us, recall first the symmetry properties of the free Green's function

$$G^{(+)}_{o\ell}(k;r,s) = G_{o\ell}(k;s,r) \tag{4.1}$$

and

$$G^{(+)}_{o\ell}(k;r,s) = G^{(+)*}_{o\ell}(-k^*;s,r) \tag{4.2}$$

from these relations, and the hermiticity of the potential, it follows that

$$K_\ell^{(+)}(k;r,s) = K_\ell^{(+)*}(-k^*;s,r) \tag{4.3}$$

which implies that the Fredholm determinant and the function $\eta(k)$ satisfy a similar relation

$$\Delta_\ell^{(+)}(k) = \Delta_\ell^{(+)*}(-k^*) \tag{4.4}$$

$$\eta(k) = \eta^*(-k^*) \tag{4.5}$$

These relations suggest taking the complex conjugate in both sides of (2.12), changing k into $-k^*$ and making use of (3.3), we get

$$u^*_{n\ell}(-k^*_n;r) = \int_0^\infty u^*_{n\ell}(-k^*,s)\; K_\ell^{(+)}(k_n;s,r)\, ds \tag{4.6}$$

showing that $u^*_{n\ell}(-k^*_n,r)$ is a left eigenfunction of $K_\ell^{(+)}(k;s,r)$. Therefore, the adjoint $\tilde{u}_{n\ell}(k_n,r)$ of the Gamow function is given by

$$\tilde{u}_{n\ell}(k_n,r) = u^*_{n\ell}(-k^*_n,r) \tag{4.7}$$

In the case of spinless particles, the potential is a real function, symmetric in its arguments, and $G^+_{o\ell}(k;r,s)$ is also symmetric, then (3.6) may be rearranged to give

$$\tilde{u}_{n\ell}(k_n,r) = \int_0^\infty K_\ell^{(+)}(k;r,s)\; \tilde{u}_{n\ell}(k_n,s)ds \tag{4.8}$$

which is the same equation as (2.9). Therefore, for

spinless particles, the Gamow function is proportional to its adjoint. Without loss of generality, we may write

$$\tilde{u}_{n\ell}(k_n,r) = u_{n\ell}(k_n,r) \tag{4.9}$$

5.NORMALIZATION OF GAMOW STATES

In what follows it will be shown that the residue of the Green's function of the radial equation (2.1) at a simple pole located in the fourth quadrant of the k plane is the product of the Gamow functions $u_{n\ell}(k_n,r)u_{n\ell}(k_n,r')$times a constant which is equal to the inverse of the normalization integral of the Gamow function.

Before proceeding with our argument, and in order to avoid a clumsy notation we will introduce the shorthand notation

$$|f><g| = f(k,r)\,\tilde{g}(k,s)$$

$$<f|g> = \int_0^\infty \tilde{f}(k,r)\,g(k,r)dr$$

$$\underset{\sim}{K}^+_\ell|f> = \int_0^\infty ds\ K^{(+)}_\ell(k;r,s)f(s)$$

$$<f|\underset{\sim}{K}^{(+)}_\ell = \int_0^\infty ds\tilde{f}(k,s)\ K^{(+)}_\ell\ (k;s,r)$$

Now, the complete Green's function or resolvent of the radial integrodifferential equation (2.1) satisfies the inhomogeneous Lippmann-Schwinger equations

$$\underset{\sim}{G}^{(+)}_\ell(k) = \underset{\sim}{G}^{(+)}_{o\ell}\ (k) + \underset{\sim}{K}^+_\ell(k)\ \underset{\sim}{G}^{(+)}_\ell\ (k\ ) \tag{5.1}$$

and

$$\underset{\sim}{G}^{(+)}_\ell\ (k) = \underset{\sim}{G}^{(+)}_{o\ell}\ (k)\ + \underset{\sim}{G}^{(+)}_\ell\ (k)\ \underset{\sim}{K}^{(+)}_\ell \tag{5.2}$$

N. von der Heydt[5] has shown that, when the non-local potential U_ℓ satisfies the conditions (2.15) to (2.19), $\underset{\sim}{G}^{(+)}_\ell$ (k), as function of k, is a meromorphic function in the half plane Im $k \geq -\alpha$. When α is finite, $\underset{\sim}{G}^{(+)}_\ell$ (k) has, at most, a finite number of poles in that half plane, which are located precisaly where the zeroes of $\Delta_\ell^{(+)}(k)$ are.

Near a complex pole, located at $k=k_n$, $\underset{\sim}{G}_\ell^{(+)}(k)$ may be written as

$$\underset{\sim}{G}_\ell^{(+)}(k) = \frac{R_{n\ell}(k_n)}{k-k_n} + F_n(k) \qquad (5.5)$$

the function $F_n(k)$, as function of k, is regular at $k=k_n$.

In order to derive an integral equation for R_n, we substitute (5.3) for $\underset{\sim}{G}_\ell^{(+)}(k)$ in (5.1), we multiply both sides of the resulting equation by $(k-k_n)$ and take the limit when k goes to k_n, in this way we get

$$\underset{\sim}{R}_n(k_n) = \underset{\sim}{G}_{o\ell}^{(+)}(k_n)\underset{\sim}{U}_\ell\underset{\sim}{R}_n(k_n) \qquad (5.4)$$

This is just the homogeneous integral equation (2.16) satisfied by the Gamow state $|u_{n\ell}(k_n)\rangle$. Therefore, $\underset{\sim}{R}_n(k_n)$ is proportional to the Gamow state.

$$\underset{\sim}{R}_n(k_n)=|u_{n\ell}(k_n)\rangle\ \langle f_n| \qquad (5.5)$$

To derive an integral equation for $\langle f_n|$ we proceed in a similar way. We substitute (5.3) and (5.5) for $\underset{\sim}{G}_\ell^{(+)}(k)$ in (5.2), then we multiply both sides of the resulting equation by $(k-k_n)$ and take the limit when k goes to k_n

$$\langle f_n| = \langle f_n(k_n)|K_\ell^{(+)}(k_n) \qquad (5.6)$$

Therefore, $\langle f_n(k_n)|$ is proportional to the adjoint of the Gamow state, but since the Gamow state is equal to its adjoint, $\langle f_n(k_n)|$ is equal to the Gamow state. With these results, near one of its complex poles, the Green's function may be written as

$$G_\ell^{(+)}(k;r,r') = u_{n\ell}(k_n,r)\,\frac{C_n}{k-k_n}\,u_{n\ell}(k_n,r') + F_n(k;r,r') \qquad (5.7)$$

In order to derive an expression for the constant C_n, it is convenient to extend the validity of the expression (5.7) in such a way that the functions in the numerator of the singular term are defined for all values of the wave number k in the half plane Im k$\geq$-α, and not just at the point k_n. This may be achieved replacing in (5.7) the Gamow state $|u_n(k_n)>$ by the Fredholm eigenfunction which is equal to this state when k is equal to k_n. To simplify the notation we will call this Fredholm eigenfunction $|u_{n\ell}(k)>$, keeping in mind that, when another Gamow state $|u_{m\ell}(k_m)>$ belongs to the same branch of $\eta(k)$ $|u_{n\ell}(k)>$ and $|u_{m\ell}(k)>$ are actually the same function.

$$G_\ell^{(+)}(k) = |u_{n\ell}(k)> \frac{C_n}{k-k_n} <u_{n\ell}(k)| + F_n(k) \quad (5.8)$$

The function $F_n(k)$ which, by definition, is the difference of the Green's function $G_\ell^{(+)}(k)$ and the singular term in (4.8), reduces to $F_n(k)$ when the wave number k in the argument of the Fredholm eigenfunctions is equal to k_n.

From the Lippmann Schwinger equation (5.2) we obtain the following relation

$$\underset{\sim}{G}_\ell^{+}(k)\underset{\sim}{U}_\ell \,|\, u_{n\ell}(k)> = \underset{\sim}{G}_\ell^{(+)}(k)\underset{\sim}{U}_\ell \,|u_{n\ell}(k)> +$$

$$+ \underset{\sim}{G}_\ell^{(+)}(k)\, \underset{\sim}{U}_\ell\, \underset{\sim}{G}_{o\ell}^{(+)}(k)\, |u_{n\ell}(k)> \quad (5.9)$$

recalling that $|u_{n\ell}(k)>$ is a solution of the homogeneous Fredholm equation (3.3), this expression reduces to

$$(\eta(k)-1)\underset{\sim}{G}_\ell^{(+)}(k)\underset{\sim}{U}_\ell|u_{n\ell}(k)> = |u_{n\ell}(k)> \quad (5.10)$$

Now, we substitute the expression (5.8) for $\underset{\sim}{G}_\ell^{(+)}(k)$ in this equation, and obtain

$$(\eta(k)-1) \quad [\, u_{n\ell}(k) \frac{C_n}{k-k_n} <u_{n\ell}(k) \mid U_\ell|u_{n\ell}(k)> +$$

$$+ \underset{\sim}{F}_n(k)\underset{\sim}{U}_\ell|\, u_{n\ell}(k)> \,] = |u_{n\ell}(k)> \quad (5.11)$$

Taking the limit when k goes to k_n, and recalling that $\eta(k_n)=1$, we get

$$C_n = \frac{1}{\left(\frac{d\eta}{dk}\right)_{k_n} < u_{n\ell}(k_n) \, |U_\ell| u_{n\ell}(k_n)>} \tag{5.12}$$

This expression is not yet in the desired form since it still contains the derivative of the Fredholm eigenvalue which is not written in terms of Gamow functions.

The derivative of the Fredholm eigenvalue $\eta^{-1}(k)$ may be computed from the identity

$$\eta^{-1}(k) = \frac{< u_{n\ell}(k) | U_\ell G^{(+)}_{o\ell}(k) U_\ell \, | u_{n\ell}(k) >}{< u_{n\ell}(k) |U_\ell| u_{n\ell}(k)>} \tag{5.13}$$

Taking the derivative with respect to k in both sides of this equation, we get

$$\frac{d\eta(k)}{dk} = -\,\eta(k) \frac{< u_{n\ell}(k) |U_\ell \frac{\partial G^{(+)}_{o\ell}(k)}{\partial k} U_\ell| \, u_n(k)>}{< u_{n\ell}(k) |U_\ell| u_{n\ell}(k)>} \tag{5.14}$$

The derivative of the free Green's function may be obtained from the spectral representation

$$\frac{\partial G^{(+)}_{o\ell}(k)}{\partial k} = -2k \int_0^\infty G^{(+)}_{o\ell}(k;r,s) G^{(+)}_{o\ell}(k;s,r')ds \tag{5.15}$$

When this expression is substituted for $\frac{\partial G^{(+)}_{o\ell}(k)}{\partial k}$ in (5.14) and, recalling that $|u_{n\ell}(k)>$ is the Fredholm eigenfunction belonging to the eigenvalue $\eta^{-1}(k)$, the result is simplified with the help of (3.3), we get

$$\frac{d\eta(k)}{dk} = - \frac{2k}{\eta(k)} \frac{\langle u_{n\ell}(k) \mid u_{n\ell}(k)\rangle}{\langle u_{n\ell}(k) \mid U_\ell \mid u_{n\ell}(k)\rangle} \qquad (5.16)$$

Now, we substitute this result in (5.12) and take the limit when k goes to k_n, in this way we obtain

$$C_n = \frac{1}{2k_n} \frac{1}{\lim\limits_{k \to k_n} [\int_0^\infty u^2{}_{n\ell}(k^+,r)dr]} \qquad (5.17)$$

Two remarks are in order, first, in (5.17) we have made use of the equality of the Fredholm eigenfunction and its adjoint. Second, the integral in the denominator in (4.17) is defined with k in the upper half of the k-plane and, after the integration is performed, the resulting function of k is analytically continued to k_n in the lower half of the k-plane. The limit cannot be taken inside the integration sign because the resulting integral does not exist.

When the expression (5.17) is substituted for C_n in (5.8), we obtain

$$\frac{1}{2\pi i} \oint_C G_\ell^{(+)} (k;r,r')dk$$

$$= u_{n\ell}(k_n,r) \frac{1}{\lim\limits_{k \to k_n} [\int_0^\infty u^2{}_{n\ell}(k^+ r)dr]} u_{n\ell}(k_n,r') \qquad (5.18)$$

It follows from this result that the normalization rule appropriate for Gamow states is

$$\lim_{k \to k_n} \int_0^\infty u^2{}_{n\ell}(k^+,r)dr = 1 \qquad (5.19)$$

Hernández and Mondragón[4] have shown that, when the Gamow state is defined in a cutoff potential of range R_0, (5.19) reduces to the normalization condition given by Hokkyo, Romo and García-Calderón and Peierls[2].

6. EXPANSION OF THE GREEN'S FUNCTION IN TERMS OF BOUND RESONANT AND SCATTERING STATES.

The Green's function of the Schrödinger equation with a non-local potential may be expanded in eigenfunctions of the energy belonging to negative and complex eigenvalues plus an integral over scattering states of complex wave number. A result similar to ours is well known in the case of local short ranged potentials[2].

The spectral representation of the Green's function of the radial equation with a non-local potential (2.1) is

$$G_\ell^{(+)}(k;r,r') = \sum_{\substack{\text{bound}\\\text{states}}} v_{n\ell}(r) \frac{1}{k^2 - k^2_{n\ell}} v_{n\ell}(r') + \frac{2}{\pi} \int^{\infty} \phi_\ell(k',r) \frac{1}{k^2 - k'^2} \phi^*_\ell(k',r')dk' \qquad (6.1)$$

When the non-local potential satisfies the condition stated in (2.15) and (2.16), $G_\ell^{(+)}(k;r,r')$, as a function of k, is meromorphic in the half-plane Im $k>-\alpha$ and, when α is finite, there it has, at most, a finite number of poles. In the half plane Im $k \geq 0$, it may have poles only on the positive imaginary axis or on the real axis. When $k \neq 0$, the poles are simple. In case there is a pole at $k=0$, this may be simple or double[5].

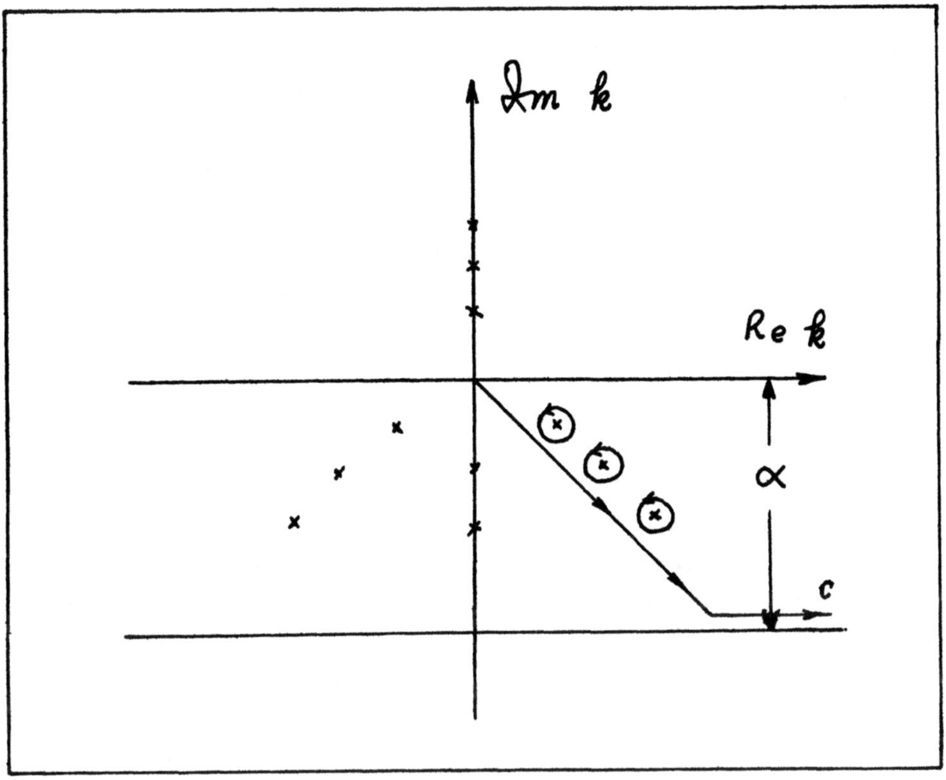

Fig. 1. The straight line with slope-1 is the integration contour C in the wave number plane k that appears in Eq.(6.2).

The integration contour of eq.(6.1) is deformed into the lower half plane, as shown in Fig. 1. When the deformed contour crosses over resonant poles, the theorem of the residue, and eq. (5.18) yield

$$G_{\ell}^{(+)}(k;r,r') = \sum_{\substack{\text{bound}\\ \text{states}}} v_{n\ell}(r) \frac{1}{k^2 + \frac{2m|E_n|}{\hbar^2}} v_{n\ell}(r') +$$

$$\sum_{\substack{\text{resonant}\\ \text{states}}} u_{n\ell}(k_n,r) \frac{1}{2k_n(k-k_n)} u_{n\ell}(k_n,r) + \int_C \phi_{\ell}(Z,r) \frac{1}{k^2 - Z^2} \phi_{\ell}(Z\ ,r')dZ \qquad (6.2)$$

where $v_{n\ell}(r)$ are the bound state eigenfunctions belonging to the negative energy eigenvalues, $u_{n\ell}(k_n,r)$ are the Gamow functions belonging to the complex enery eigenvalues and $\phi_\ell(Z,r)$ is a scattering function of complex wave number Z. Right and left solutions of the scattering equation are distinguished with a tilde on the left solution. This distinction is made because when the wave number is complex, the left solution is not equal to the Hermitian conjugate of the right solution.

7. EXPANSION OF A FUNCTION IN TERMS OF BOUND, RESONANT AND SCATTERING STATES.

In this section it will be shown that an arbitrary square integrable function may be expanded in terms of a set of functions containing bound and resonant states, and a continuum of scattering wave functions of complex wave number.

We start by recalling that the orthonormal set of bound and scattering solutions of the radial Schrödinger equation (2.1) is a complete set. This has been shown for non-local potentials by N. von der Heydt[5] under fairly general conditions. Then, for any two square integrable functions, $\Phi(r)$ and $\chi(r)$, the following relation holds

$$\int_0^\infty \Phi^*(r)\chi(r)dr = \sum_{n=0}^{N} \sum_{s=1}^{m_s} \langle\Phi|v_{n\ell s}\rangle\langle v_{n\ell s}|\chi\rangle + \int_0^\infty \langle\Phi|\phi_\ell(k)\rangle\langle\phi_\ell(k)|\chi\rangle dk \qquad (7.1)$$

where

$$\langle\Phi|v_{n\ell s}\rangle = \int_0^\infty \Phi^*(r)v_{n\ell s}(r)dr \qquad (7.2a)$$

$$\langle v_{n\ell s}|\chi\rangle = \int_0^\infty v_{n\ell s}(r)\chi(r)dr \qquad (7.2b)$$

$$<\Phi|u_\ell(k)> = \int_0^\infty \Phi(r)\phi_\ell(k,r)\,dr \qquad (7.3a)$$

and

$$<\phi_\ell(k)|\chi> = \int_0^\infty \phi_\ell^*(k,r)\chi(r)\,dr \qquad (7.3b)$$

In this expression, the function $v_{n\ell s}(r)$ is a bound state solution of the Schrödinger equation (2.1) that belongs to the negative energy eigenvalue E_n, and it is also a solution of the homogeneous Lippmann-Schwinger equation (2.9). The function $\phi(k,r)$ is a scattering partial wave radial solution of eq. (2.1) which is also a solution of an inhomogeneous Lippmann-Schwinger equation. When the nonlocal potential satisfies the conditions stated in section 2 eq. (2.15) and (2.16), N may be infinite only when $\alpha=0$.

Now, from the spectral representation of the Green's function for outgoing and for incoming particles it follows that

$$\int <\Phi|u_{n\ell}> <u_{n\ell}|\chi> =$$

$$\int_0^\infty [\frac{1}{2\pi i}\int_0^\infty\int_0^\infty \Phi^*(r)\,[G_\ell^{(+)}(k;r,r')-G_\ell^{(-)}(k;r,r')]\chi(r')drdr']dk \qquad (7.4)$$

The Green's function for incoming particles, has no singularities in the lower half of the k plane, while the Green's function for outgoing particles has poles in that part of the plane as explained in section 6. The integration contour of eq. (7.4) may be deformed into the lower half plane as shown in Fig. 1, and when the deformed contour C crosses over resonant poles, the theorem of the residue and equation (5.18), and (5.19) yield

$$\int \langle \Phi | \phi_\ell(k) \rangle \langle \phi_\ell(k) | \chi \rangle \, dk =$$

$$= \sum_{\substack{n \\ \text{(resonant states)}}} \left[\int_0^\infty \Phi^*(r) u_{n\ell}(k^+ r) dr \right]_{kn} \left[\int_0^\infty u_{n\ell}(k^+ s) \chi(s) ds \right]_{kn}$$

$$+ \int_C dZ \left[\int_0^\infty \Phi^*(r) \phi_\ell(k,r) dr \int_0^\infty \tilde{\phi}_\ell(k,s) \chi(s) ds \right]_{k=Z} \qquad (7.5)$$

Since $\Phi(r)$ and $\chi(r)$ are arbitrary functions, the above discussion justifies writing the expansion

$$\chi(r) = \sum_{\substack{m \\ \text{(bound states)}}} v_{m\ell}(r) \langle u_{n\ell} | \chi \rangle +$$

$$+ \sum_{\substack{n \\ \text{(resonant states)}}} u_{n\ell}(k_n, r) \langle u_{n\ell} | \chi \rangle +$$

$$+ \int_C \tilde{\phi}_\ell(Z,r) \langle \tilde{\phi}_\ell(Z) | \chi \rangle dZ \qquad (7.6)$$

The expansion coefficients are given by

$$\langle v_{m\ell} | \chi \rangle = \int_0^\infty v_{m\ell}(r) \chi(r) dr \qquad (7.7)$$

$$\langle u_{n\ell} | \chi \rangle = \left[\int_0^\infty u_{n\ell}(k^+, r) \chi(r) dr \right]_{k=k_n} \qquad (7.8)$$

$$\langle \phi_\ell(Z) | \chi \rangle = \left[\int_0^\infty \phi_\ell(k,r) \chi(r) dr \right]_{k=Z} \qquad (7.9)$$

8. CONCLUSIONS

We have shown that Fredholm's theory of linear integral equations is an appropriate tool for the study of the properties of resonances and resonant states in non-local potentials with compact or non-compact support. In this way we derived expansions of the wave function and the Green's function of the Schrödinger equation with a non-local potential in terms of bound states, resonant states and a continuum of scattering functions with complex wave number. These results are a generalization to the case of non-local potentials of similar expansions obtained by other authors for the case of local potentials of finite range. The rules for normalization, orthogonality and completeness satisfied by the eigenstates of the Schrödinger equation belonging to complex eigenvalues, with Im E_n 0 were also derived. The generalization of these results to the case of a non-local potential plus a Coulomb potential and its application to the analysis of a realistic example will be published elsewhere

REFERENCES

1. A. Böhm, J. Math. Phys. 22, 2813, (1981).
   M. Gadella, J. Math. Phys. 24, 1462, 2142 (1983).
   M. Gadella, Physica A (Netherlands) 124A, 317, (1984).

2. Y.B. Zel'dovich, JETP (Sov. Phys.) 12 (1961) 542.
   N. Hokkyo, Progr. Theor. Phys. 33 (1965) 1116.
   T. Berggren, Nucl. Phys. A109 (1968) 265.
   W. Romo, Nucl. Phys. A116 618 (1968).
   B. Gyarmati and T. Vertse, Nucl. Phys. A160 (1971) 523.
   G. García-Calderón and R.E. Peierls, Nucl. Phys. A265 (1976) 443.

3. E. Hernández and A. Mondragón, Phys. Rev. C29 (1983) 722.
   E. Hernández and A. Mondragón Notas Física 5 (1982)219.
   T. Berggren, Phys. Lett. 38B (1972) 1782.
   B. Gyarmati, A.T. Kruppa and Z. Papp. Phys. Rev. C.

4. E. Hernández and A. Mondragón Ibid.
   B. Gyarmati, A. T. Kruppa, Z. Papp and G. Wolf. Nucl. Phys. A417 (1984) 393.
   A.T. Kruppa and Z. Papp, Comp. Phys. Comm. 36 (1985) 59.

5. N. von der Heydt, Ann. Phys. (Leipzig) 29, 309 (1973).

INTEGRABILITY, CHAOS AND NONLINEAR SUPERPOSITION FORMULAS FOR DIFFERENTIAL MATRIX RICCATI EQUATIONS

M.A. del Olmo, M. Rodríguez* and P. Winternitz
Centre de Recherches Mathématiques
Université de Montréal
C.P. 6128, Succursale A
Montréal, Québec
H3C 3J7, Canada

* Present address:
Departamento de Métodos Matemáticos de la Fisica
Facultad de Físicas
Universidad Complutense, 28040 Madrid
Spain

Abstract. We point out that nonlinear superposition principles can be used to identify integrable systems of nonlinear ordinary differential equations among families of nonintegrable ones. A superposition formula is then obtained for a class of integrable equations, namely the matrix Riccati equations.

1. INTEGRABLE AND NONINTEGRABLE SYSTEMS RICCATI EQUATIONS.

We shall call a system of first order nonlinear ordinary differential equations (ODE's) a system of Riccati equations if all the nonlinearities are quadratic

$$y'^{\mu} = a^{\mu} + b^{\mu}{}_{\alpha} y^{\alpha} + c^{\mu}{}_{\alpha\beta} y^{\alpha} y^{\beta}, \qquad \mu = 1,\dots,n. \tag{1}$$

The coefficients a^{μ}, $b^{\mu}{}_{\alpha}$, $c^{\mu}{}_{\alpha\beta}$ are, in general, arbitrary functions of t, the prime denotes differentiation with respect to time t and summation over repeated indices is to be understood.

Equations of the type (1) can serve as prototypes of systems demonstrating chaotic behavior. Examples of such systems with extremely sensitive dependence on the initial conditions are the Lorenz equations, various versions of the Volterra-Lotke equations, or the Henon-Heiles equations[1-3].

On the other hand, many particular cases of equations of type (1) exist which are integrable and hence have stable solutions with regular long term (global) behavior. Clearly it is of interest to identify such systems and to solve them. These systems can then serve as tools for studying "nearby" nonintegrable systems, obtained by perturbing the coefficients in the integrable systems. Several methods are commonly used for identifying integrable systems among families of nonintegrable ones. In addition to direct searches for first integrals, or for linearizing transformations, we mention the method of Painlevé analysis[4]. This is an investigation of the singularity structure of the solutions of the equations, the purpose of which is to find equations for which the solutions have no moving critical points.

In this contribution we present a different integrability test[5], based on the fact that certain systems of nonlinear ODE's admit **superposition formulas**. We use this term to indicate that the general solution of such a system of n ODE's can be expressed as a function of a finite number m of particular solutions and of n constants[6-10]. This approach makes use of the following theorem, due to S. Lie[11].

Theorem 1. The necessary and sufficient condition for a system of first order ODE's

$$y' = \eta(y, t) \tag{2}$$

to admit a superposition formula

$$y = F(y_1,\dots,y_m, c_1,\dots,c_n) \tag{3}$$

is that :

(i) The system (2) have the form

$$y' = \sum_{k=1}^{s} Z_k(t)\, \xi_k(y) \tag{4}$$

(ii) The vector fields

$$\hat{X}_k = \sum_{\mu=1}^{n} \xi_k^{\mu}(y)\frac{\partial}{\partial y^{\mu}}, \quad \xi_k = \left(\xi_k^1, \dots, \xi_k^n\right) \tag{5}$$

generate a finite dimensional Lie algebra.

All indecomposable systems of equations satisfying Lie's criteria have been recently classified, making use of the theory of transitive primitive Lie algebras[9]. All systems of n such equations with $n \leq 3$ have recently been integrated[5].

Equations satisfying Lie's criterion do not necessarily have polynomial nonlinearities; examples have been given of equations with rational and other nonlinearities[9,10]. On the other hand, families of Riccati type equations have been obtained that do admit superposition formulas and still have coefficients that are arbitrary functions of t. Among these we mention :

1. Projective Riccati equations[6,7]

$$y'^{\mu} = a^{\mu} + b^{\mu}{}_{\alpha} y^{\alpha} + y^{\mu} c_{\alpha} y^{\alpha} \qquad (6)$$

2. Conformal Riccati equations[7]

$$y'^{\mu} = a^{\mu} + \Lambda^{\mu}{}_{\nu} y^{\nu} + s\, y^{\mu} + y^{\mu} c_{\alpha} y^{\alpha} - c^{\mu} y_{\alpha} y^{\alpha}/2 \qquad (7)$$

$$\Lambda^{\mu}{}_{\nu} + \Lambda^{\nu}{}_{\mu} = 0$$

3. Matrix Riccati equations

$$W' = A + WB + CW + WDW \qquad (8)$$

$$W, A \in R^{n \times k},\ B \in R^{k \times k},\ C \in R^{n \times n},\ D \in R^{k \times n}$$

(equations (6) are a special case of the MRE (8) with $k = 1$).

In order to show the qualitative difference between Riccati equations with and without superposition formulas, consider a simple example[5]

$$x' = y + x(c_1 x + c_2 y) \qquad y' = -(2 + Q\cos 2t)y + y(c_1 x + c_3 y) \qquad (9)$$

where c_1, c_2, c_3 and Q are constants. For $c_3 = c_1$ (9) is a special case of projective Riccati equations (6), for $c_3 \neq c_1$ the conditions of Lie's theorem are not satisfied. For $c_3 = c_2$ we can integrate analytically, for $c_3 \neq c_2$ numerical integration is required. On Fig. 1 we show Poincaré surfaces of section for two different cases, namely $Q = 0.2$ and $c_3 = c_2$ and $Q = 0.2$, $c_3 = c_2 - 0.1 = 0.6$. In both cases the orbits are plotted for $t = k\pi$ (k = integer).

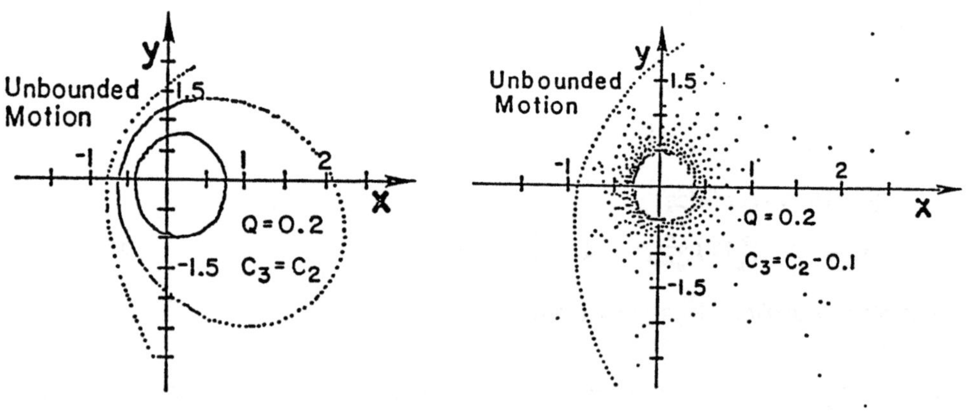

Fig. 1

In both cases we can have unbounded motion (for initial conditions to the left of the dotted line) or bounded motion (to the right of that boundary). In the case with a superposition formula we obtain regular closed orbits. For $c_3 \neq c_2$ we start the integration at a point ($x_0 \sim -1$, $y_0 = 0$, $t = 0$) close to the boundary. The trajectory starts out in a complicated manner that depends very sensitively on the exact value of x_0; it then settles into a spiralling motion towards the origin. We observe (for many different values of x_0, Q, and $c_3 - c_2 \neq 0$) a phenomenon of "transitory" chaos in a non Hamiltonian system.

In order to show how the nonlinear superposition formulas provide solutions, we consider the matrix Riccati equations (8).

2. SUPERPOSITION FORMULAS FOR THE MATRIX RICCATI EQUATION.

The case we present here corresponds to the MRE (8) based on the Lie group $SL(N, \mathbf{C})$ and a maximal parabolic subgroup $P(k)$, i.e. we consider $N = n+k$, and $n = rk$, where $n \geq k$ and $r \in \mathbf{Z}$. It has been shown that we only need 5 particular solutions to construct the general solution of $r=1$ (square MRE's[8]) and $n+2$ if $k=1$ (projective MRE's[6,7]).

Let $P(k)$ be a maximal subgroup of the Lie group $SL(n+k, \mathbf{C})$. We construct the homogeneous space $SL(n+k, \mathbf{C})/P(k) \simeq G_k(\mathbf{C}^{n+k})$, the Grassmannian of k-planes in $\mathbf{C}^{n+k}$. We introduce homogeneous coordinates on $G_k(\mathbf{C}^{n+k})$ as the components of a matrix ξ of rank k

$$\xi = \begin{pmatrix} X \\ Y \end{pmatrix} \sim \begin{pmatrix} X' \\ Y' \end{pmatrix} = \begin{pmatrix} XG \\ YG \end{pmatrix}, \quad X \in \mathbf{C}^{n\times k},\ Y \in \mathbf{C}^{k\times k},\ G \in GL(k, \mathbf{C}). \qquad (10)$$

The inherent redundancy in homogeneous coordinates is removed by introducing affine coordinates $W = XY^{-1}$ on $G_k(\mathbf{C}^{n+k})$ (for $\det Y \neq 0$).

The system of ODE's related to the action of $SL(n+k, \mathbf{C})$ on $G_k(\mathbf{C}^{n+k})$ is precisely the rectangular MRE (8) with $W, A \in \mathbf{C}^{n\times k}$, $B \in \mathbf{C}^{k\times k}$, $C \in \mathbf{C}^{n\times n}$, $D \in \mathbf{C}^{k\times n}$, where $A,\dots,D$ are matrix functions of (time) t. The right hand side of (8) corresponds to a curve in the Lie algebra $sl(n+k, \mathbf{C})$. The general solution of this system is given by the action of $SL(n+k, \mathbf{C})$ on $G_k(\mathbf{C}^{n+k})$, i.e.

$$W(t) = (G_{11}(t)U + G_{12}(t))\,(G_{21}(t)U + G_{22}(t))^{-1}, \quad G = \begin{pmatrix} G_{11} & G_{12} \\ G_{21} & G_{22} \end{pmatrix} \in SL(n+k, \mathbf{C}). \quad (11)$$

Here $G(t)$ is a curve in $SL(n+k, \mathbf{C})$, and $U \in \mathbf{C}^{n\times k}$ is a constant matrix, specifying the initial conditions for $W(t)$. The problem is now to determine $G_{ik}(t)$ as functions of a (minimal) sufficient number of particular solutions $W_i(t)$ of the equation (8). These solutions are called a **fundamental set of solutions.**

In the following we will consider the case $n \geq k$ (the case $n < k$ is reduced to this one by transposition). The minimal number m of known particular solutions needed verifies $mnk \geq (n+k)^2 - 1$, where the right hand term is the number of independent matrix elements of $G(t)$. Then m particular solutions $W_1(t),\dots,W_m(t)$ will determine $G(t)$, at least locally, if the stability group of m initial values $W_i(t_0)$ on the product of m copies of $G_k(\mathbf{C}^{n+k})$ is contained in the center of $SL(n+k, \mathbf{C})$. It is possible to choose a fundamental set of solutions genericly and to transform it into a "standard" particularly convenient set. These results can be expressed in homogeneous or in affine coordinates (see ref. 12 for all details and proofs of statements).

The points of $G_k(C^{n+k})$ are expressed as

$$\xi^T = (X_1{}^T,\dots,X_r{}^T, Y^T),\ W^T = (W_1{}^T,\dots,W_r{}^T),\ X_i,Y,W_i \in C^{k\times k},\ i = 1,\dots,r \qquad (12)$$

in homogeneous or affine coordinates, respectively (the superscript T denotes transposition). Correspondingly, we shall write the elements of $G \in SL(n+k, C)$ of (11) as

$$G_{11} = \begin{bmatrix} M_{11},\dots, M_{1r} \\ \cdot\ \cdot\ \cdot\ \cdot\ \cdot \\ M_{r1},\dots, M_{rr} \end{bmatrix},\ G_{12}^T = (N_1^T,\dots,N_2^T),\ G_{21} = (P_1,\dots, P_r),\ G_{22} = Q. \quad (13)$$

Theorem 2. The following "standard set" of $r+3$ initial conditions of the MRE (8), given in homogeneous coordinates, has only the center of $SL(n+k, C)$ as its isotropy group :

$$\{\xi_1^S,\dots,\xi_{r+3}^S\} = \left\{ \begin{bmatrix} I_k \\ 0 \\ \vdots \\ 0 \\ - \\ 0 \end{bmatrix} \begin{bmatrix} 0 \\ I_k \\ \vdots \\ 0 \\ - \\ 0 \end{bmatrix},\dots, \begin{bmatrix} 0 \\ 0 \\ \vdots \\ I_k \\ - \\ 0 \end{bmatrix} \begin{bmatrix} 0 \\ 0 \\ \vdots \\ 0 \\ - \\ I_k \end{bmatrix} \begin{bmatrix} I_k \\ I_k \\ \vdots \\ I_k \\ - \\ I_k \end{bmatrix} \begin{bmatrix} \Lambda_1 \\ \Lambda_2 \\ \vdots \\ \Lambda_r \\ - \\ I_k \end{bmatrix} \right\}. \quad (14)$$

The blocks $\Lambda_i \in C^{k\times k}$ are such that one of them, say Λ_1 satisfies $\Lambda_1 = \mathrm{diag}(\lambda_1, \lambda_2,\dots, \lambda_k)$, with $\lambda_i \in C$, $\lambda_i \neq \lambda_j$ for $i \neq j$ and another one, say Λ_2, has no irreducible invariant subspaces in common with Λ_1.

Theorem 3. Given a set of $r+3$ initial conditions of the MRE (8) in affine coordinates $\{U_1,\dots,U_{r+3}\} \subset C^{rk\times k}$ satisfying the conditions

(i) $$\det \begin{bmatrix} U_1 & \dots & U_r & U_{r+1} \\ I_k & \dots & I_k & I_k \end{bmatrix} \neq 0,$$

(ii) $$\det \begin{bmatrix} U_1 & \dots & U_{j-1} & U_{j+1} & \dots & U_{r+1} & U_{r+2} \\ I_k & \dots & I_k & I_k & \dots & I_k & I_k \end{bmatrix} \neq 0,\ j = 2,\dots,r$$

(iii) $$\det \begin{bmatrix} U_1 & \dots & U_r & U_{r+3} \\ I_k & \dots & I_k & I_k \end{bmatrix} \neq 0.$$

(iv) The matrices

$$T_i = S_i R_i (S_{r+1} R_{r+1})^{-1} \in C^{k\times k}, \quad i = 1,2$$

have no common nontrivial irreducible eigenspaces and one of them, say T_1 has k distinct eigenvalues, where S_i and R_i are defined by

$$U = \begin{bmatrix} U_1 & \dots & U_{r+1} \\ I_k & \dots & I_k \end{bmatrix},$$

$$\begin{bmatrix} (S_1)^{-1} \\ \vdots \\ (S_{r+1})^{-1} \end{bmatrix} = U^{-1} \begin{bmatrix} U_{r+2} \\ I_k \end{bmatrix}, \quad \begin{bmatrix} R_1 \\ \vdots \\ R_{r+1} \end{bmatrix} = U^{-1} \begin{bmatrix} U_{r+3} \\ I_k \end{bmatrix}.$$

Then, there exists a transformation $G \in SL(n+k, C)$ transforming the set

$$\{\xi_i\} = \left\{ \begin{bmatrix} U_i \\ I_k \end{bmatrix} \right\} \quad i = 1,\dots,r+3$$

into the standard set ξ^S_i of (14)

The superposition formula can be obtained by reconstructing the group element $G(t)$ in terms of $r+3$ particular solutions.

We parametrize the group element $G(t)$ as in (11) and (13). Writing (11) for the first $r+1$ "standard" solutions $W_i(t)$ we obtain

$$M_{ij} = W_{ji} P_j, \quad N_i = W_{r+1,i} Q, \quad i,j = 1,\dots,r$$

where we put

$$W_i(t) = \begin{bmatrix} W_{i1}(t) \\ \vdots \\ W_{ir}(t) \end{bmatrix}, \quad i = 1,\dots,r+3.$$

Using $W_{r+2}(t)$ we obtain a system of inhomogeneous linear equations for P_i in terms of the known solutions $W_j(t)$ $(j = 1,\dots,r+2)$ and the still unknown matrix $Q(t) \in C^{k\times k}$:

$$\tilde{W} \begin{bmatrix} P_1 \\ \vdots \\ P_r \end{bmatrix} = (W_{r+1} - W_{r+2}) Q, \quad \tilde{W} = \begin{bmatrix} W_{r+2,1} - W_{11}, \dots, W_{r+2,1} - W_{r1} \\ \dots \\ W_{r+2,r} - W_{1r}, \dots, W_{r+2,r} - W_{rr} \end{bmatrix}.$$

The solution exists and is unique as long as $\det \tilde{W} \neq 0$. Finally, to determine Q we use the remaining solution $W_{r+3}(t)$:

$$\tilde{\tilde{W}} \begin{pmatrix} F_1 Q \Lambda_1 \\ \vdots \\ F_r Q \Lambda_r \end{pmatrix} = (W_{r+1} - W_{r+3}) Q \tag{15}$$

where

$$\tilde{W} \begin{pmatrix} F_1 \\ \vdots \\ F_r \end{pmatrix} = (W_{r+1} - W_{r+2}) Q, \quad \tilde{\tilde{W}} = \begin{pmatrix} W_{r+3,1} - W_{11}, \dots, W_{r+3,1} - W_{r1} \\ \cdots\cdots\cdots\cdots\cdots\cdots\cdots\cdots \\ W_{r+3,r} - W_{1r}, \dots, W_{r+3,r} - W_{rr} \end{pmatrix}.$$

and $\Lambda_1, \dots, \Lambda_r$ are defined in Theorem 2.

Using (15) and

$$\begin{pmatrix} H_1 \\ \vdots \\ H_r \end{pmatrix} = \tilde{\tilde{W}}^{-1} (W_{r+1} - W_{r+3}),$$

we can write the following equations

$$Q \Lambda_i Q^{-1} = (F_i)^{-1} H_i, \; i = 1, \dots, r$$

which determine Q. Note that the matrices $F^{-1}{}_i H_i$ are conjugate to constant matrices. The existence of W^{-1} is assured by the conditions imposed in Theorem 3. For the same reasons $F^{-1}{}_i$ exists, $i = 1, \dots, r$. (Theorem 3, (ii).)

REFERENCES

1. E.N. Lorenz, J. Atmos. Sci. **20**, 130 (1963).
2 C. Sparrow, The Lorenz Equations, Bifurcations, Chaos and Strange Attractors, Springer, New York, 1982.
3. H. Hénon and C. Heiles, Astron. J. **69**, 73 (1964).
4. A. Ramani, B. Dorizzi, B. Grammaticos, J. Math. Phys. **24**, 2282 (1983).
5. T.C. Bountis, V. Papageorgiou and P. Winternitz, J. Math. Phys. **27**, xxx (1986).
6. R.L. Anderson, Lett. Math. Phys. **4**, 1 (1980).
7. R.L. Anderson, J. Harnad and P. Winternitz, Physica **D4**, 164 (1982).
8. J. Harnad, P. Winternitz and R.L. Anderson, J. Math. Phys. **24**, 1062 (1983).
9. S. Shnider and P. Winternitz, Lett. Math. Phys. **8**, 69 (1984); J. Math. Phys. **25**, 3155 (1984).
10. M.A. del Olmo, M.A. Rodríguez and P. Winternitz, J. Math. Phys. **27**, 14 (1986).
11. S. Lie and G. Scheffers, Vorlesungen über continuierlichen Gruppen mit geometrischen und anderen Anwendungen. Teubner. Leipzig. 1893 (Reprinted by Chelsea, New York, 1967).
12. M.A. del Olmo, M.A. Rodríguez, Preprint CRM-1309, Montréal, 1985.

LIST OF PARTICIPANTS

Michael Baranger
M.I.T. Room 6-312
Cambridge, Massachusetts 02139
U.S.A.

Peter Bauer
Laurentian University
1364 Rideau St.
Sudbury, Ontario
Canada

Manuel Berrondo
Physics Institute-UNAM
Apdo. Postal 139-B
Cuernavaca, Morelos 62190
Mexico

Michael Berry
H.H. Wills Physics Laboratory
Tyndall Avenue
Bristol BS8 1TL
England

Estela Blaisten
Physics Institue-UNAM
Apdo. Postal 20-364
01000 Mexico, D.F.

Ma. Esther Brandan
Physics Institute-UNAM
Apdo. Postal 20-364
01000 Mexico, D.F.

Oriol Bohigas
Institut Physique Nucleaire
B.P. #1
91406 Orsay
France

Giulio Casati
Dipartimento de Fisica
Via Celoria, 16
20133 Milano, Italy

Boris V. Chirikov
USSR Academy of Sciences
Siberian Division
Institute of Nuclear Physics
Novosibirsk 90, USSR

Angel Dacal
Physics Institute-UNAM
Apdo. Postal 20-364
01000 Mexico, D.F.

Till von Egidy
Physik-Department, E18
Technische Universitat Munchen
James Franck Str.
D-8046 Garching bei Munchen
Germany

Daniel Espinosa Wolf
Physics Institute-UNAM
Apdo. Postal 20-364
01000 Mexico, D.F.

Gustavo Espinosa
Physics Institute-UNAM
Apdo. Postal 139-B
Cuernavaca, Morelos 62190
Mexico

Silvia Espinosa Wolf
Physics Institute-UNAM
Apdo. Postal 20-364
01000 Mexico, D.F.

Jorge Flores
Physics Institute-UNAM
Apdo. Postal 20-364
01000 Mexico, D.F.

J.B. French
Physics Department
University of Rochester
Rochester, N.Y. 14627
U.S.A.

William A. Friedman
Department of Physics
University of Wisconsin
Madison, Wisconsin 53706
U.S.A.

C. Konrad Gelbke
Natl. Superconducting Cyclotron Lab
Michigan State University
East Lansing, MI 48824
U.S.A.

Halo Guarneri
Dip. di Fisica Teorica
Pavia, Italy

R. Haq
Laurentian University
1364 Rideau St.
Sudbury, Ontario, Canada

William Harter
Physics Department
University of Arkansas
Fayetteville, AR 72701
U.S.A.

Eric Heller
Department of Chemistry-BG10
University of Washington
Seattle, Washington 98195
U.S.A.

Helmut Hofmann
Physics Department
Technical University of Munich
Munich, West Germany

Jorge V. José
Dept. Physics
Northeastern University
360 Huntington Avenue
Boston, Massachusetts 02115
U.S.A.

Remy Jost
C.N.R.S. - S.N.C.I.
25, Avenue des Martyrs
38042 Grenoble Cedex
France

Sigurd Larsen
Physics Dept. 009-00
Temple University
Philadelphia, Pennsylvania 19122
U.S.A.

Bernhard Lesche
Physics Institute-UNAM
Apdo. Postal 139-B
Cuernavaca, Morelos 62190
Mexico

Gustavo Martínez Meckler
Physics Institute-UNAM
Apdo. Postal 20-364
01000 Mexico, D.F.

Pier A. Mello
Physics Institute-UNAM
Apdo. Postal 20-364
01000 Mexico, D.F.

Arturo Menchaca
Physics Institute-UNAM
Apdo. Postal 20-364
01000 Mexico, D.F.

Dawn Meredith
Calif. Institute of Technology
Kellogg Radiation Laboratory 106-38
Pasadena, CA 91125
U.S.A.

Jesús Miranda Hernández
Physics Institute-UNAM
Apdo. Postal 20-364
01000 Mexico, D.F.

Gary E. Mitchell
Department of Physics
North Carolina State University
Box 8202
Raleigh, N.C. 27695
U.S.A.

Alfonso Mondragón
Physics Institute-UNAM
Apdo. Postal 20-364
01000 Mexico, D.F.

Raúl J. Mondragón
H.H. Wills Laboratory
Tyndall Avenue
Bristol BS8 1TL
England

Luciano G. Moretto
Lawrence Berkeley Laboratory
Berkeley, CA 94720
U.S.A.

Marcos Moshinski
Physics Institute-UNAM
Apdo. Postal 20-364
01000 Mexico, D.F.

Hide Nishioka
Max-Planck-Institut fur Kernphysik
Postfach 10 39 80
D-6900 Heidelberg 1
West Germany

Mariano A. del Olmo
C.R.M. - Universite de Montreal
Case postale 6128, Succursale A
Montreal, Quebec H3C 3J7
Canada

Ma. Esther Ortiz
Physics Institute-UNAM
Apdo. Postal 20-364
01000 Mexico, D.F.

Alfredo Ozorio de Almeida
Instituto de Fisica
Universidade Estadual de Campinas
Caixa Postal 1170
13100 Campinas, SP Brazil

Akhilesh Pandey
Dept. of Physics and Astronomy
University of Rochester
Rochester, N.Y. 14627
U.S.A.

Jitendra C. Parikh
Cyclotron Institute
Texas A&M University
College Station, TX 77843
U.S.A.

Jerzy Plebanski
CINESTAV
Tabasco 204, Col. Roma
Mexico, D.F.

Diego Rappaport
Departamento de Matemáticas
UAM - Ixtapalapa
Ixtapalapa
Mexico, D.F.

José Récamier
Physics Institute-UNAM
Apdo. Postal 139-B
Cuernavaca, Morelos 62190
Mexico

Linda E. Reichl
Physics Dept.- Stat. Mech. Center
University of Texas
Austin, TX 78712
U.S.A.

Jean Richert
Centre de Recherches Nucleaires
B.P. 20
67037 Strasbourg Cedex
France

Marko Robnik
H.H. Wills Physics Laboratory
University of Bristol
Tyndall Avenue
Bristol BS8 1TL, England

Miguel Rodríguez
C.R.M. - Universite de Montreal
Case postale 6128, Succursale A.
Montreal, Quebec H3C 3J7
Canada

Carlos Román
Physics Institute-UNAM
Apdo. Postal 139-B
Cuernavaca, Morelos 62190
Mexico

Humberto Saint-Martin
Physics Institute-UNAM
Apdo. Postal 139-B
Cuernavaca, Morelos 62190
Mexico

Thomas H. Seligman
Physics Institute-UNAM
Apdo. Postal 139-B
Cuernavaca, Morelos 62190
Mexico

V.B. Sheorey
Physical Research Lab., Room 461
Navrangpura, Ahmedabad 380 009
India

Kurt E. Shuler
Dept. Chemistry (B-014)
Univ. Calif. San Diego
La Jolla, CA 92093
U.S.A.

Uzy Smilansky
Dept. of Nuclear Physics
The Weizmann Institute
Rehovot, Israel 76100

Ellen B. Stechel
Organization 1151
Sandia National Laboratories
P. O. Box 5800
Albuquerque, NM 87185
U.S.A.

Steven Tomsovik
Physics Department
University of Rochester
Rochester, N.Y. 14620
U.S.A.

Gabriel Vázquez
Physics Institute-UNAM
Apdo. Postal 139-B
Cuernavaca, Morelos 62190
Mexico

Jacobus Verbaarschot
Department of Physics
Urbana University
Urbana, Illinois 61801
U.S.A.

William Visscher
Theoretical Division
Los Alamos National Laboratory
Los Alamos, New Mexico 87545
U.S.A.

Hans A. Weidenmuller
Max-Planck-Institut fur Kernphysik
D-6900 Heidelberg 1
West Germany

Michael Wilkinson
Physics Dept. - 405-47
CALTECH
Pasadena, CA 91125
U.S.A.

Pavel Winternitz
Centre de Recherches Mathematiques
Universite de Montreal
Montreal H3C 3J7, Quebec
Canada

Kurt Wolf
Inst. de Investigaciones en Matemáticas
Aplicada y en Sistemas
Apdo. Postal 484-A
Cuernavaca, Morelos
Mexico

Martin Zirnbauer
Kellogg Laboratory 106-38
CALTECH
Pasadena, CA 91107
U.S.A.